California Earth's System

"Or why the Golden State is really golden"

Custom Edition for CSU-Fresno-EES168

James S. Monroe | Reed Wicander | Richard Hazlett
Bernard W. Pipkin | Dee D. Trent | Paul Bierman
Frank DeCourten

CENGAGE Learning™

Australia • Brazil • Japan • Korea • Mexico • Singapore • Spain • United Kingdom • United States

California Earth's System: "Or why the Golden State is really golden", Custom Edition for CSU-Fresno-EES168

James S. Monroe | Reed Wicander
Richard Hazlett | Bernard W. Pipkin
Dee D. Trent | Paul Bierman
Frank DeCourten

Executive Editors:
Maureen Staudt
Michael Stranz

Senior Project Development Manager:
Linda deStefano

Marketing Specialist:
Courtney Sheldon

Senior Production/Manufacturing Manager:
Donna M. Brown

PreMedia Manager:
Joel Brennecke

Sr. Rights Acquisition Account Manager:
Todd Osborne

Cover Image:
Getty Images*

*Unless otherwise noted, all cover images used by Custom Solutions, a part of Cengage Learning, have been supplied courtesy of Getty Images with the exception of the Earthview cover image, which has been supplied by the National Aeronautics and Space Administration (NASA).

ISBN-13: 978-1-111-46435-6

ISBN-10: 1-111-46435-9

Cengage Learning
5191 Natorp Boulevard
Mason, Ohio 45040
USA

Cengage Learning is a leading provider of customized learning solutions with office locations around the globe, including Singapore, the United Kingdom, Australia, Mexico, Brazil, and Japan. Locate your local office at:
international.cengage.com/region.
Cengage Learning products are represented in Canada by Nelson Education, Ltd.
For your lifelong learning solutions, visit **custom.cengage.com.**
Visit our corporate website at **cengage.com.**

Printed in the United States of America

Acknowledgements

The content of this text has been adapted from the following product(s):

Physical Geology: Exploring the Earth (with ThomsonNOW# Printed Access Card)
Monroe/Wicander/Hazlett ISBN-10: (0-495-01148-7)
ISBN-13: (978-0-495-01148-4)

Geology and the Environment
Pipkin/Trent/Hazlett/Bierman ISBN-10: (0-495-11497-9)
ISBN-13: (978-0-495-11497-0)

Custom Table of Contents

Understanding Earth: A Dynamic and Evolving Planet

1

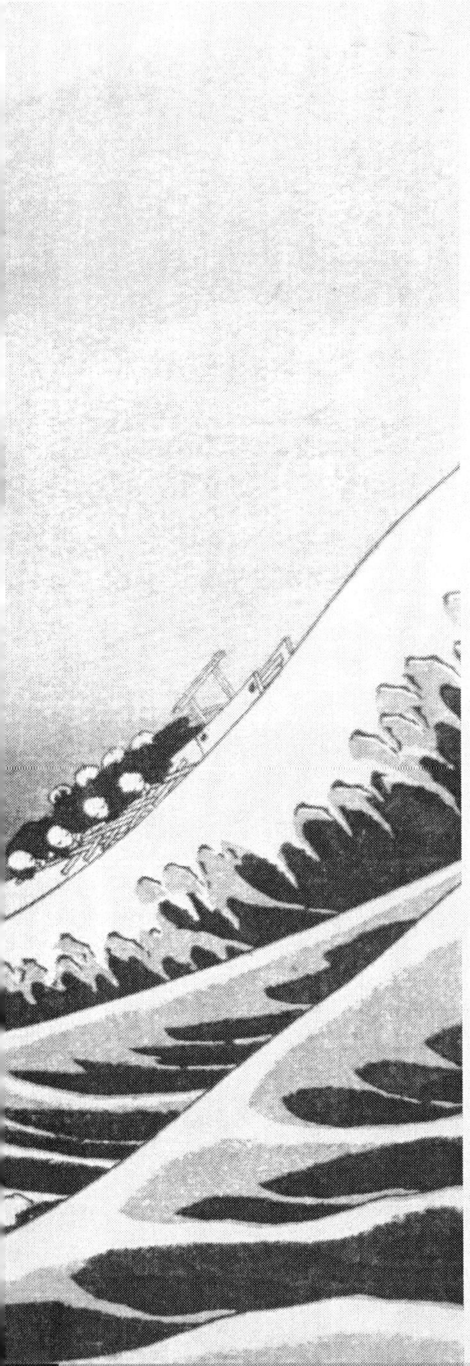

Katsusika Hokusai (1760–1849)/Art Resource, NY

The Great Wave of Kanagawa, by Katsusika Hokusai (1760–1849), color woodblock print.

Tsunami
Fingers of the sea
Pull back their waves, then release
Cerulean blue
—Alyssa White

ESSENTIAL QUESTIONS TO ASK

1.1 Introduction

- *What do natural disasters have to do with geology?*
- *What is a system?*

1.2 Geology in Our Everyday Lives

- *What is geology and what do geologists do?*
- *How does geology relate to the human experience?*
- *How does geology affect our everyday lives?*

1.3 Global Geologic and Environmental Issues Facing Humankind

- *What is considered the greatest environmental problem facing the world today?*
- *How is the greenhouse effect related to global warming?*

1.4 Origin of the Universe and Solar System, and Earth's Place in Them

- *How did the universe begin?*
- *What is our solar system composed of, and what is its origin and history?*
- *How did Earth form?*

1.5 Earth as a Dynamic and Evolving Planet

- *Why is Earth a dynamic planet?*
- *What are Earth's three concentric layers, and what are their characteristics?*

1.6 Geology and the Formulation of Theories

- *What is a theory?*

1.7 Plate Tectonic Theory

- *What is plate tectonic theory?*
- *Why is plate tectonic theory important in geology?*

1.8 The Rock Cycle

- *What is a rock?*
- *What is the rock cycle?*
- *What are the characteristics of each of the three major rock groups?*
- *How are the rock cycle and plate tectonics related?*

1.9 Geologic Time and Uniformitarianism

- *Why is it important to understand how geologic time differs from the human perspective of time?*
- *What is the principle of uniformitarianism, and why is it important in geology?*

GEOLOGY MATTERS

GEOLOGY IN UNEXPECTED PLACES:
A Bit of Egypt in Central Park, New York City, and London, England

GEOLOGY IN FOCUS:
The Kyoto Protocol—An Update

GEOLOGY AT WORK:
Forensic Geology and Palynology

GEOLOGY IN YOUR LIFE:
Global Warming and Climate Change, and How They Affect You

PHYSICAL Geology Now™ This icon, which appears throughout the book, indicates an opportunity to explore interactive tutorials, animations, or practice problems available on the Physical GeologyNow website at **http://now.brookscole.com/phygeo6.**

1.1 Introduction

Would you know what to do if suddenly the ground beneath you started shaking violently, or you were at the beach and the sea unexpectedly retreated to expose a large area that was formerly underwater, or you were on vacation somewhere near a volcano when it erupted? These are no longer hypothetical questions that you can dismiss because you think they don't apply to you. In this day and age of global business and travel, your knowledge of what to do in such situations may mean the difference between life and death. Every day we read about natural disasters and the toll they take in human suffering. The December 26, 2004, magnitude-9.0 earthquake in the Indian Ocean produced a tsunami that killed, by conservative estimates, more than 220,000 people is just the latest in a long line of natural disasters.

In a recent report published by the Earth Institute at Columbia University and the World Bank, more than half of the world's population is at risk from floods, droughts, cyclones and hurricanes, landslides, volcanoes, or earthquakes. According to the report, "Natural Disaster Hotspots: A Global Risk Analysis," more than a third of the United States population lives in hazard-prone areas, and 82% of the world's population lives in areas subject to flooding! These natural disasters not only result in human suffering but also exact an economic toll on nations. They hit particularly hard the poorest and least developed countries, which also have the fewest resources for recovery.

■ *What do natural disasters have to do with geology?*

Geology is an integral part of our lives. Natural disasters such as earthquakes, volcanic eruptions, tsunami, landslides, and floods are all components of geology that affect everyone. The more we understand how these phenomena work, the better prepared we can be to lessen their destruction and minimize the loss of life resulting from them.

Our standard of living depends directly on our consumption of natural resources, resources that formed millions and billions of years ago. However, the way we consume natural resources and interact with the environment, both as individuals and as a society, also determines our ability to pass on this standard of living to the next generation.

The major theme of this book is that Earth is a complex, dynamic planet that has changed continuously since its origin some 4.6 billion years ago. These changes and the present-day features we observe result from the interactions among Earth's various internal and external systems, subsystems, and cycles. Earth is unique among the planets of our solar system in that it supports life and has oceans of water, a hospitable atmosphere, and a variety of climates. It is ideally suited for life as we know it because of a combination of factors, including its distance from the Sun and the evolution of its interior, crust, oceans, and atmosphere. Life processes have, over time, influenced the evolution of Earth's atmosphere, oceans, and to some extent its crust. In turn, these physical changes have affected the evolution of life.

By viewing Earth as a whole—that is, thinking of it as a system—we not only see how its various components are interconnected but also better appreciate its complex and dynamic nature. The system concept makes it easier for us to study a complex subject such as Earth because it divides the whole into smaller components we can easily understand without losing sight of how the components all fit together as a whole.

■ *What is a system?*

A **system** is a combination of related parts that interact in an organized fashion. An automobile is a good example of a system. Its various components or subsystems, such as the engine, transmission, steering, and brakes, are all interconnected in such a way that a change in any one of them affects the others.

We can examine Earth in the same way we view an automobile—that is, as a system of interconnected components that interact and affect each other in many ways. The principal subsystems of Earth are the *atmosphere, biosphere, hydrosphere, lithosphere, mantle,* and *core* (▶ Figure 1.1). The complex interactions among these subsystems result in a dynamically changing planet in which matter and energy are continuously recycled into different forms (Table 1.1).

Table 1.1

Interactions among Earth's Principal Subsystems

	Atmosphere	Hydrosphere	Biosphere	Lithosphere
Atmosphere	Interaction among various air masses	Surface currents driven by wind Evaporation	Gases for respiration Dispersal of spores, pollen, and seed by wind	Weathering by wind erosion Transport of water vapor for precipitation of rain and snow
Hydrosphere	Input of water vapor and stored solar heat	Hydrologic cycle	Water for life	Precipitation Weathering and erosion
Biosphere	Gases from respiration	Removal of dissolved materials by organisms	Global ecosystems Food cycles	Modification of weathering and erosion processes Formation of soil
Lithosphere	Input of stored solar heat Landscapes affect air movements	Source of solid and dissolved materials	Source of mineral nutrients Modification of ecosystems by plate movements	Plate tectonics

Figure 1.1 Subsystems of Earth The atmosphere, hydrosphere, biosphere, lithosphere, mantle, and core are all subsystems of Earth. This simplified diagram shows how these subsystems interact, with some examples of how materials and energy are cycled throughout the Earth system. The interactions between these subsystems make Earth a dynamic planet that has evolved and changed since its origin 4.6 billion years ago.

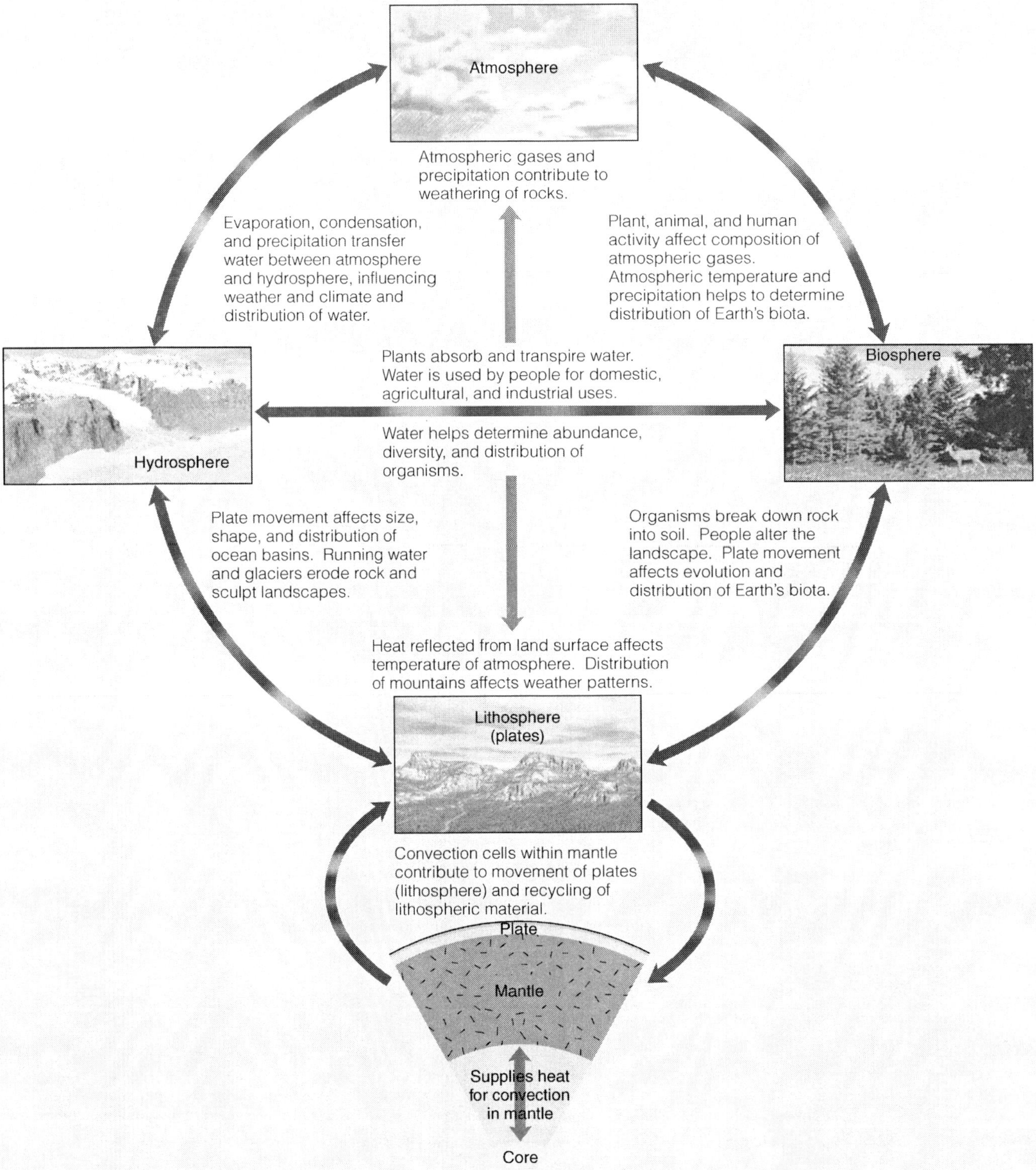

For example, the movement of plates has profoundly affected the formation of landscapes, the distribution of mineral resources, and atmospheric and oceanic circulation patterns, which in turn have affected global climate changes.

We must also not forget that humans are part of the Earth system, and our activities can produce changes with potentially wide-ranging consequences. When people discuss and debate such environmental issues as acid rain, the greenhouse effect and global warming, and the depleted ozone layer, it is important to remember that these are not isolated events but part of the larger Earth system. Furthermore, remember that Earth goes through time cycles that are much longer than humans are used to. Although they may have disastrous short-term effects on the human species, global warming and cooling are part of a longer-term cycle that has resulted in many glacial advances and retreats during the past 1.8 million years.

As you study the various topics covered in this book, keep in mind the themes discussed in this chapter and how, like the parts of a system, they are interrelated. By relating each topic to its place in the entire Earth system, you will gain a greater appreciation of why geology is so integral to our lives.

Section 1.1 Summary

- More than half of the world's population is at risk from various natural disasters. The more we understand how these phenomena work, the better prepared we can be to lessen their destructive effects and minimize loss of life.

- Earth can be viewed as a system of interconnected components that interact and affect each other, resulting in a dynamically changing planet.

- Humans are part of the Earth system, and our activities can produce changes that have potentially wide-ranging consequences.

1.2 Geology in Our Everyday Lives

■ *What is geology and what do geologists do?*

Geology, from the Greek *geo* and *logos*, is defined as the study of Earth, but now it must also include the study of the planets and moons in our solar system. Geology is generally divided into two broad areas—physical geology and historical geology. *Physical geology* is the study of Earth materials, such as minerals and rocks, as well as the processes operating within Earth and on its surface. *Historical geology* examines the origin and evolution of Earth, its continents, oceans, atmosphere, and life.

The discipline of geology is so broad that it is subdivided into numerous fields or specialties. Table 1.2 lists many of the diverse fields of geology and their relationship to the sciences of astronomy, biology, chemistry, and physics.

Nearly every aspect of geology has some economic or environmental relevance. Many geologists are involved in exploration for mineral and energy resources, using their specialized knowledge to locate the natural resources on which our industrialized society is based. As the demand for these nonrenewable resources increases, geologists apply the basic principles of geology in increasingly sophisticated ways to help focus their attention on areas with a high potential for economic success.

Whereas some geologists work on locating mineral and energy resources, other geologists use their expertise in solving environmental problems. Finding adequate sources of groundwater for the ever-burgeoning needs of communities and industries is becoming increasingly important, as is monitoring surface and underground water pollution and cleaning it up. Geologic engineers help find safe locations for dams, waste-disposal sites, and power plants and also design earthquake-resistant buildings.

Geologists are also engaged in making short- and long-range predictions about earthquakes and volcanic eruptions and the potential destruction that may result. Following the tragic events in Indonesia at the end of 2004, geologists are now more involved than ever in working with governmental agencies and civil defense planners to ensure that timely

Table 1.2

Specialties of Geology and Their Broad Relationship to the Other Sciences

Specialty	Area of Study	Related Science
Geochronology	Time and history of Earth	Astronomy
Planetary geology	Geology of the planets	
Paleontology	Fossils	Biology
Economic geology	Mineral and energy resources	
Environmental geology	Environment	Chemistry, Biology
Geochemistry	Chemistry of Earth	Chemistry
Hydrogeology	Water resources	
Mineralogy	Minerals	
Petrology	Rocks	
Geophysics	Earth's interior	Physics
Structural geology	Rock deformation	
Seismology	Earthquakes	
Geomorphology	Landforms	
Oceanography	Oceans	
Paleogeography	Ancient geographic features and locations	
Stratigraphy/ sedimentology	Layered rocks and sediments	

warnings are given to potentially affected regions and contingency plans are in place when such natural disasters as tsunami occur.

■ *How does geology relate to the human experience?*

You would probably be surprised at the extent to which geology pervades our everyday lives and at the numerous references to geology in the arts, music, and literature. Many sketches and paintings depict rocks and landscapes realistically. Leonardo da Vinci's *Virgin of the Rocks* and *Virgin and Child with Saint Anne*, Giovanni Bellini's *Saint Francis in Ecstasy* and *Saint Jerome*, and Asher Brown Durand's *Kindred Spirits* are just a few examples by famous painters (▶ Figure 1.2). In addition, we know that glaciers extended farther than they do today during Europe's Little Ice Age (from about 1500 to the middle or late 1800s) from such landscape paintings as Samuel Birmann's *Unterer Grindelwald* painted in 1826 (see Figure 17.1a).

In the field of music, Ferde Grofé's *Grand Canyon Suite* was no doubt inspired by the grandeur and timelessness of Arizona's Grand Canyon and its vast rock exposures. The rocks on the island of Staffa in the Inner Hebrides provided the inspiration for Felix Mendelssohn's famous *Hebrides Overture*.

References to geology abound in *The German Legends of the Brothers Grimm*. Jules Verne's *Journey to the Center of the Earth* describes an expedition into Earth's interior. There is even a forensic geology mystery series featuring fictional geologist Em Hansen, who uses her knowledge of geology to solve crimes. On one level, the poem "Ozymandias" by Percy B. Shelley deals with the fact that nothing lasts forever and even solid rock eventually disintegrates under the ravages of time and weathering. And popular comic strips contain references to geology; two of the best known are *B.C.* by Johnny Hart and *The Far Side* by Gary Larson.

Geology has also played an important role in the history and culture of humankind. Empires throughout history have risen and fallen on the distribution and exploitation of natural resources. Wars have been fought for the control of such natural resources as oil, gas, gold, silver, diamonds, and other valuable minerals. The Earth's surface, or its *topography*, which is shaped by geologic agents, has been a crucial factor in military strategy. Napoleon included two geologists in his expeditionary forces when he invaded Egypt in 1798, and the Russians used geologists as advisors in selecting sites for fortifications during the Russo-Japanese War of 1904–1905. Natural barriers such as mountain ranges and rivers frequently serve as political boundaries, and the shifting of river channels has sparked numerous border disputes. Deserts, which most people think of as inhospitable areas, have been the home to many people, such as the Bedouin, and have shaped their cultures over the centuries.

▶ **Figure 1.2 Layered Rocks around Gorges** *Kindred Spirits* by Asher Brown Durand (1849) realistically depicts the layered rocks along gorges in the Catskill Mountains of New York State. Durand was one of numerous artists of the 19th-century Hudson River School, which was known for realistic landscapes. This painting shows Durand conversing with the recently deceased Thomas Cole, the original founding force of the Hudson River School.

■ *How does geology affect our everyday lives?*

We observe the most obvious connection between geology and our everyday lives when natural disasters strike. Less apparent, but equally significant, are the connections between geology and economic, social, and political issues. Although most readers of this book will not become professional geologists, everyone should have a basic understanding of the geologic processes that ultimately affect all of us.

Natural Events Events such as destructive volcanic eruptions, devastating earthquakes, disastrous landslides, tsunami, floods, and droughts make headlines and affect people in obvious ways. Although we cannot prevent most of these natural disasters from happening, the more we learn about what causes them, the better we will be able to predict them and mitigate the severity of their impact.

Economics and Politics Equally important, but not always as well understood or appreciated, is the connection between geology and economic and political power. Mineral and energy resources are not equally distributed and no country is self-sufficient in all of them. Throughout history, people have fought wars to secure these resources. The United States was involved in the 1990–1991 Gulf War largely because it needed to protect its oil interests in that region. Many foreign policies and treaties develop from the need to acquire and maintain adequate supplies of mineral and energy resources.

Our Role as Decision Makers You may become involved in geologic decisions in various ways—for instance, as a member of a planning board or as a property owner with mineral rights. In such cases, you must have a basic knowledge of geology to make informed decisions. Many professionals must also deal with geologic issues as part of their jobs. Lawyers, for example, are becoming more involved in issues ranging from ownership of natural resources to how development activities affect the environment. As government plays a greater role in environmental issues and regulations, members of Congress have increased the number of staff devoted to studying issues related to the environment and geology.

Consumers and Citizens If issues like nonrenewable energy resources, waste disposal, and pollution seem simply too far removed or too complex to be fully appreciated, consider for a moment just how dependent we are on geology in our daily routines (▶ Figure 1.3).

Much of the electricity for our appliances comes from the burning of coal, oil, or natural gas or from uranium consumed in nuclear-generating plants. It is geologists who locate the coal, petroleum, natural gas, and uranium. The copper or other metal wires through which electricity travels are manufactured from materials found as the result of mineral exploration. The buildings we live and work in owe their existence to geologic resources. The concrete in the foundations is a mixture of clay, sand, or gravel, and limestone; drywall is made largely from the mineral gypsum; and windows have the mineral quartz as the principal ingredient in their manufacture.

When we go to work, the car or public transportation we use is powered and lubricated by some type of petroleum by-product and is constructed of metal alloys and plastics. And the roads or rails we ride over come from geologic materials, such as gravel, asphalt, concrete, or steel. All these items are the result of processing geologic resources.

As individuals and societies, we enjoy a standard of living that is obviously directly dependent on the consumption of geologic materials. We therefore need to be aware of how our use and misuse of geologic resources may affect the environment. We must develop policies that not only encourage the wise management of our natural resources but also allow for continuing economic development among all the world's nations. Geologists will continue to play an important role in meeting these demands by locating the needed resources and ensuring that the environment is protected for the benefit of future generations.

▶ **Figure 1.3 Life Expectancy** According to the Mineral Information Institute in Golden, Colorado, the average American born in 2004 has a life expectancy of 77.3 years and will need 1,620,000 kg of minerals, metals, and fuels to sustain his or her standard of living over a lifetime. That is an average of 20,886 kg of mineral and energy resources per year for every man, woman, and child in the United States.

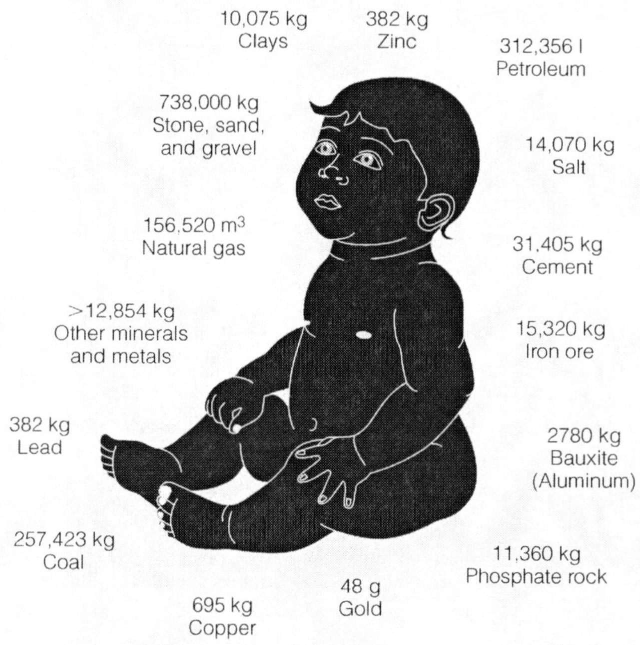

Section 1.2 Summary

- Geology is the study of Earth. It is divided into two broad areas—physical geology (the study of Earth materials) and historical geology (the study of the origin and evolution of Earth and life).
- Geology pervades the human experience in ways most people are not even aware of, such as in art, music, and literature. Geology has also played an important role in the history and culture of humankind.
- Geology affects all of us as it relates to natural disasters, our economic well-being, politics, and our responsibility as global citizens.

1.3 Global Geologic and Environmental Issues Facing Humankind

■ *What is considered the greatest environmental problem facing the world today?*

Most scientists would argue that overpopulation is the greatest environmental problem facing the world today (▶ Figure 1.4). The world's population reached 6.4 billion in 2004, and projections indicate that this number will grow by at least another billion during the next two decades, bringing Earth's human population to more than 7.5 billion. Although this may not seem to be a geologic problem, remember that these people must be fed, housed, and clothed, and all with a minimal impact on the environment. Much of this population growth will be in areas that are already at risk from such natural hazards as earthquakes, tsunami, volcanic eruptions, and floods. Adequate water supplies must be found and kept from being polluted. Additional energy resources will be needed to help fuel the economies of nations with ever-increasing populations. New techniques must be developed to reduce the use of our dwindling nonrenewable resource base and to increase our recycling efforts so that we can decrease our dependence on new sources of these materials.

▶ **Figure 1.4 Overpopulation** Overpopulation is the greatest environmental problem facing the world today. Until the world's increasing population is brought under control, people will continue to strain Earth's limited resources. Shown here is the Oshodi Market in Lagos, Nigeria. It is located at the crossroads of several bus, rail, and highway junctions in one of the world's most populous cities. Upwards of 4 million people buy and sell goods in this large market daily.

James Marshall/The Image Works

The problems of overpopulation and how it affects the global ecosystem vary from country to country. For many poor and nonindustrialized countries, the problem is too many people and not enough food. For the more developed and industrialized countries, it is too many people rapidly depleting both the nonrenewable and renewable natural resource base. And in the most industrially developed countries, it is people producing more pollutants than the environment can safely recycle on a human time scale. The common thread tying these varied situations together is an environmental imbalance caused by a human population that is exceeding Earth's short-term carrying capacity.

■ *How is the greenhouse effect related to global warming?*

An excellent example of how Earth's various subsystems are interrelated is the relationship between the greenhouse effect and global warming. As a by-product of respiration and the burning of organic material, carbon dioxide is a component of the global ecosystem and is constantly being recycled as part of the carbon cycle. The concern in recent years over the increase in atmospheric carbon dioxide levels relates to its role in the greenhouse effect.

The recycling of carbon dioxide between Earth's crust and the atmosphere is an important climate regulator because carbon dioxide as well as other gases, such as methane, nitrous oxide, chlorofluorocarbons, and water vapor, allow sunlight to pass through them but trap the heat reflected back from Earth's surface. This retention of heat is called the *greenhouse effect.* It results in an increase in the temperature of Earth's surface and, more important, its atmosphere, thus producing global warming (▶ Figure 1.5).

With industrialization and its accompanying burning of tremendous amounts of fossil fuels, carbon dioxide levels in the atmosphere have been steadily increasing since about 1880, causing many scientists to conclude that a global warming trend has already begun and will result in severe global climate shifts. Most computer models based on the current rate of increase in greenhouse gases show Earth warming as a whole by as much as 5°C during the next hundred years. Such a temperature change will be uneven, however, with the greatest warming occurring in the higher latitudes. As a consequence of this warming, rainfall patterns will shift dramatically, which will have a major effect on the largest grain-producing areas of the world, such as the American Midwest. Drier and hotter conditions will intensify the severity and frequency of droughts, leading to increased crop failure and higher food prices. With such shifts in climate, Earth's deserts may expand, with a resulting decrease in the amount of valuable crop and grazing lands.

Continued global warming will result in a rise in mean sea level as ice caps and glaciers melt and contribute their water to the world's oceans. It is predicted that at the current rate of glacial melting, sea level will rise 21 cm by the 2050s, thus increasing the number of people at risk from flooding in coastal areas by approximately 20 million!

We would be remiss, however, if we did not point out that many other scientists are not convinced that the global warming trend is the direct result of increased human activity related to industrialization. They point out that although the amount of greenhouse gases has increased, we are still uncertain about their rate of generation and rate of removal, and whether the rise in global temperature during the past century resulted from normal climate variations through time or from human activity. Furthermore, these scientists point out that even if there is a general global warming during the next

Figure 1.5 The Greenhouse Effect and Global Warming

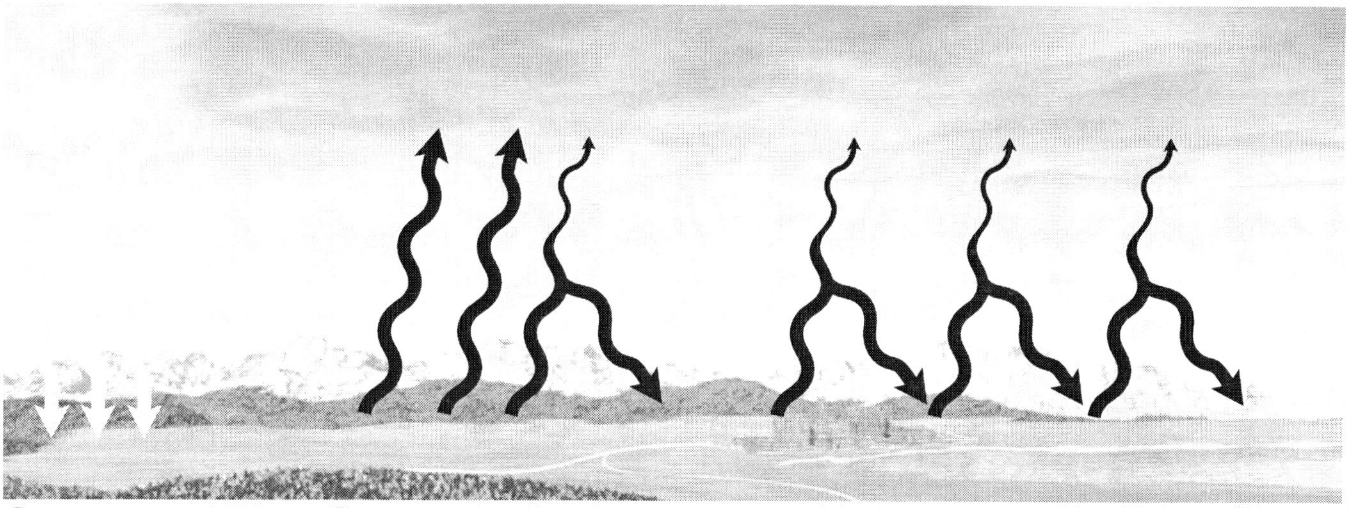

1 Short-wavelength radiation from the Sun that is not reflected back into space penetrates the atmosphere and warms Earth's surface.

2 Earth's surface radiates heat in the form of long-wavelength radiation back into the atmosphere, where some of it escapes into space. The rest is absorbed by greenhouse gases and water vapor and reradiated back toward Earth.

3 Increased concentrations of greenhouse gases trap more heat near Earth's surface, causing a general increase in surface and atmospheric temperatures, which leads to global warming.

hundred years, it is not certain that the dire predictions made by proponents of global warming will come true.

Earth, as we know, is a remarkably complex system, with many feedback mechanisms and interconnections throughout its various subsystems and cycles. It is very difficult to predict all of the consequences that global warming would have for atmospheric and oceanic circulation patterns and its ultimate effect on Earth's biota.

Section 1.3 Summary

- Most scientists would argue that the greatest environmental problem facing the world today is overpopulation. This is because a way must be found to feed, clothe, and house the world's increasing population with a minimal negative effect on the environment.

- Carbon dioxide is a by-product of respiration and the burning of organic materials, and it is constantly being recycled as part of the global carbon cycle. As such, carbon dioxide is an important climate regulator.

- The greenhouse effect is the phenomenon in which carbon dioxide and other gases allow sunlight to pass through them in the atmosphere but trap the heat reflected back from Earth's surface, thus increasing the temperature of Earth's surface and atmosphere.

- Global warming is an increase in the temperature of the atmosphere over time due to the greenhouse effect. As a consequence, global weather patterns will shift and mean sea level will rise, resulting in environmental changes that affect the world's biota.

1.4 Origin of the Universe and Solar System, and Earth's Place in Them

■ *How did the universe begin?*

Most scientists think that the universe originated about 15 billion years ago in what is popularly called the **Big Bang**. The Big Bang is a model for the evolution of the universe in which a dense, hot state was followed by expansion, cooling, and a less dense state. In a region infinitely smaller than an atom, both time and space were set at zero. Therefore there is no "before the Big Bang," only what occurred after it. The reason is that space and time are unalterably linked to form a space–time continuum demonstrated by Einstein's theory of relativity. Without space, there can be no time.

How do we know the Big Bang took place approximately 15 billion years ago? Why couldn't the universe have always existed as we know it today? Two fundamental phenomena

An important environmental issue facing the world today is global warming. How can this problem be approached from a global systems perspective? What are the possible consequences of global warming, and can we really do anything about it? Are there ways to tell whether global warming occurred in the geologic past?

indicate that the Big Bang occurred. First, the universe is expanding. When astronomers look beyond our own solar system, they observe that everywhere in the universe galaxies are moving away from each other at tremendous speeds. By measuring this expansion rate, astronomers can calculate how long ago the galaxies were all together at a single point. Second, everywhere in the universe there is a pervasive background radiation of 2.7 Kelvin (K) above absolute zero (absolute zero equals −273°C; 2.7 K = −270.3°C). This background radiation is thought to be the faint afterglow of the Big Bang.

According to the currently accepted theory, matter as we know it did not exist at the moment of the Big Bang and the universe consisted of pure energy. During the first second following the Big Bang, the four basic forces—*gravity* (the attraction of one body toward another), *electromagnetic force* (combines electricity and magnetism into one force and binds atoms into molecules), *strong nuclear force* (binds protons and neutrons together), and *weak nuclear force* (responsible for the breakdown of an atom's nucleus, producing radioactive decay)—separated and the universe experienced enormous expansion. About 300,000 years later, the universe was cool enough for complete atoms of hydrogen and helium to form; photons (the energetic particles of light) separated from matter and light burst forth for the first time.

During the next 200 million years, as the universe continued expanding and cooling, stars and galaxies began to form and the chemical makeup of the universe changed. Initially the universe was 100% hydrogen and helium, whereas today it is 98% hydrogen and helium and 2% all other elements by weight. How did such a change in the universe's composition occur? Throughout their life cycle, stars undergo many nuclear reactions in which lighter elements are converted into heavier elements by nuclear fusion. When a star dies, often explosively, the heavier elements that were formed in its core are returned to interstellar space and are available for inclusion in new stars. In this way, the composition of the universe is gradually enhanced in heavier elements.

■ *What is our solar system composed of, and what is its origin and history?*

Our solar system, which is part of the Milky Way Galaxy, consists of the Sun, 9 planets, 101 known moons or satellites (although this number keeps changing with the discovery of new moons and satellites surrounding the Jovian planets), a tremendous number of asteroids—most of which orbit the Sun in a zone between Mars and Jupiter—and millions of comets and meteorites as well as interplanetary dust and gases (▶ Figure 1.6). Any theory formulated to explain the origin and evolution of our solar system must therefore take into account its various features and characteristics.

Many scientific theories for the origin of the solar system have been proposed, modified, and discarded since the French scientist and philosopher René Descartes first proposed, in 1644, that the solar system formed from a gigantic whirlpool within a universal fluid. Today the **solar nebula theory** for the origin of our solar system involves the condensation and collapse of interstellar material in a spiral arm of the Milky Way Galaxy.

The collapse of this cloud of gases and small grains into a counterclockwise-rotating disk concentrated about 90% of the material in the central part of the disk and formed an embryonic Sun, around which swirled a rotating cloud of material called a *solar nebula*. Within this solar nebula were localized eddies in which gases and solid particles condensed. During the condensation process, gaseous, liquid, and solid particles began to accrete into ever-larger masses

▶ **Figure 1.6 Diagrammatic Representation of the Solar System** This representation of the solar system shows the planets and their orbits around the Sun.

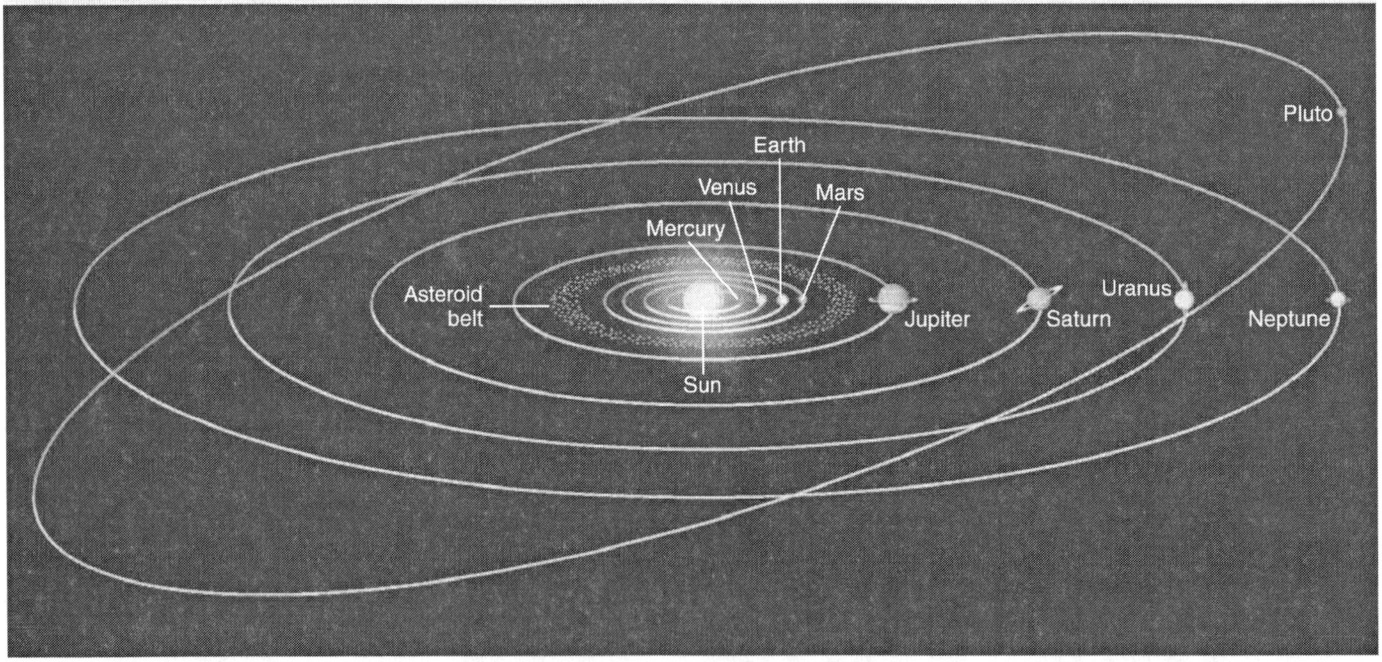

Figure 1.7 Planetesimals At the stage of development shown here, planetesimals have formed in the inner solar system, and large eddies of gas and dust remain at great distances from the embryonic Sun.

Courtesy of Dana Berry/NASA/GFSC

called *planetesimals* (Figure 1.7), which collided and grew in size and mass until they eventually became planets.

The composition and evolutionary history of the planets are a consequence, in part, of their distance from the Sun (see "The Terrestrial and Jovian Planets" on pages 14 and 15). The **terrestrial planets**—Mercury, Venus, Earth, and Mars—so named because they are similar to *terra,* Latin for "earth," are all small and composed of rock and metallic elements that condensed at the high temperatures of the inner nebula. The **Jovian planets**—Jupiter, Saturn, Uranus, and Neptune—so named because they resemble Jupiter (the Roman god was also named Jove), all have small rocky cores compared to their overall size, and are composed mostly of hydrogen, helium, ammonia, and methane, which condense at low temperatures.

While the planets were accreting, material that had been pulled into the center of the nebula also condensed, collapsed, and was heated to several million degrees by gravitational compression. The result was the birth of a star, our Sun.

During the early accretionary phase of the solar system's history, collisions between various bodies were common, as indicated by the craters on many planets and moons. Asteroids probably formed as planetesimals in a localized eddy between what eventually became Mars and Jupiter in much the same way that other planetesimals formed the terrestrial planets. The tremendous gravitational field of Jupiter, however, prevented this material from ever accreting into a planet. Comets, which are interplanetary bodies composed of loosely bound rocky and icy material, are thought to have condensed near the orbits of Uranus and Neptune.

The solar nebula theory of the formation of the solar system thus accounts for most of the characteristics of the planets and their moons, the differences in composition between the terrestrial and Jovian planets, and the presence of the asteroid belt. Based on the available data, the solar nebula theory best explains the features of the solar system and provides a logical explanation for its evolutionary history.

■ *How did Earth form?*

Some 4.6 billion years ago, various planetesimals in our solar system gathered enough material together to form Earth and eight other planets. Scientists think that this early Earth was probably cool, of generally uniform composition and density throughout, and composed mostly of silicates (compounds of silicon and oxygen), iron and magnesium oxides, and small amounts of all the other chemical elements. Subsequently, when the combination

Figure 1.8 Homogeneous Accretion Theory for the Formation of a Differentiated Earth

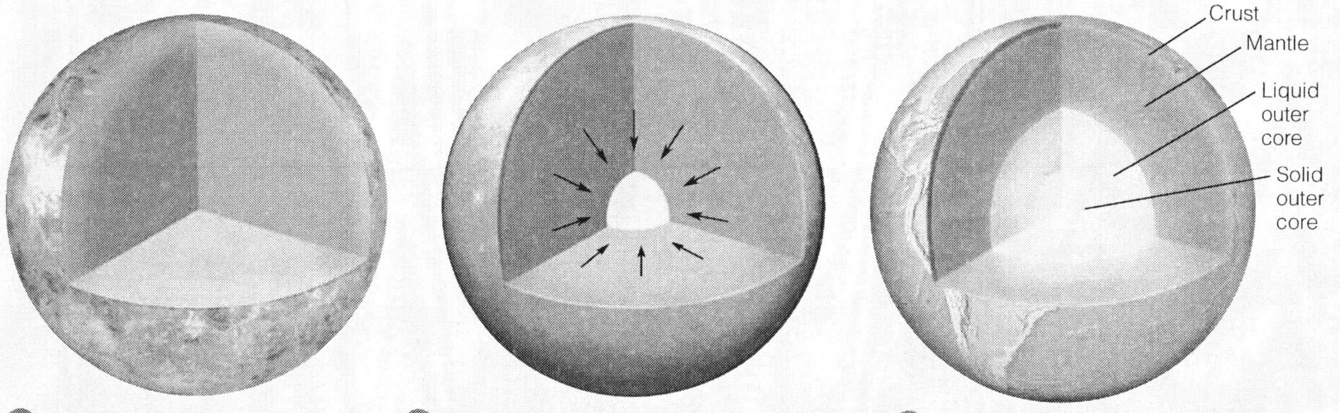

1 Early Earth probably had a uniform composition and density throughout.

2 The temperature of early Earth reached the melting point of iron and nickel, which, being denser than silicate minerals, settled to Earth's center. At the same time, the lighter silicates flowed upward to form the mantle and the crust.

3 In this way, a differentiated Earth formed, consisting of a dense iron–nickel core, an iron-rich silicate mantle, and a silicate crust with continents and ocean basins.

of meteorite impacts, gravitational compression, and heat from radioactive decay increased the temperature of Earth enough to melt iron and nickel, this homogeneous composition disappeared and was replaced by a series of concentric layers of differing com position and density, resulting in a differentiated planet (▶ Figure 1.8).

This differentiation into a layered planet is probably the most significant event in Earth history. Not only did it lead to the formation of a crust and eventually continents, but it also was probably responsible for the emission of gases from the interior that eventually led to the formation of the oceans and atmosphere.

Section 1.4 Summary

- The universe originated some 15 billion years ago in what is popularly called the Big Bang. The evidence for the Big Bang is that the universe is expanding and has a pervasive background radiation of 2.7 K above absolute zero (absolute zero equals −273°C; 2.7 K = −270.3°C).
- Our solar system began with the condensation and collapse of interstellar matter in a spiral arm of the Milky Way Galaxy. The collapse formed a counterclockwise-rotating disk in which a rotating cloud of gases and particles swirled around an embryonic Sun. This material eventually condensed into larger bodies and formed the nine planets of our solar system.
- The planets are classified as either terrestrial (Mercury, Venus, Earth, and Mars), which are small and composed of rocky and metallic elements that condense at high temperatures, or Jovian (Jupiter, Saturn, Uranus, and Neptune), which have small rocky cores compared to their overall size and are composed mostly of hydrogen, helium, ammonia, and methane, which condense at low temperatures.
- When Earth formed, it was probably cool and of generally uniform composition and density throughout. As a result of meteorite impacts, gravitational compression, and heat from radioactive decay, the temperature of Earth increased, which caused its various elements and compounds to form a series of concentric layers of differing composition and density, resulting in a differentiated planet.
- This differentiation not only led to the formation of a crust and continents but also contributed to the eventual formation of Earth's oceans and atmosphere.

1.5 Earth as a Dynamic and Evolving Planet

■ *Why is Earth a dynamic planet?*

Earth is a dynamic planet that has continuously changed during its 4.6-billion-year existence. The size, shape, and geographic distribution of continents and ocean basins have changed through time, the composition of the atmosphere has evolved, and life-forms existing today differ from those that lived during the past. Mountains and hills have been worn away by erosion, and the forces of wind, water, and ice have sculpted a diversity of landscapes. Volcanic eruptions and earthquakes reveal an active interior, and folded and fractured rocks are testimony to the tremendous power of Earth's internal forces.

■ *What are Earth's three concentric layers, and what are their characteristics?*

Earth consists of three concentric layers: the core, the mantle, and the crust (▶ Figure 1.9). This orderly division results from density differences between the layers as a function of variations in composition, temperature, and pressure.

The **core** has a calculated density of 10–13 grams per cubic centimeter (g/cm^3) and occupies about 16% of Earth's total volume. Seismic (earthquake) data indicate that the core consists of a small, solid inner region and a larger, apparently liquid, outer portion. Both are thought to consist largely of iron and a small amount of nickel.

The **mantle** surrounds the core and comprises about 83% of Earth's volume. It is less dense than the core (3.3–5.7 g/cm^3) and is thought to be composed largely of *peridotite*, a dark, dense igneous rock containing abundant iron and magnesium. The mantle can be divided into three distinct zones based on physical characteristics. The lower mantle is solid and forms most of the volume of Earth's interior. The **asthenosphere**

▶ **Figure 1.9 Cross Section of Earth Illustrating the Core, Mantle, and Crust** The enlarged portion shows the relationship between the lithosphere (composed of the continental crust, oceanic crust, and solid upper mantle) and the underlying asthenosphere and lower mantle.

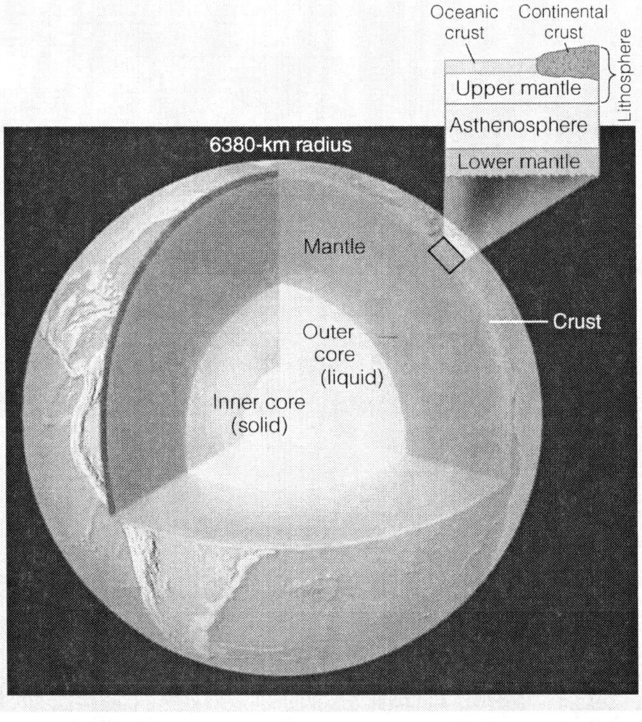

The Terrestrial and Jovian Planets

The planets of our solar system are divided into two major groups that are quite different, indicating that the two underwent very different evolutionary histories. The four inner planets—Mercury, Venus, Earth, and Mars—are the terrestrial planets; they are small and dense (composed of a metallic core and silicate mantle-crust), ranging from no atmosphere (Mercury) to an oppressively thick one (Venus). The outer four planets (excluding Pluto, which some astronomers don't regard as a planet at all)—Jupiter, Saturn, Uranus, and Neptune—are the Jovian planets; they are large, ringed, low-density planets with rocky cores surrounded by thick atmospheres.

1. The relative sizes of the planets and the Sun. (Distances between planets are not to scale.)

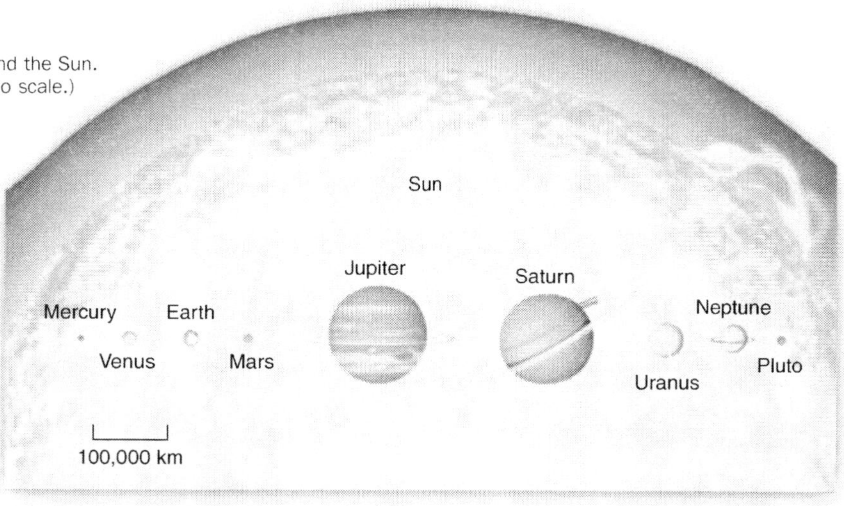

JPL/NASA

JPL/NASA

2. The **Moon** is one-fourth the diameter of Earth, has a low density relative to the terrestrial planets, and is extremely dry. Its surface is divided into low-lying dark colored plains and light-colored highlands that are heavily cratered, attesting to a period of massive meteorite bombardment in our solar system more than 4 billion years ago. The hypothesis that best accounts for the origin of the Moon has a giant planetesimal, the size of Mars or larger, crashing into Earth 4.6 to 4.4 billion years ago, causing ejection of a large quantity of hot material that cooled and formed the Moon.

3. Venus is surrounded by an oppressively thick atmosphere that completely obscures its surface. However, radar images from orbiting spacecraft reveal a wide variety of terrains, including volcanic features, folded mountain ranges, and a complex network of faults.

JPL/NASA

JPL/NASA

4. Mercury has a heavily cratered surface that has changed very little since its early history. Because Mercury is so small, its gravitation attraction is insufficient to retain atmospheric gases; any atmosphere that it may have held when it formed probably escaped into space quickly.

JPL/NASA

▶ **5. Earth** is unique among our solar system's planets in that is has a hospitable atmosphere, oceans of water, a variety of climates, and it supports life.

▼ **6. Mars** has a thin atmosphere, little water, and distinct seasons. Its southern hemisphere is heavily cratered like the surfaces of Mercury and the Moon. The northern hemisphere has large, smooth plains, fewer craters, and evidence of extensive volcanism. The largest volcano in the solar system is found in the northern hemisphere, as are huge canyons, the largest of which, if present on Earth, would stretch from San Francisco to New York!

JPL/NASA

JPL/NASA

◀ **7. Jupiter** is the largest of the Jovian planets. With its moons, rings, strong magnetic field, and intense radiation belts, Jupiter is the most complex and varied planet in our solar system. Jupiter's cloudy and violent atmosphere is divided into a series of different colored bands and a variety of spots (the Great Red Spot) that interact in incredibly complex motions.

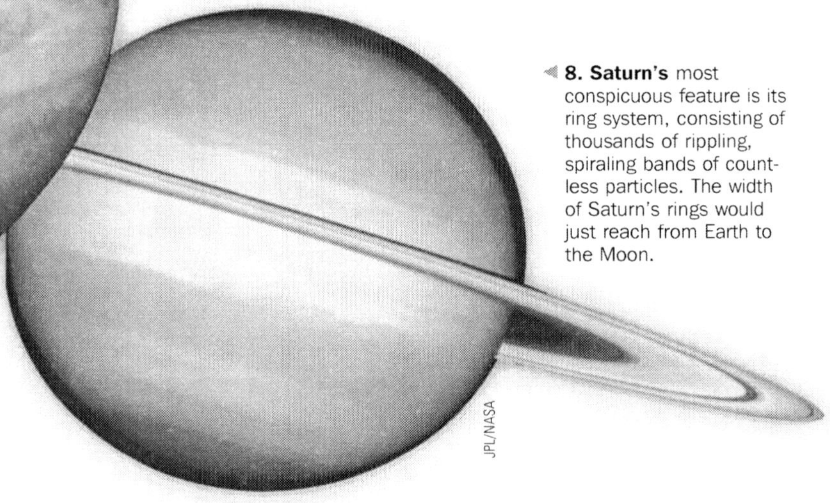

JPL/NASA

◀ **8. Saturn's** most conspicuous feature is its ring system, consisting of thousands of rippling, spiraling bands of countless particles. The width of Saturn's rings would just reach from Earth to the Moon.

JPL/NASA

▲ **9. Uranus** is the only planet that lies on its side, that is, its axis of rotation nearly parallels the plane of the ecliptic. Some scientists think that a collision with an Earth-sized body early in its history may have knocked Uranus on its side. Like the other Jovian planets, Uranus has a ring system, albeit a faint one.

▶ **10. Neptune** is a dynamic stormy planet with an atmosphere similar to those of the other Jovian planets. Winds up to 2000 km/h blow over the planet, creating tremendous storms, the largest of which, the Great Dark Spot, seen in the center, is nearly as big as Earth and is similar to the Great Red Spot on Jupiter.

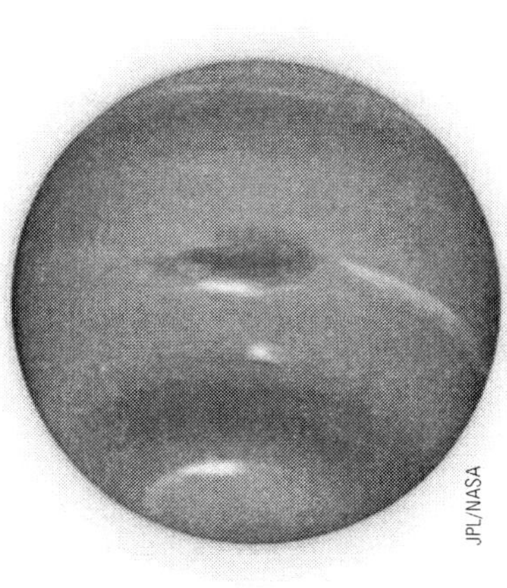

JPL/NASA

surrounds the lower mantle. It has the same composition as the lower mantle but behaves plastically and flows slowly. Partial melting within the asthenosphere generates *magma* (molten material), some of which rises to the surface because it is less dense than the rock from which it was derived. The upper mantle surrounds the asthenosphere. The solid upper mantle and the overlying crust constitute the **lithosphere**, which is broken into numerous individual pieces called **plates** that move over the asthenosphere, partially as a result of underlying *convection cells* (▶ Figure 1.10). Interactions of these plates are responsible for such phenomena as earthquakes, volcanic eruptions, and the formation of mountain ranges and ocean basins.

PHYSICAL GeologyNow™ ▶ **Active Figure 1.10 Movement of Earth's Plates** Earth's plates are thought to move partially as a result of underlying mantle convection cells in which warm material from deep within Earth rises toward the surface, cools, and then upon losing heat descends back into the interior, as shown in this diagrammatic cross section.

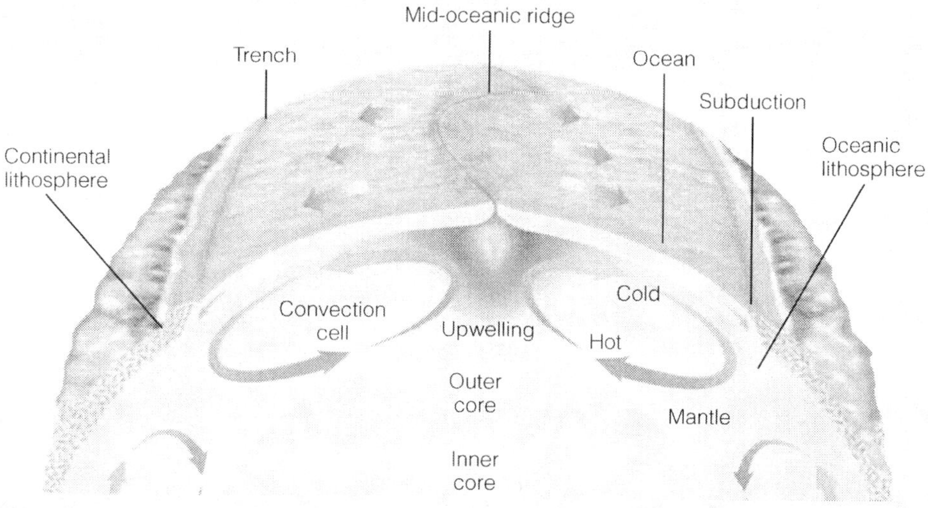

The **crust**, Earth's outermost layer, consists of two types. *Continental crust* is thick (20–90 km), has an average density of 2.7 g/cm^3, and contains considerable silicon and aluminum. *Oceanic crust* is thin (5–10 km), denser than continental crust (3.0 g/cm^3), and is composed of the dark igneous rock *basalt* and *gabbro*.

PHYSICAL GeologyNow™ Log into GeologyNow and select this chapter to work through a **Geology Interactive** activity on "Core Studies" (click Earth's Layers⟶Core Studies).

Section 1.5 Summary

- Earth is a dynamic planet that has changed continuously during its 4.6-billion-year history.
- Earth is differentiated into three concentric layers: core, mantle, and crust.
- The core occupies about 16% of Earth's volume, has a calculated density of 10–13 g/cm^3, is composed largely of iron with some nickel, and is divided into a small solid portion surrounded by an apparently liquid larger portion.
- The mantle surrounds the core, occupies about 83% of Earth's volume, has a density of 3.3–5.7 g/cm^3, and is composed largely of peridotite. It is divided into a solid lower portion, followed by a plastically behaving asthenosphere, and overlain by a solid upper portion.
- Earth's outermost layer is its crust. It is divided into thick continental crust that has an average density of 2.7 g/cm^3 and contains considerable silicon and aluminum, and thin oceanic crust that has a density of 3.0 g/cm^3 and is composed of basalt and gabbro.

1.6 Geology and the Formulation of Theories

■ *What is a theory?*

The term **theory** has various meanings. In colloquial usage, it means a speculative or conjectural view of something—hence, the widespread belief that scientific theories are little more than unsubstantiated wild guesses. In scientific usage, however, a theory is a coherent explanation for one or several related natural phenomena supported by a large body of objective evidence. From a theory scientists derive predictive statements that can be tested by observations and/or experiments so that their validity can be assessed. The law of universal gravitation is an example of a theory that describes the attraction between masses (an apple and Earth in the popularized account of Newton and his discovery).

Theories are formulated through the process known as the **scientific method**. This method is an orderly, logical approach that involves gathering and analyzing facts or data about the problem under consideration. Tentative explanations, or **hypotheses**, are then formulated to explain the observed phenomena. Next the hypotheses are tested to see whether what was predicted actually occurs in a given situation. Finally, if one of the hypotheses is found, after repeated tests, to explain the phenomena, then the hypothesis is proposed as a theory. Remember, however, that in science even a theory is still subject to further testing and refinement as new data become available.

The fact that a scientific theory can be tested and is subject to such testing separates science from other forms of human inquiry. Because scientific theories can be tested, they have the potential of being supported or even proved wrong. Accordingly, science must proceed without any appeal to beliefs or supernatural explanations, not because such beliefs or explanations are necessarily untrue but because we have no way to investigate them. For this reason, science makes no claim about the existence or nonexistence of a supernatural or spiritual realm.

Each scientific discipline has certain theories that are of particular importance. In geology, plate tectonic theory has changed the way geologists view Earth. Geologists now look at Earth from a global perspective in which all its subsystems and cycles are interconnected, and Earth history is seen as a continuum of interrelated events that are part of a global pattern of change.

Section 1.6 Summary

- The scientific method is a logical, orderly approach that involves gathering data, formulating and testing hypotheses, and proposing theories.
- A hypothesis is a provisional explanation for observations that is subject to repeated testing.
- A theory is a coherent explanation for natural phenomena that is supported by a large body of objective evidence.

1.7 Plate Tectonic Theory

■ *What is plate tectonic theory?*

The recognition that the lithosphere is divided into rigid plates that move over the asthenosphere forms the foundation of **plate tectonic theory** (▶ Figure 1.11). Zones of volcanic activity, earthquakes, or both mark most plate boundaries. Along these boundaries plates separate (diverge), collide (converge), or slide sideways past each other (▶ Figure 1.12).

The acceptance of plate tectonic theory is recognized as a major milestone in the geologic sciences, comparable to the revolution Darwin's theory of evolution caused in biology. Plate tectonics has provided a framework for interpreting the composition, structure, and internal processes of Earth on a global scale. It has led to the realization that the continents and ocean basins are part of a lithosphere–atmosphere–hydrosphere system that evolved together with Earth's interior (Table 1.3).

■ *Why is plate tectonic theory important in geology?*

A revolutionary concept when it was proposed in the 1960s, plate tectonic theory has had far-reaching consequences in all fields of geology because it provides the basis for relating many seemingly unrelated phenomena. Besides being responsible for the major features of Earth's crust, plate

▶ **Figure 1.11 Earth's Plates** Earth's lithosphere is divided into rigid plates of various sizes that move over the asthenosphere.

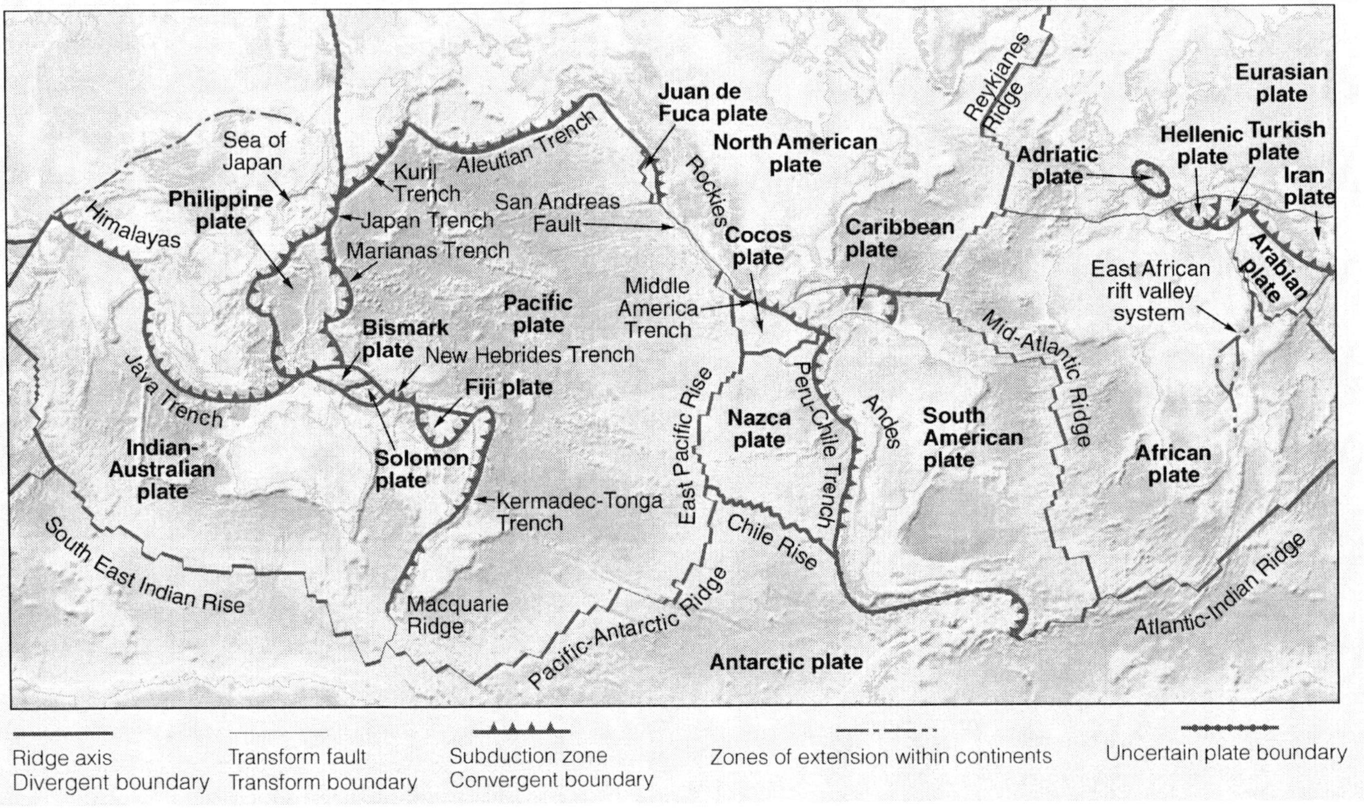

PHYSICAL GeologyNow™ ▶ **Geo-focus Figure 1.12 Relationship Between Lithosphere, Asthenosphere, and Plate Boundaries** An idealized cross section illustrating the relationship between the lithosphere and the underlying asthenosphere and the three principal types of plate boundaries: divergent, convergent, and transform.

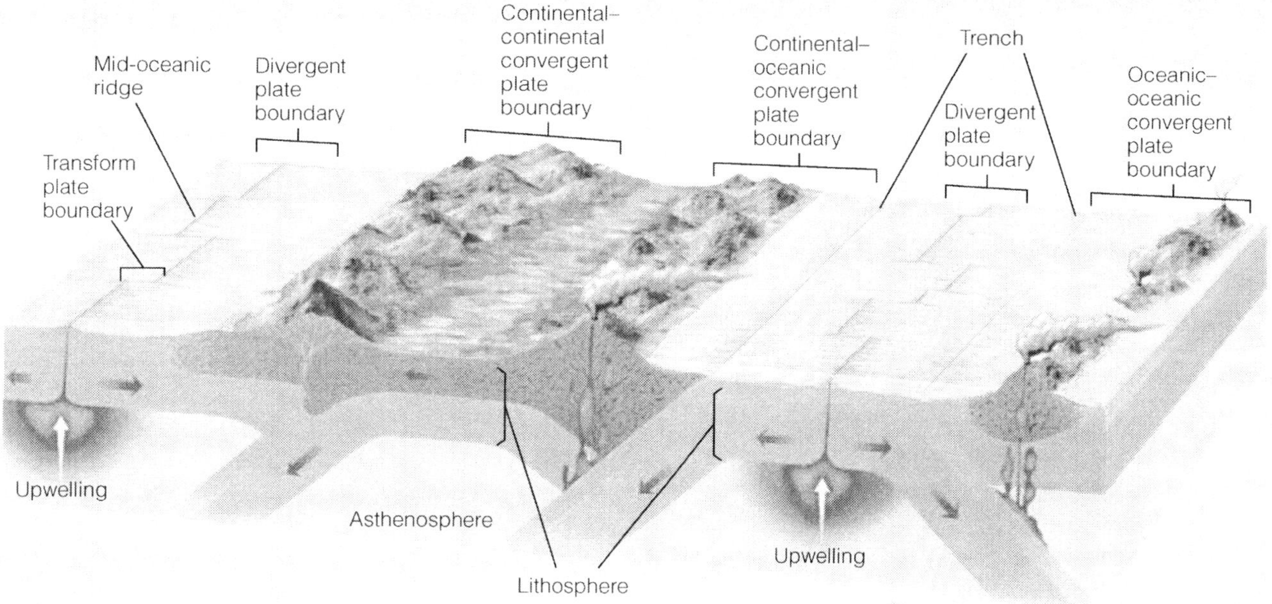

movements also affect the formation and occurrence of Earth's natural resources as well as the distribution and evolution of the world's biota.

The impact of plate tectonic theory has been particularly notable in the interpretation of Earth's history. For example, the Appalachian Mountains in eastern North America and the mountain ranges of Greenland, Scotland, Norway, and Sweden are not the result of unrelated mountain-building episodes but, rather, are part of a larger mountain-building event that involved the closing of an ancient "Atlantic Ocean" and the formation of the supercontinent Pangaea about 251 million years ago.

PHYSICAL GeologyNow™ Log into GeologyNow and select this chapter to work through a **Geology Interactive** activity on "Plate Locations" (click Plate Tectonics→Plate Locations).

Table 1.3

Plate Tectonics and Earth Systems

Solid Earth
Plate tectonics is driven by convection in the mantle and in turn drives mountain-building and associated igneous and metamorphic activity.
Atmosphere
Arrangement of continents affects solar heating and cooling, and thus winds and weather systems. Rapid plate spreading and hot-spot activity may release volcanic carbon dioxide and affect global climate.
Hydrosphere
Continental arrangement affects ocean currents. Rate of spreading affects volume of mid-oceanic ridges and hence sea level. Placement of continents may contribute to onset of ice ages.
Biosphere
Movement of continents creates corridors or barriers to migration, the creation of ecological niches, and transport of habitats into more or less favorable climates.
Extraterrestrial
Arrangement of continents affects free circulation of ocean tides and influences tidal slowing of Earth's rotation.

Source: Adapted by permission from Stephen Dutch, James S. Monroe, and Joseph Moran, *Earth Science* (Minneapolis/St. Paul: West Publishing Co., 1997).

Section 1.7 Summary

- Plate tectonic theory holds that the Earth's lithosphere is divided into rigid plates that move over the asthenosphere and diverge, converge, and slide past each other.
- Plate movement results in volcanic activity, earthquakes, and mountain building. It also affects the formation and distribution of natural resources and influences the distribution and evolution of Earth's biota.

1.8 The Rock Cycle

What is a rock?

A **rock** is an aggregate of **minerals**, which are naturally occurring, inorganic, crystalline solids that have definite physical and chemical properties. Minerals are composed of elements such as oxygen, silicon, and aluminum, and elements are made up of atoms, the smallest particles of matter that retain the characteristics of an element. More than 3500 minerals have been identified and described, but only about a dozen make up the bulk of the rocks in Earth's crust (see Table 3.3).

Geologists recognize three major groups of rocks—*igneous*, *sedimentary*, and *metamorphic*—each of which is characterized by its mode of formation. Each group contains a variety of individual rock types that differ from one another on the basis of their composition or texture (the size, shape, and arrangement of mineral grains).

What is the rock cycle?

The **rock cycle** provides a way of viewing the interrelationships between Earth's internal and external processes (▶ Figure 1.13). It relates the three rock groups to each other; to surficial processes such as weathering, transportation, and deposition; and to internal processes such as magma generation and metamorphism.

What are the characteristics of each of the three major rock groups?

Igneous rocks result when magma crystallizes or volcanic ejecta such as ash accumulate and consolidate. As magma cools, minerals crystallize, and the resulting rock is characterized by interlocking mineral grains. Magma that cools slowly beneath the surface produces *intrusive igneous rocks* (▶ Figure 1.14a); magma that cools at the surface produces *extrusive igneous rocks* (▶ Figure 1.14b).

Rocks exposed at Earth's surface are broken into particles and dissolved by various weathering processes. The particles and dissolved materials may be transported by wind, water, or ice and eventually deposited as *sediment*. This sediment may then be compacted or cemented (lithified) into sedimentary rock.

Sedimentary rocks form in one of three ways: consolidation of rock fragments, precipitation of mineral matter from solution, or compaction of plant or animal remains (▶ Figure 1.14c, d). Because sedimentary rocks form at or near Earth's surface, geologists can make inferences about the environment in which they were deposited, the transporting agent, and perhaps even something about the source from which the sediments were derived (see Chapter 7). Accordingly, sedimentary rocks are especially useful for interpreting Earth history.

Metamorphic rocks result from the alteration of other rocks, usually beneath the surface, by heat, pressure, and the chemical activity of fluids. For example, marble, a rock preferred by many sculptors and builders, is a metamorphic rock produced when the agents of metamorphism are applied to the sedimentary rocks limestone or dolostone. Metamorphic rocks are either *foliated* (▶ Figure 1.14e) or *nonfoliated* (▶ Figure 1.14f). Foliation, the parallel alignment of minerals due to pressure, gives the rock a layered or banded appearance.

How are the rock cycle and plate tectonics related?

Interactions between plates determine, to some extent, which of the three rock groups will form (▶ Figure 1.15). For example, when plates converge, heat and pressure generated along the plate boundary may lead to igneous activity and metamorphism within the descending oceanic plate, thus producing various igneous and metamorphic rocks.

Some of the sediments and sedimentary rocks on the descending plate are melted, whereas other sediments and sedimentary rocks along the boundary of the nondescending plate are metamorphosed by the heat and pressure generated along the converging plate boundary. Later, the mountain range or chain of volcanic islands formed along the convergent plate boundary will be weathered and eroded, and the new sediments will be transported to the ocean to begin yet another cycle.

The interrelationship between the rock cycle and plate tectonics is just one example of how Earth's subsystems and cycles are all interrelated. Heating within Earth's interior results in convection cells that power the movement of plates, and also in magma, which forms intrusive and extrusive igneous rocks. Movement along plate boundaries may result in volcanic activity, earthquakes, and in some cases mountain building. The interaction between the atmosphere, hydrosphere, and biosphere contributes to the weathering of rocks exposed on Earth's surface. Plates descending back into Earth's interior are subjected to increasing heat and pressure, which may lead to metamorphism as well as the generation of magma and yet another recycling of materials.

PHYSICAL Geology Now™ Log into GeologyNow and select this chapter to work through a **Geology Interactive** activity on "The Rock Cycle" (click Rocks and the Rock Cycle→Rock Cycle).

Figure 1.13 The Rock Cycle This cycle shows the interrelationships between Earth's internal and external processes and how the three major rock groups are related.

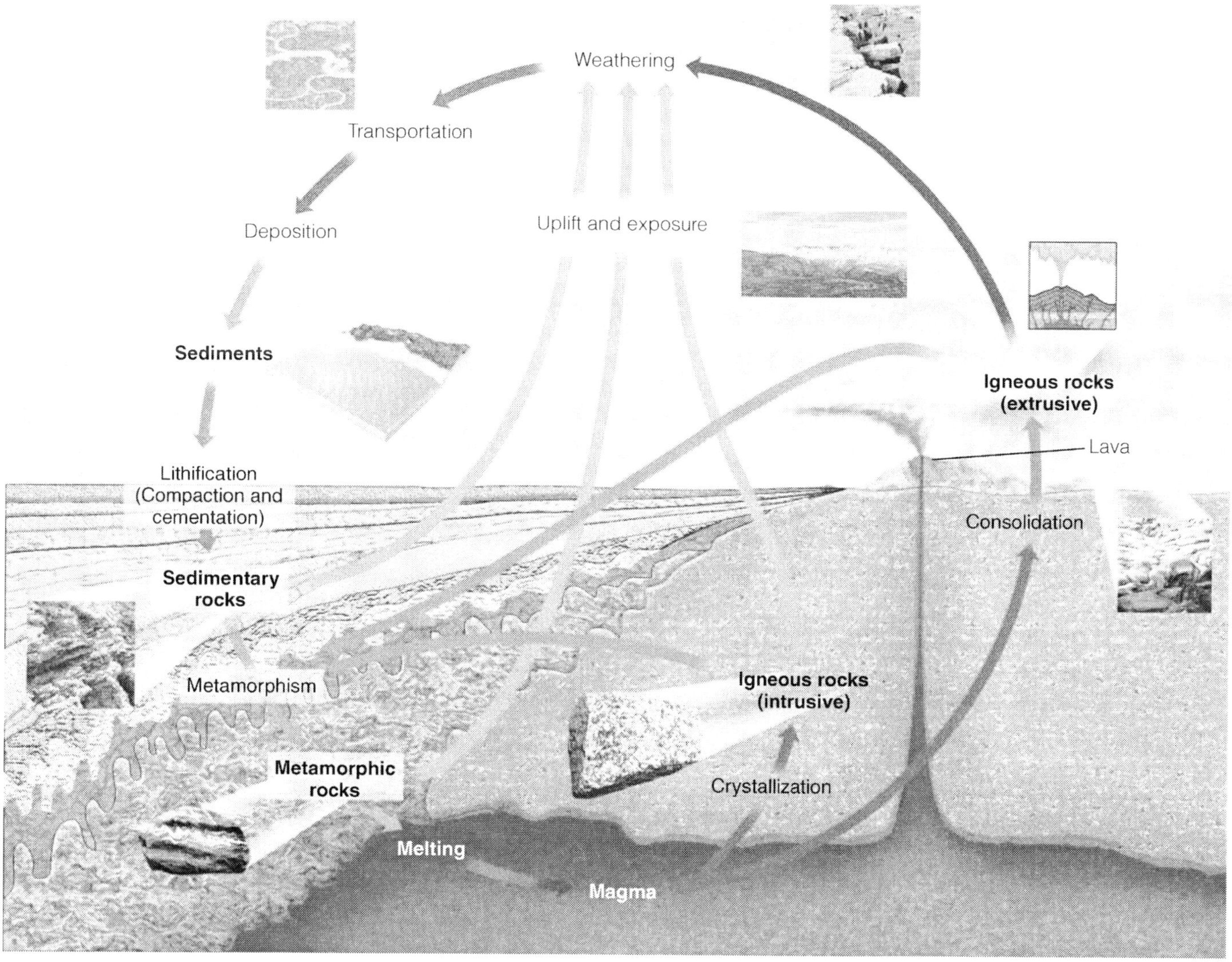

Section 1.8 Summary

- A rock is an aggregate of minerals.
- There are three major groups of rocks, each characterized by its mode of formation.
- Igneous rocks result from the cooling and crystallization of magma or the accumulation and consolidation of volcanic ejecta such as ash.
- Sedimentary rocks form at Earth's surface from the consolidation of rock fragments, precipitation of mineral matter from solution, or compaction of plant or animal remains.
- Metamorphic rocks result from the alteration of other rocks, typically beneath Earth's surface, by heat, pressure, and the chemical activity of fluids.
- The rock cycle relates the three rock groups to each other; to Earth's surficial processes such as weathering, transportation, and deposition; and to its internal processes such as magma generation and metamorphism.
- Plate tectonics is the mechanism that drives the rock cycle and recycles the three rock groups between Earth's interior and its surface.

Figure 1.13 The Rock Cycle (continued)

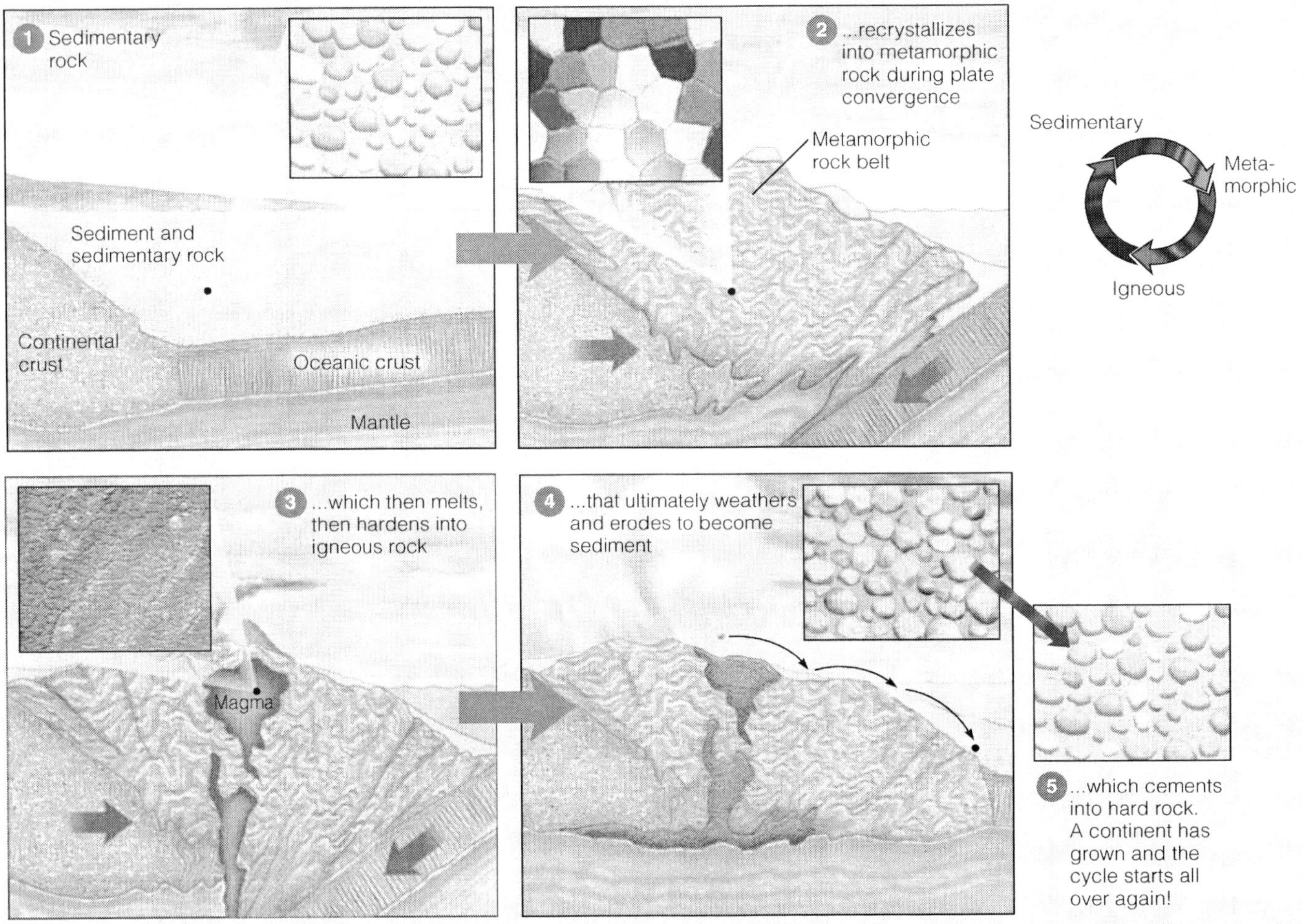

1.9 Geologic Time and Uniformitarianism

■ *Why is it important to understand how geologic time differs from the human perspective of time?*

An appreciation of the immensity of geologic time is central to understanding the evolution of Earth and its biota. Indeed, time is one of the main aspects that sets geology apart from the other sciences, except astronomy. Most people have difficulty comprehending geologic time because they tend to think in terms of the human perspective—seconds, hours, days, and years. Ancient history is what occurred hundreds or even thousands of years ago. When geologists talk of ancient geologic history, however, they are referring to events that happened hundreds of millions or even billions of years ago. To a geologist, recent geologic events are those that occurred within the last million years or so.

It is also important to remember that Earth goes through cycles of much longer duration than the human perspective of time. Although they may have disastrous effects on the human species, global warming and cooling are part of a larger cycle that has resulted in numerous glacial advances and retreats during the past 1.8 million years. Because of their geologic perspective on time and how the various Earth subsystems and cycles are interrelated, geologists can make valuable contributions to many of the current environmental debates, such as those involving global warming and sea-level changes.

The **geologic time scale** subdivides geologic time into a hierarchy of increasingly shorter time intervals; each time subdivision has a specific name. The geologic time scale resulted from the work of many 19th-century geologists who pieced together information from numerous rock exposures and constructed a chronology based on changes in Earth's biota through time. Subsequently, with the discovery of radioactivity in 1895 and the development of various

Figure 1.14 Hand Specimens of Common Igneous, Sedimentary, and Metamorphic Rocks

Andrew J. Martinez/Photo Researchers, Inc.

(a) **Granite,** an intrusive igneous rock.

Joyce Photographics/Photo Researchers, Inc.

(b) **Basalt,** an extrusive igneous rock.

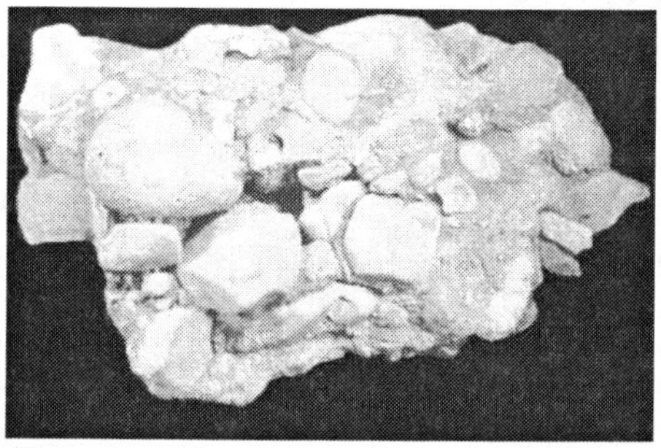

Sue Monroe

(c) **Conglomerate,** a sedimentary rock formed by the consolidation of rounded rock fragments.

Charles D. Winters/Photo Researchers, Inc.

(d) **Limestone,** a sedimentary rock formed by the extraction of mineral matter from seawater by organisms or by the inorganic precipitation of the mineral calcite from seawater.

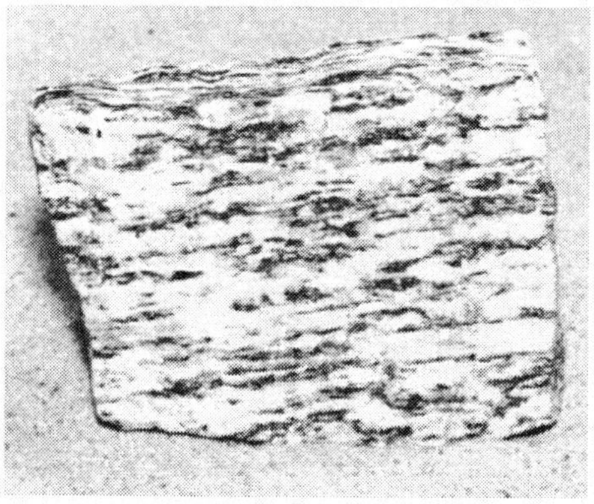

Andrew J. Martinez/Photo Researchers, Inc.

(e) **Gneiss,** a foliated metamorphic rock.

A. J. Copley/Visuals Unlimited

(f) **Quartzite,** a nonfoliated metamorphic rock.

radiometric dating techniques, geologists have been able to assign numerical ages (also known as absolute ages) in years to the subdivisions of the geologic time scale (▶ Figure 1.16).

■ *What is the principle of uniformitarianism, and why is it important in geology?*

One of the cornerstones of geology is the **principle of uniformitarianism,** which is based on the premise that present-day processes have operated throughout geologic time. Therefore, to understand and interpret geologic events from evidence preserved in rocks, we must first understand present-day processes and their results. In fact, uniformitarianism fits in completely with the system approach we are following for the study of Earth.

Uniformitarianism is a powerful principle that allows us to use present-day processes as the basis for interpreting the past and for predicting potential future events. We should keep in mind, however, that uniformitarianism does not exclude sudden or catastrophic events such as volcanic erup-

PHYSICAL GeologyNow™ ▸ **Geo-focus Figure 1.15 Plate Tectonics and the Rock Cycle** Plate movement provides the driving mechanism that recycles Earth materials. The cross section shows how the three major rock groups—igneous, metamorphic, and sedimentary—are recycled through both the continental and oceanic regions. Subducting plates are partially melted to produce magma, which rises and either crystallizes beneath Earth's surface as intrusive igneous rock or spills out on the surface, solidifying as extrusive igneous rock. Rocks exposed at the surface are weathered and eroded to produce sediments that are transported and eventually lithified into sedimentary rocks. Metamorphic rocks result from pressure generated along converging plates or adjacent to rising magma.

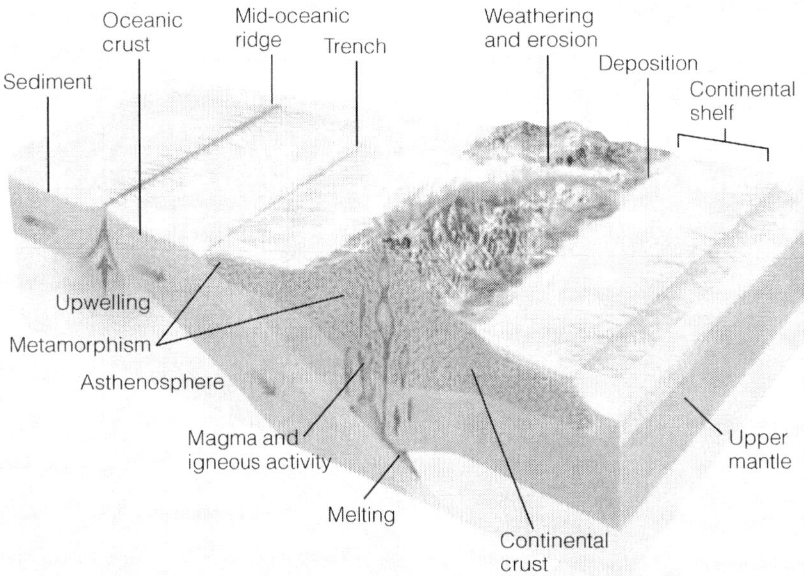

tions, earthquakes, tsunami, landslides, or floods. These are processes that shape our modern world, and some geologists view Earth history as a series of such short-term or punctuated events. This view is certainly in keeping with the modern principle of uniformitarianism.

Furthermore, uniformitarianism does not require that the rates and intensities of geologic processes be constant through time. We know that volcanic activity was more intense in North America 5 to 10 million years ago than it is today, and that glaciation has been more prevalent during the last several million years than in the preceding 300 million years.

What uniformitarianism means is that even though the rates and intensities of geologic processes have varied during the past, the physical and chemical laws of nature have remained the same. Although Earth is in a dynamic state of change and has been ever since it formed, the processes that have shaped it during the past are the same ones operating today.

What Would You Do?

Because of budget shortfalls and in an effort to save money, your local school board is considering eliminating the required geology course all students must pass to graduate from high school. As a concerned parent and citizen, you have rallied community support to keep geology in the curriculum, and you will be making your presentation to the school board next week. What will be your arguments to keep geology as part of the basic body of knowledge all students should have when they graduate from high school? Why should students have a working knowledge of geology? What arguments would you expect from the school board in favor of eliminating geology from the curriculum, and how will you counter their reasons?

Section 1.9 Summary

- Time is one aspect that sets geology apart from most other sciences, and an appreciation of the immensity of geologic time is crucial to understanding the evolution of Earth and its biota.

- The geologic time scale subdivides geologic time into intervals of various duration, with names and numerical age dates in years assigned to each interval.

- The principle of uniformitarianism states that present-day processes have operated throughout geologic time. It is one of the cornerstones of geology because it provides us with the basis for interpreting the past and predicting possible future events.

PHYSICAL Geology Now™ ▶ **Geo-focus Figure 1.16 Geologic Time Scale** Numbers to the right of the columns are ages in millions of years before the present. Photographs show characteristic life-forms for the indicated time interval. Dates are from Gradstein, F., Ogg, J., and Smith, A., *A Geologic Time Scale 2004* (Cambridge, UK: Cambridge University Press, 2005), Figure 1.2.

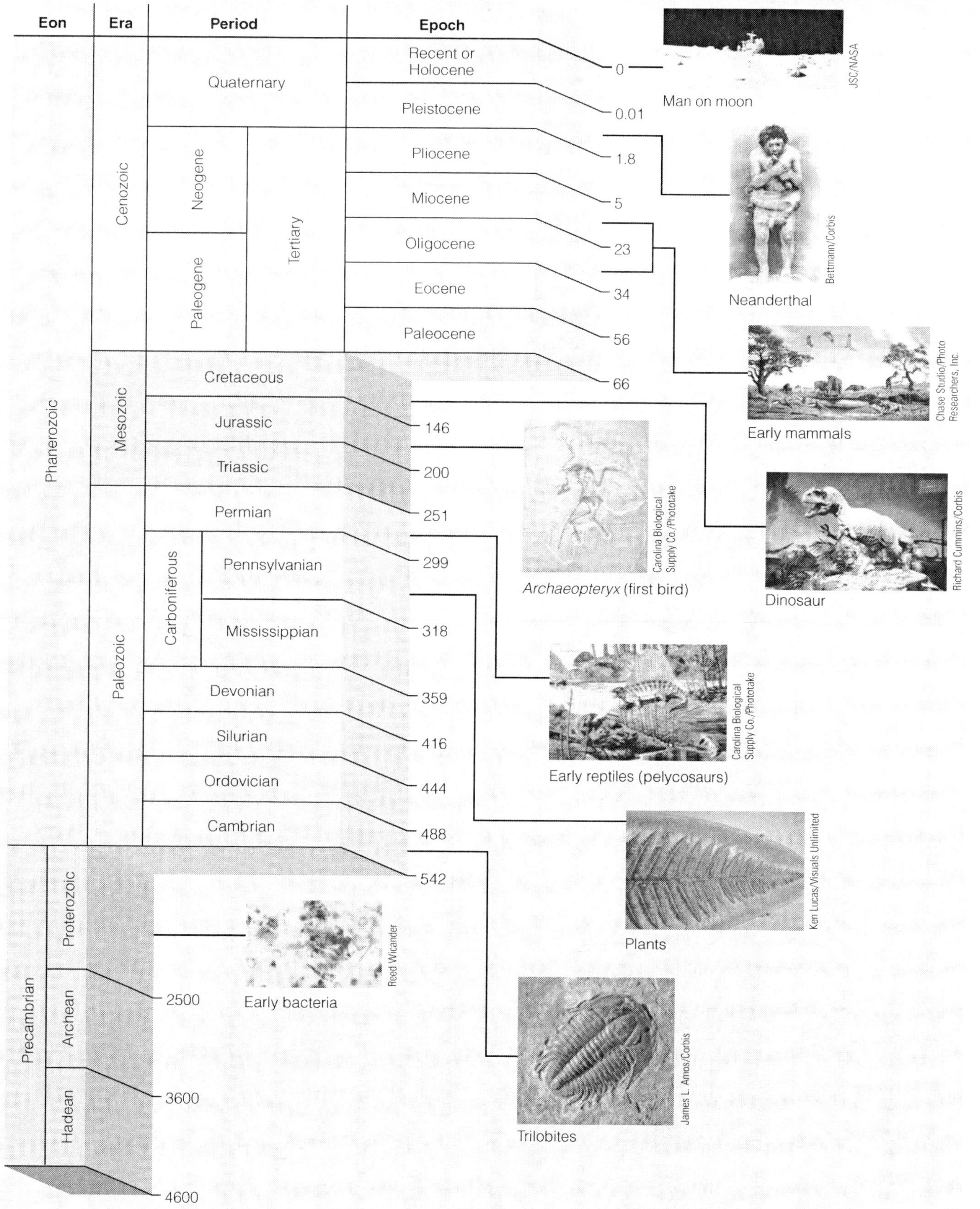

Review Workbook

ESSENTIAL QUESTIONS SUMMARY

1.1 Introduction

■ *What do natural disasters have to do with geology?*

Geology is an integral part of our lives, and natural disasters such as earthquakes, volcanic eruptions, tsunami, landslides, and floods are geologic phenomena that affect everyone. The more we understand how these phenomena work, the better prepared we can be to minimize their destructive effects and lessen the loss of life.

■ *What is a system?*

A system is a combination of related parts, components, or subsystems that interact in an organized fashion. Earth can be viewed as a complex system whose principal subsystems are the atmosphere, biosphere, hydrosphere, lithosphere, mantle, and core. The interaction of these subsystems has resulted in a dynamically changing planet in which matter and energy are continuously recycled into different forms.

1.2 Geology in Our Everyday Lives

■ *What is geology and what do geologists do?*

Geology, the study of Earth, is divided into two broad areas—physical geology (the study of Earth materials) and historical geology (the study of the origin and evolution of Earth). Geologists engage in a variety of occupations, such as exploration for mineral and energy resources, to solve various environmental problems.

■ *How does geology relate to the human experience?*

Geology pervades our everyday lives, with references to it in the arts (Asher Brown Durand's painting *Kindred Spirits*), music (Ferde Grofé's *Grand Canyon Suite*), and literature (Percy B. Shelley's poem "Ozymandias"). Earth's surface features, or topography (such as mountain ranges and rivers), are a result of the interplay among its various subsystems and have dictated military tactics as well as political boundaries.

■ *How does geology affect our everyday lives?*

Geology affects all of us when natural disasters occur. Our economic and political well-being depends, in part, on the distribution and concentration of natural resources. And government plays an increasingly important role in environmental issues related to geology.

1.3 Global Geologic and Environmental Issues Facing Humankind

■ *What is considered the greatest environmental problem facing the world today?*

Most scientists would argue that overpopulation is the greatest problem facing the world today. Increasingly large numbers of people must be fed, housed, and clothed, with a minimal impact on the environment.

■ *How is the greenhouse effect related to global warming?*

The greenhouse effect is the retention of heat in the atmosphere, which results in an increase in the temperature of Earth's surface and atmosphere, thus producing global warming.

1.4 Origin of the Universe and Solar System, and Earth's Place in Them

■ *How did the universe begin?*

The universe began with a Big Bang approximately 15 billion years ago. Evidence for the Big Bang is that the universe is expanding and that the entire universe has a pervasive background radiation of 2.7 K above absolute zero (2.7 K = −270.3°C), which is thought to be the faint afterglow of the Big Bang.

■ *What is our solar system composed of, and what is its origin and history?*

Our solar system consists of the Sun, 9 planets, 101 known moons, asteroids, comets, meteorites, and interplanetary dust and gases. It formed from a rotating cloud of interstellar matter that condensed, collapsed under the influence of gravity, and flattened into a rotating disk from which the Sun, planets, and moons formed.

■ *How did Earth form?*

Earth formed from a swirling eddy of nebular material 4.6 billion years ago, accreting as a solid body and soon thereafter differentiating into a layered planet during a period of internal heating.

1.5 Earth as a Dynamic and Evolving Planet

■ *Why is Earth a dynamic planet?*

Earth has continuously changed during its 4.6-billion-year existence as a result of the interactions between its various subsystems and cycles.

■ *What are Earth's three concentric layers, and what are their characteristics?*

Earth's three concentric layers are the core, mantle, and crust. The core consists of a small, solid inner region and a larger, seemingly liquid, outer portion, both of which are thought to be composed of iron and a small amount of nickel. The mantle, composed largely of peridotite, surrounds the core and is divided into a solid lower mantle, an asthenosphere that behaves plastically and flows slowly, and a solid upper mantle. The outermost layer, the crust, is divided into thick continental crust and thin oceanic crust.

1.6 Geology and the Formulation of Theories

■ *What is a theory?*

A theory is a testable explanation for some natural phenomenon that has a large body of supporting evidence. It is arrived at through the scientific method, which involves gathering and analyzing facts, formulating hypotheses to explain the phenomenon, testing the hypotheses, and finally proposing a theory.

1.7 Plate Tectonic Theory

■ *What is plate tectonic theory?*

Plate tectonic theory states that the lithosphere is divided into rigid plates that diverge, converge, or slide past each other as they move over the asthenosphere.

■ *Why is plate tectonic theory important in geology?*
Plate tectonic theory is a unifying explanation for many geologic features and events, and provides a framework for interpreting the composition, structure, and internal processes of Earth on a global scale.

1.8 The Rock Cycle

■ *What is a rock?*
A rock is an aggregate of minerals.

■ *What is the rock cycle?*
The rock cycle illustrates the interactions between Earth's internal and external processes and how the three rock groups are interrelated.

■ *What are the characteristics of each of the three major rock groups?*
Igneous rocks result from the crystallization of magma or the consolidation of volcanic ejecta. Sedimentary rocks are typically formed by the consolidation of rock fragments, precipitation of mineral matter from solution, or compaction of plant or animal remains. Metamorphic rocks result from the alteration of other rocks, usually beneath Earth's surface, by heat, pressure, and chemically active fluids.

■ *How are the rock cycle and plate tectonics related?*
Plate movement is the driving mechanism of the rock cycle. Plate interaction determines, to some extent, which of the three rock groups will form.

1.9 Geologic Time and Uniformitarianism

■ *Why is it important to understand how geologic time differs from the human perspective of time?*
An appreciation of geologic time is essential to understanding Earth's evolution. Earth goes through cycles of much longer duration than the human perspective of time, and thus we must adjust our concept of time to incorporate the fact that Earth is more than 20,000 times as old as the human species.

■ *What is the principle of uniformitarianism, and why is it important in geology?*
This principle holds that the laws of nature have been constant through time and that the same processes operating today have operated in the past, though at different rates. Therefore, to understand and interpret geologic events from evidence preserved in rocks, geologists must first understand present-day processes and their results.

ESSENTIAL TERMS TO KNOW

asthenosphere (p. 13)
Big Bang (p. 10)
core (p. 13)
crust (p. 16)
geologic time scale (p. 21)
geology (p. 6)
hypothesis (p. 16)
igneous rock (p. 19)
Jovian planets (p. 12)
lithosphere (p. 16)
mantle (p. 13)
metamorphic rock (p. 19)
mineral (p. 19)
plate (p. 16)
plate tectonic theory (p. 17)
principle of uniformitarianism (p. 22)
rock (p. 19)
rock cycle (p. 19)
scientific method (p. 16)
sedimentary rock (p. 19)
solar nebula theory (p. 11)
system (p. 4)
terrestrial planets (p. 12)
theory (p. 16)

REVIEW QUESTIONS

1. Why is viewing the Earth as a system a good way to study Earth? Are humans a part of the Earth system? If so, what role, if any, do we play in Earth's evolution?
2. What is geology? Why is it important that everyone have a basic understanding of geology, even if they aren't going to become geologists?
3. In what ways does geology affect our everyday lives at the individual, local, and nation-state levels?
4. Why do most scientists think overpopulation is the greatest environmental problem facing the world today? Do you agree? If not, what do you think is the greatest threat to our existence as a species?
5. What is the Big Bang? What evidence do we have that the universe began approximately 15 billion years ago?
6. How does the solar nebula theory account for the formation of our solar system, its features, and evolutionary history?
7. Why is Earth considered a dynamic planet? What are its three concentric layers and their characteristics?
8. Why is plate tectonic theory so important to geology? How does it fit into a systems approach to the study of Earth?
9. Using plate movement as the driving mechanism of the rock cycle, explain how the three rock groups are related and how each rock group can be converted into a different rock group.
10. Explain how the principle of uniformitarianism allows for catastrophic events.

APPLY YOUR KNOWLEDGE

1. Discuss why an accurate geologic time scale is particularly important for geologists in examining global temperature changes during the past, and how an understanding of geologic time is crucial to the current debate on global warming and its consequences.
2. Based on what you know about landforms and topography, do you think an understanding of geology would be helpful in planning a military campaign against another country? What geologic factors or information would be useful to know? Can you think of any examples from your knowledge of history where geology played a role in the planning or execution of a military campaign? After you have completed your geology course, answer this question again and see whether your answer is any different from what it is now.

GEOLOGY MATTERS

GEOLOGY IN UNEXPECTED PLACES

A Bit of Egypt in Central Park, New York City, and London, England

Most tourists visiting Central Park in New York City probably don't expect to see a relic of Egypt's past rising above the trees. But that is exactly what you will find if you visit Cleopatra's Needle, located behind the Metropolitan Museum of Art (▶ Figure 1).

The ruler of Egypt, the Khedive Ismail Pasha, promised the United States in 1869 an obelisk in appreciation for its aid in the construction of the Suez Canal. The obelisk and its companion, which is now in London, were originally erected in front of the great temple of Heliopolis during the reign of Pharaoh Tuthmosis III (1504–1450 B.C.). They were later moved to Alexandria, where they remained until one was transported to London in 1878 and erected along the Thames on the Victoria Embankment (▶ Figure 2) and the other was transported to New York in 1880 and erected at its present site in 1881. Both are dubbed "Cleopatra's Needle," even though they have nothing to do with Cleopatra and their inscriptions celebrate the pharaohs of ancient Egypt.

Henry Gorridge, a lieutenant commander of the U.S. Navy, was assigned the task of transporting the second of the twin obelisks and its pedestal to New York in 1880. The obelisk arrived at the Quarantine Station in New York in July 1880, and then it took 112 days to transport it to its final destination in Central Park, where it was raised in early 1881 in front of more than 10,000 cheering New Yorkers.

What of the geology of the obelisks? Both are made from the red granite of Syene and have suffered the effects of weathering. The New York obelisk is 21.4 m tall and weighs 193 tons, whereas the London obelisk is 21.1 m high and weighs 187 tons. Because of air pollution, the New York obelisk is not as well preserved as the one in London, and the inscriptions on the New York "Needle" are readable on only two sides.

Figure 1
Cleopatra's needle is located in New York's Central Park behind the Metropolitan Museum of Art. This obelisk, a gift from the Khedive Ismail, shows columns of hieroglyphic inscriptions on its sides. Chemical weathering, resulting from air pollution, has visibly damaged the obelisk. Also note more weathering and damage to the dower portion of the obelisk.

Lauree Feldman/Index Stock Imagery

Figure 2
The London obelisk was presented to Britain by the Viceroy of Egypt in 1819 and erected at its present location on the Victoria Embankment along the Thames River in London in 1878.

Michael Nicholson/Corbis

GEOLOGY IN FOCUS The Kyoto Protocol—An Update

At the 1997 Climate Treaty Meeting in Kyoto, Japan, a group of nations took the first steps to reduce greenhouse gas emissions by negotiating a treaty calling for the United States and 37 other industrialized nations to reduce their emissions by an average of 5.2% below their 1990 levels between 2008 and 2012. Although this might not seem like a significant decrease, it should be kept in mind that the industrialized nations, which make up only 20% of the world's population, have to date emitted 90% of all human-produced carbon emissions.

The Kyoto Protocol requires the largest polluters to make the greatest sacrifices. The United States must cut its greenhouse gas emissions to 7% below 1990 levels, and the European Union and Japan must reduce their emissions by 8% and 6%, respectively. The greenhouse gases that must be reduced are carbon dioxide, methane, nitrous oxide, hydrofluorocarbons, perfluorocarbons, and sulfur hexafluoride. It was also agreed that further reductions will be phased in after 2012.

Developing countries are not required to reduce their emissions of greenhouse gases unless they want to, and countries with extensive forests and jungles can receive credit toward their emission-reduction goals. Research indicates that the deforestation of large areas, particularly in the tropics, is also a cause of increased levels of carbon dioxide. This is because trees and plants use carbon dioxide in photosynthesis and thus act as carbon dioxide "sinks" by removing it from the atmosphere. With a decrease in the global vegetation cover, less carbon dioxide is removed from the atmosphere by photosynthesis, and carbon dioxide levels can increase.

Whereas a reduction in greenhouse gas emissions is certainly desirable, the economic impact on industrialized countries must also be considered. It is estimated that meeting the Kyoto Protocol reductions could cause as much as a 2% drop in the U.S. gross domestic production, which is equivalent to about $227 billion. Furthermore, the economic impact would not be spread evenly throughout the population, and there could be devastating effects on many industries, particularly those that produce and use large quantities of energy. Thus the possible benefits to be gained by reducing greenhouse gas emissions must be weighed against the possible economic hardships that meeting such reductions would entail (▶ Figure 1).

On February 16, 2005, the Kyoto Protocol went into effect. One hundred forty nations that produce 55% of the world's greenhouse gas emissions have ratified the Kyoto Protocol and pledged to reduce overall emissions by an average of 5.2% by 2012. Although the protocol officially ends in 2012, most observers think that it is the first step toward controlling global warming and that it will be extended with modifications, based on how successful it is in reducing overall greenhouse gas emissions.

Even though all the European Community members and Japan ratified the agreement, the world's largest polluter, the United States, refused to ratify the protocol, citing that it could cost millions of U.S. jobs. According to the United States, the Kyoto Protocol is flawed because it exempts large, developing countries, such as China (which also refused to ratify the protocol and is the world's second largest greenhouse gas producer), from having to reduce their pollution. Many of these developing and advanced developing countries, such as South Korea, Brazil, and India, are already concerned that if the protocol is extended beyond 2012, their economies may suffer when they must comply with emissions reductions.

Only time will tell how successful the Kyoto Protocol is in reducing greenhouse gas emissions, and what effect it will have on global warming and the global economy. It is a start, however, and as we know, Earth's climate is a complex system with many variables contributing to its evolution over a long period of time.

Figure 1
The scene of smokestacks at an oil refinery serves to focus on some of the issues relating to the recently ratified Kyoto Protocol and the debate concerning global warming. All industrial activity pollutes the environment to some extent, and the debate is, in part, how much human pollution is contributing to global warming. By requiring nations to reduce their greenhouse gas emissions, particularly those that produce and use large quantities of energy, it is hoped that such reductions will lessen the rate of global warming and its consequences. However, several industrialized as well as developing countries have refused to ratify the Kyoto Protocol because of the perceived potentially crippling effects it may have on their economies. Only time will tell whether the Kyoto Protocol is successful in stemming the increasing levels of greenhouse gas emissions resulting from industrialization.

Royalty-Free/Corbis

Geology AT WORK

Forensic Geology and Palynology

Although Sherlock Holmes, created by Sir Arthur Conan Doyle, is probably the best-known detective in the world, many people would be surprised to know that he also was a forensic geologist of sorts. In Doyle's novel *A Study in Scarlet,* we are introduced to Sherlock Holmes by his future biographer and companion, Dr. John H. Watson. Soon after moving into 221B Baker Street, Dr. Watson was still trying to figure out just what Sherlock Holmes did for a living. He knew Holmes was very knowledgeable in such subjects as botany, chemistry, and anatomy, and yet "of contemporary literature, philosophy and politics he appeared to know next to nothing." According to Dr. Watson, Sherlock Holmes's "knowledge of Geology [is]—Practical, but limited. Tells at a glance different soils from each other. After walks has shown me splashes upon his trousers, and told me by their colour and consistence in what part of London he had received them."

About a century later, Sarah Andrews, a geologist and author, has written 10 mysteries featuring fictional forensic geologist Em Hansen, who uses her knowledge of geology to solve mysteries and crimes. Do real-life forensic geologists exist, or are they only fictional characters? The answer is yes; they exist and are employed in a number of public and private laboratories in the United States as well as many other countries. In fact, the Federal Bureau of Investigation laboratory in Quantico, Virginia, employs several geologist/forensic scientists. In its Materials Analysis Unit, specialists perform a variety of geologic/mineralogic, metallurgic, and elemental analyses on evidence associated with various crimes and disasters. In the Mineralogy Subunit, tests can be done on virtually all naturally occurring geologic materials such as minerals, rocks, soil, and gemstones. In addition, human-made products like glass, abrasives, and building materials are analyzed because these products are derived from geologic materials and have characteristic signatures that may reveal their origin.

What do forensic geologists do? A forensic geologist identifies, analyzes, and compares such earth materials as soil, minerals, rocks, and fossils from a crime scene to determine where a particular material came from, when an incident might have taken place, and even what the cause of the incident was or who might be responsible. A solid grounding in the geologic sciences is essential, and a forensic geologist must know how to identify minerals, rocks, soils, and fossils; how to read and interpret geologic and topographic maps; and how to put all this information together so that it can be used in a court of law.

Just as Sherlock Holmes could tell where a person had been by the soil adhering to his shoes or clothing, the identification of soil found on a victim or suspect is probably the most common type of forensic geology practiced. There are numerous instances in the crime annals of soil or rock chips being used to help solve a case.

A good example involved the apparent accidental shooting death of John Dodson in the Uncompahgre Mountains of western Colorado on October 15, 1995. According to her testimony, his wife, Jane, had returned to their hunting camp to find that John had been shot three times. Jane's ex-husband, J. C. Lee, was also camped near the Dodson hunting camp at the time. He naturally became a prime suspect, but he reported that he had been hunting with his boss far from the camp when the shooting occurred, and furthermore someone had stolen his .308-caliber rifle and a box of cartridges from his tent while he was hunting.

Figure 1
Bentonite, a clay brought in to stop water from seeping out of this cattle pond in western Colorado's Uncompahgre Mountains, provided critical geologic evidence that led to the conviction of Janice Dodson for the murder of her husband.

Bill Booth/Office of the District Attorney/Grand Junction, CO

A cattle pond near the campsite of J. C. Lee provided the evidence that eventually convicted Jane Dodson in the shooting death of her husband (▶ Figure 1). The pond was lined with bentonite, a clay not found in the area but brought in to stop water from seeping out of the bottom of the pond. In her initial statement to the police, Jane reported that she had stepped into a mud bog near her camp and had removed her coveralls, which were caked with mud. Investigators tested the mud from the cattle pond near J. C. Lee's camp, the mud in the bog near the Dodson's camp, and the mud on Jane's clothing. The forensic evidence indicated the mud on Jane's coveralls was consistent with mud from the cattle pond, and not consistent with mud near the Dodson's camp. This meant that Jane Dodson was at her ex-husband's camp around the time his rifle (which was the same caliber as a shell casing and bullet found near the body) was stolen. A jury later convicted Jane Dodson of the murder of her husband, and she is serving a life sentence in Colorado's state prison for women.

Another intriguing story, but this time involving fraud, is the collapse of Bre-X, a Canadian mining company that supposedly discovered the world's largest gold deposit at Busang, Indonesia. According to company releases in 1995, the reserves were estimated to be 200 million ounces of gold (approximately 8% of the world's gold), worth more than $70 billion at that time! Unfortunately, very little gold actually existed at Busang, and with that revelation, Bre-X collapsed in 1997. Instead of millions of ounces of gold, the core samples on which projections were based had been "salted" with outside gold.

How did such a massive fraud happen? From the beginning, no outside analyses of core samples at the mine were allowed. Only the analyses performed by the company hired by Bre-X were released to the public. Furthermore, standard practices of core analysis were not followed. Typically, a core is cut lengthwise and half the sample is analyzed, leaving the other half as a backup to verify the initial results should that be necessary. In the case of Bre-X, the entire core was pulverized and handled only by company geologists and the company hired to do the analysis. Thus, no material was available for independent analysis. Only when another company brought in at the insistence of the Indonesian government to run the mine drilled its own test holes was the fraud discovered. Test results showed that the original samples had been salted with gold dust from elsewhere. If standard procedures had been followed in splitting the original samples, the fraud might never have happened. The collapse of Bre-X is one of the largest stock scandals in Canadian history, and it resulted in numerous lawsuits from investors who lost billions of dollars.

A field related to forensic geology is forensic palynology, which is the study of pollen and spores to help solve criminal cases. Pollen is the microscopic grains produced by the male element of a flowering plant; it is what causes allergy sufferers so much discomfort. Pollen also can help crime investigators determine where someone or something has been; trace the origin of items like packages, letters, antiques, and drugs; and corroborate or discredit a suspect's alibi. Forensic palynologists generally work as independent consultants or for private laboratories.

Despite its potential, forensic palynology is still underutilized by law enforcement. An example of how palynology can be useful is a case worked on by Dr. Lynne Milne, who teaches at the Centre for Forensic Science at the University of Western Australia. During a period of several years, a number of rapes and assaults had occurred in the Perth area of Western Australia, and the police believed the crimes were committed by the same person, possibly an itinerant worker. To help the police narrow down their search area, Dr. Milne collected pollen and other samples from slits in the shoe soles the rapist had allegedly left near the scene of an assault in 1995. In addition to pollen that was common to the area, pollen grains produced by cereals were also found, leading police to theorize that the suspect had been walking on the stubble left from cutting hay. Hay of the type that produces the rare pollen grains found in Dr. Milne's samples is very common in the chaff-cutting businesses located near a particular town in Western Australia. Because this industry employs itinerant workers, police concentrated their efforts in this area and several weeks later arrested a suspect who confessed to the rapes and assaults. Though not directly linking the suspect to the crime, the analysis of pollen from the crime scene enabled police to narrow their search for the perpetrator.

GEOLOGY IN YOUR LIFE

Global Warming and Climate Change, and How They Affect You

The greenhouse effect, global warming, climate change: These headlines and topics are in the news all the time, global issues that affect us all and the planet we live on. But just how will global warming and the resultant climate change personally affect you? Are they something that you should really be concerned about? After all, there are exams to worry about, graduation, finding a job, and that doesn't even include the everyday issues we all must deal with, not to mention your personal life. Yet part of the college experience is examining and debating the "big picture" and issues facing society today. So what about global warming and you?

Figure 1
(a) Map showing the location of the Maldives Islands in the Indian Ocean. **(b)** Aerial view of the Maldives Islands. A rise in sea level due to global warming could easily submerge these islands.

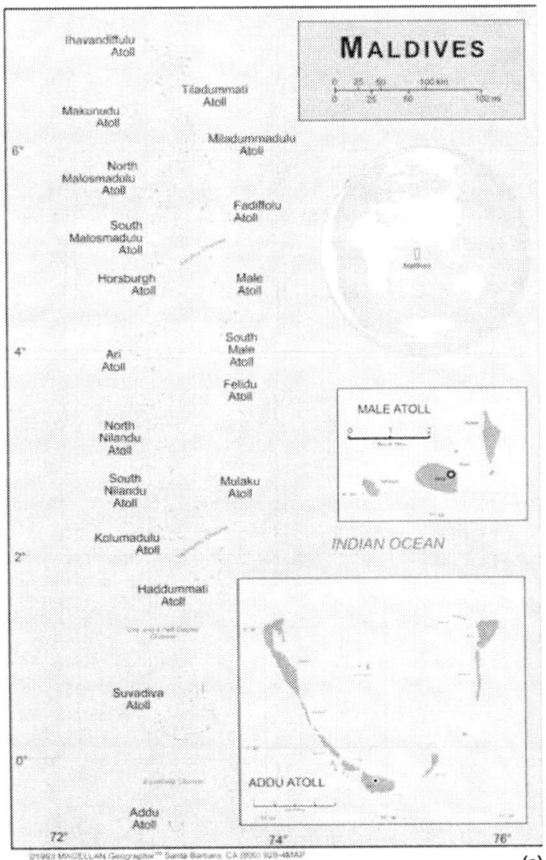

MAPS.com/Corbis

(a)

Volvox/Index Stock Imagery

(b)

You may recall that we talked about the greenhouse effect and its relationship to global warming. The greenhouse effect helps regulate Earth's temperature because as sunlight passes through the atmosphere, some of the heat is trapped in the lower atmosphere and not radiated back into space, thus effectively warming Earth's surface and atmosphere. The issue is not whether we have a greenhouse effect, because we do, but the degree to which human activity, such as the burning of fossil fuels, is increasing the greenhouse effect and thus causing global warming.

Based on many studies using a variety of techniques, it is clear that carbon dioxide (one of the greenhouse gases that allows short-wavelength solar radiation to pass through it but traps some of the long-wavelength radiation that is reflected back from Earth's surface) levels have increased since the Industrial Revolution in the 19th century. Furthermore, global surface temperatures have increased about 0.6°C since the late 1800s and about 0.4°C during the past 25 years. However, this warming trend has not been uniform around the globe, and some areas, such as the southwestern part of the United States, have actually cooled during this time period.

One thing we must be careful about is mistaking regional trends for global trends. For example, there is compelling evidence of climate variability or extremes on a regional scale, but on a global scale, there is currently little evidence of a sustained trend in climate variability or extremes. That doesn't mean we can ignore the overall increase in average global surface temperatures because, if left unchecked, such changes can have significant environmental, ecological, and economic effects.

So what are some of the effects we should be worried about? For starters, there is the problem of rising sea level. During the past 100 years, global sea level has been rising at an average rate of 1 to 2 mm per year, and the projected rate of increase by 2100 is anywhere from 9 to 88 cm, depending on which climate model is used. What this means is that low-lying coastal areas will experience flooding and increased erosion along the coastline, endangering housing and communities. For instance, about 17 million people live less than 1 m above sea level in Bangladesh, and they are certainly at risk due to rising sea level. Furthermore, many major cities are just above sea level and could also suffer from rising sea level. A rise of sea level at the upper end could completely submerge such island nations as the Maldives in the Indian Ocean (▶ Figure 1).

Based on various climate computer models and taking into account the complexity and variability of the atmospheric–oceanic system, most predictions show Earth's average surface temperature increasing by 1.4° to 5.8°C during the period 1990–2100. This increase will result in widely varying regional responses; land areas will warm more and faster than ocean areas, particularly in the high latitudes of the Northern Hemisphere. Expect to see more hot days and heat waves over nearly all the land areas, with more droughts in all continental interiors during the summer (▶ Figure 2). There will also be increased precipitation during the 21st century, particularly in the northern middle to high latitudes. And the glaciers and ice caps will continue to retreat, with a decrease in the Northern Hemisphere's snow cover and sea ice.

What do all these predictions mean to you? With increasingly hot summers and more frequent droughts, expect to see higher food prices as crop yields decrease. There will also be an increased risk of wildfires. Deadly heat waves will result in more heat-related deaths, such as occurred in Europe during the summer of 2003. As climates change, diseases such as malaria will spread easily to areas with warmer, wetter climates. Disease-carrying mosquitoes will extend their reach as climate changes allow them to survive in formerly inhospitable regions. Higher temperatures will affect regional water supplies, creating potential water crises in the western United States within the next 20 years as well as in other areas such as Peru and western China. Just as many regions will experience longer and hotter summers, other areas will suffer from intense and increased rainfall, which will result in severe floods and landslides.

Everyone is vulnerable to weather-related disasters, but large-scale changes brought about by climate change will affect people in poor countries more than those in the more industrialized countries. However, whether these climate changes are part of a natural global cycle taking place over thousands or hundreds of thousands of years—that is, on a geologic time scale—or are driven, in part, by human activities is immaterial. The bottom line is that we already are, or eventually will be, affected in some way, be it economic or social, by the climate changes that are taking place.

Figure 2
Withered corn crop as a result of drought conditions.

Richard Hamilton Smith/Corbis

Getting Around in Geology

2

About five billion years ago an interstellar cloud of gas and dust began to collapse, probably under the force of its own gravity. As it shrank it started spinning, which caused the cloud to flatten into a disc shape. Eventually, the compression accompanying the collapse created temperatures and pressures in the center of the cloud sufficient to initiate nuclear reactions. Hydrogen atoms fused, forming helium and releasing enormous energy. Our primordial sun began to shine. Around the sun a cooler disc, the solar nebula, contained the elemental leftovers of the solar implosion that eventually would form the planets. Low-boiling-point compounds such as methane, ammonia, and various ices condensed in the cold far reaches of the nebula to form the outer planets, the very large *Jovian planets.* Closer to the sun, metallic and rocky substances with high melting points condensed and collided to form the four dense inner planets, the *terrestrial planets.* Of that group, the third and fourth planets from the sun are most interesting to geologists.

On January 3 and January 24, 2004 two spacecraft landed on the surface of Mars, each carrying a remote-controlled vehicle, the Martian Exploration Rovers "Opportunity" and "Spirit." Directed remotely by scientists at Cornell University and California's Jet Propulsion Laboratory, few thought that the vehicles would last much longer than the three months they were designed

NASA/JPL/Cornell

◆ FIGURE 1 Panorama taken by Spirit Rover on April 18, 2006 looking toward Husband Hill on the horizon. The smooth patch in the middle distance in front of Husband Hill is Home Plate. Note the Rover's vehicle tracks closer by. The Rover was positioned in its winter station at this time, patiently recharging its solar batteries in the low light of the long Martian winter.

to survive under the harsh conditions of the Martian surface. But over two years later they are still operating and have become one of the most successful space missions ever undertaken. Each of these robotic "mechanical geologists" has traveled over 7 km (4.2 mi) across the planet's stony desert landscape, with special attention given to outcrops of *sedimentary rock*—layered rock formed by deposition of eroded debris carried by wind and water or by evaporation and biologic activity. The controllers of Opportunity and Spirit are looking for definite evidence that Mars once had persistent bodies of water, such as we find on Earth today. Where there was once standing water, there once was likely to have been life—a tantalizing possibility!

Navigating the Martian landscape has required the naming of numerous geographical features, including "Home Plate," a patch of level ground that looks like part of a giant baseball diamond from Mars orbit (◆ Figure 1). Following up on the baseball theme, scientists have named the most spectacular outcrops near Home Plate after famous African-American baseball players including David Barnhill, a former pitcher for the New York Cubans, and "Bullet Joe" Rogan. These outcrops consist of layers formed by falling ash and jagged blocks of ejected rock tossed through the air. They indicate that an ancient volcanic eruption or large meteorite strike occurred nearby.

The evidence looks strong that Mars once had lakes, streams, even small seas. At Erebus Crater, Opportunity discovered sandstone showing cross-beds and drying (desiccation) cracks. From research on Earth we know that such sedimentary features indicate the presence of an ancient water body big enough to have had a current flowing through it. The water body later evaporated.

Elsewhere, Spirit has plowed up Martian soil in two places rich in crystals of iron sulfate, a salt mineral, which would most easily have accumulated by evaporation of salty water (◆ Figure 2). Many of the sedimentary rocks seen so far also contain tiny granules of iron, which probably formed as water soaked through the rocks. The most abundant form of iron is the mineral hematite, which crystallizes under oxygen-rich conditions.

The emerging picture provided by careful examination of rocks and minerals is that Mars once was a much warmer world with lots of violent geological processes and a much denser atmosphere. Water was abundant in many places; but then the Martian air thinned out and grew cold and dry. Where did all of the water go? What became of any life that might have existed? The environmental history of Mars is a story of extreme climate change taking place on a world in some ways once very similar to our own.

To answer all of the questions raised will require future geologic study of our neighbor by explorations yet to be planned. It is likely that we'll develop a clearer understanding of Earth's own long history in the process.

◆ FIGURE 2 A bed of iron-rich salt accidentally trenched up beneath a thin layer of red, wind-blown Martian dust by Spirit Rover on January 20, 2006. The trench is about 30 cm wide

The Earth System

Earth functions like a finely tuned machine, a natural, self-maintaining spaceship for its inhabitants, constantly rotating about its polar axis at 1670 km/hr and revolving around the Sun at 107,000 km/hr. Just as the elements of human physiological systems are connected and interdependent, so on Earth are there linkages and interactions between its surface of rock and soil (*lithosphere*), the air (*atmosphere*), bodies of water and ice (*hydrosphere*), and its living realm (*biosphere*). Just as a healthy human body depends upon the proper circulation of blood and other fluids and solids within, so does a "healthy Earth"—one habitable for life at least—depend upon a proper circulation of matter and energy between each of its major natural components; lithosphere, atmosphere, hydrosphere, and biosphere.

One example of an important but little-appreciated interconnection is the one that exists between dust storms in the Gobi and Takla Makan deserts in China, life in the North Pacific Ocean, and Earth's climate. Airborne silt from those deserts is seasonally carried all the way to the North Pacific where it increases the nutrient content and fertility of this otherwise barren stretch of ocean. The blooming of marine plankton in response to the dust fertilization subtracts a large amount of carbon dioxide from the atmosphere, thanks to photosynthesis, which in turn keeps the air from getting as warm as it might, because carbon dioxide is a greenhouse gas. Average atmospheric temperature has a significant impact on climate worldwide, and it turns out that floating microscopic marine life such as plankton is important for making our climate comfortable.

A *system* is simply a set of components or parts that work together to perform a particular function. And all systems, including the Earth System, require energy to operate. The human body acquires its energy from eating food. An automobile requires fuel, or electricity to function. A transportation system—the network of streets and interchanges required to move automotive traffic around a city—requires routine street maintenance and the power for traffic signals to work.

The Earth System needs energy to work as well, and this comes from two sources: one *internal*, the other *external* (◆ Figure 2.1). Inside the planet's deep interior, uranium, thorium, and to a lesser extent potassium atoms are continuously decaying into more stable, lighter-weight elements. A by-product of this *radiogenic decay* is heat, which gradually escapes to Earth's surface. We see manifestations of this escaping internal heat directly expressed as volcanoes and hot springs. Some of this heat converts to mechanical energy, causing earthquakes, mountain building, and the slow shifting of massive tectonic plates (see Chapter 3).

By far more noticeable to us at the surface, however, is the energy coming from Earth's external heat source—the Sun. Without the daily input of solar radiation, our planet would be a frozen, stony ice ball, despite all of the heat slowly escaping from its interior. There would be no wind, rainfall, or circulating surface water (see Chapter 8). Since the biosphere critically depends upon sunlight for photosynthesis and various other vital processes, so too most life as we know it would be nonexistent.

The interface between the external and internal sources of energy is Earth's surface (◆ Figure 2.2). Soil—an essential factor in human carrying capacity—is a result of the conflicting processes, one set internally driven, the other set externally driven, which come together at the surface. Wind, rain, ice, snow, and chemical corrosion break down the land that

D. D. Trent

◆ FIGURE 2.1 Lone Pine Canyon in Sequoia National Park, California. This landscape is the product of mountain uplift powered by Earth's internal heat engine, interacting with erosion by glaciers, weathering, and running water powered by Earth's external heat engine, the Sun. Contrast this image with Figure 2.1. How generally do these landscapes differ? (Mars has a less active internal heat engine than Earth, is farther from the Sun, and as far as we know, has no life.)

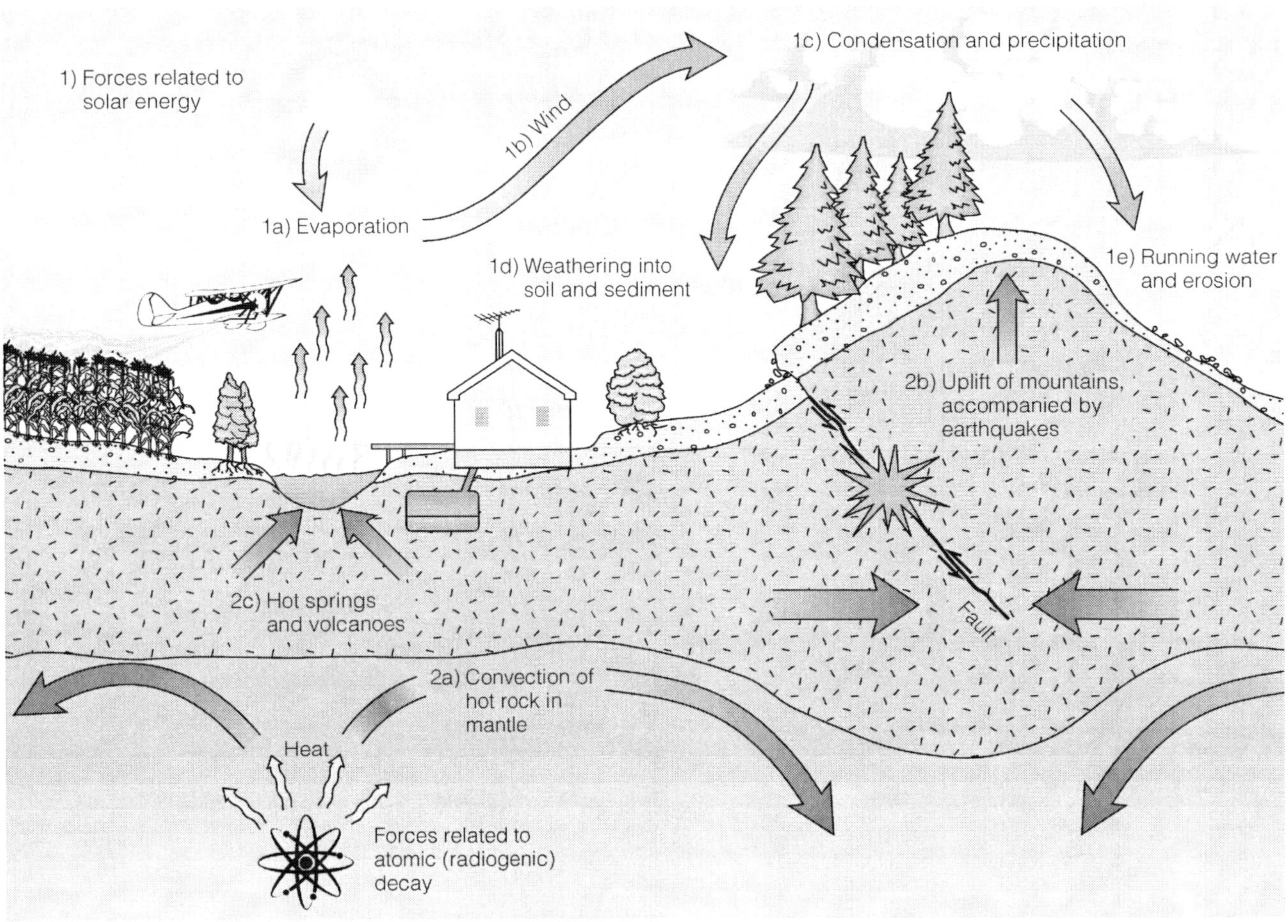

◆ FIGURE 2.2 Earth's surface is a resolution of two sets of forces, modified by the activity of life.

earthquakes, volcanic eruptions, sedimentation, and mountain building gradually construct. The form of the landscape on which we live may be thought of as a compromise in the never-ending tug-of-war of opposing forces. Biological activity modifies the ways in which these forces operate. For example, a dense cover of vegetation will inhibit erosion and sedimentation, while a landscape denuded by logging and overgrazing quickly wears way, as Aldo Leopold observed (see Chapter 1).

In this chapter we're going to take a close look at the raw material that makes up Earth's lithosphere. This is essential background for all the chapters to follow, and will provide some insight into the framework of knowledge all geologists hold in common. As the Martian example above illustrates, rocks and minerals can tell us much about a planet's changing environment. Moreover, particular kinds of rocks and minerals can provide us great environmental benefits in some cases while causing great harm in others. We begin by looking at the smallest building blocks of the Earth System: atoms, ions, and isotopes. Then we'll progress to higher states of organization and put it all together in discussing Earth's age.

Earth Materials

Elements, Atoms, and Atomic Structure

Elements are substances that cannot be changed into other substances by normal chemical methods. They are composed of infinitesimal particles called **atoms.** In 1870 Lothar Meyer published a table of the 57 then-known elements arranged in the order of their atomic weights. Meyer left blank spaces in the table wherever elements of particular weights were not known. About the same time Dmitry Mendeleyev developed a similar table, but his table was based upon the similar chemical properties of the particular known elements. Where an element's placement based on weight did not group it with elements of similar properties, Mendeleyev did not hesitate to suggest that its weight had been measured incorrectly. He also predicted the properties of the "missing" elements, and very shortly his predictions proved correct with the discovery of the elements gallium, scandium, and germanium (Appendix 1).

The weight of an atom of an element is contained almost entirely in its **nucleus,** which contains protons (atomic weight = 1, electrical charge = 1^+) and neutrons (atomic weight = 1, electrical charge = 0). An element's **atomic number** is the number of protons in its nucleus, and this number is unique to that element. The weight of an atom—its **atomic mass**—is the sum of its nuclear protons and neutrons. For example, $^{4}_{2}He$ denotes the element helium, which has 2 protons (denoted by the subscript) and 2 neutrons, for an atomic mass of 4 (superscript). Electrical neutrality of the atom is provided by balancing the positive proton charges by an equal number of negatively charged electrons orbiting in shells around the nucleus (◈ Figure 2.3). **Ions** are atoms that are positively or negatively charged due to a loss or gain of electrons (e^-) in the outer electron shell. Thus, sodium (Na) may lose an electron in its outer shell to become sodium ion (Na^+, a *cation*), and chlorine (Cl) may gain an electron to become chloride ion (Cl^-, an *anion*). *Valence* refers to ionic charge, and sodium and chlorine are said to have valences of 1^+ and 1^-, respectively. Elemental sodium is an unstable metallic solid, and chlorine is a poisonous gas. When ionized, they may combine in an orderly fashion to form the mineral halite (NaCl), common table salt, a substance necessary for human existence (◈ Figure 2.4).

Isotopes are forms (or species) of an element that have different atomic masses. The element uranium, for instance, always has 92 protons in its nucleus, but varying numbers of neutrons define its isotopes. For example, Uranium-238 (^{238}U) is the common naturally occurring isotope of uranium. It weighs 238 atomic mass units and contains 92 protons and 146 (238 *minus* 92) neutrons. Uranium-235 is the rare isotope (0.7% of all uranium) that has 143 neutrons. Similarly, ^{12}C is the common isotope of carbon, but ^{13}C and ^{14}C also exist. A modern periodic table of the elements that shows their atomic number and mass appears in Appendix 1.

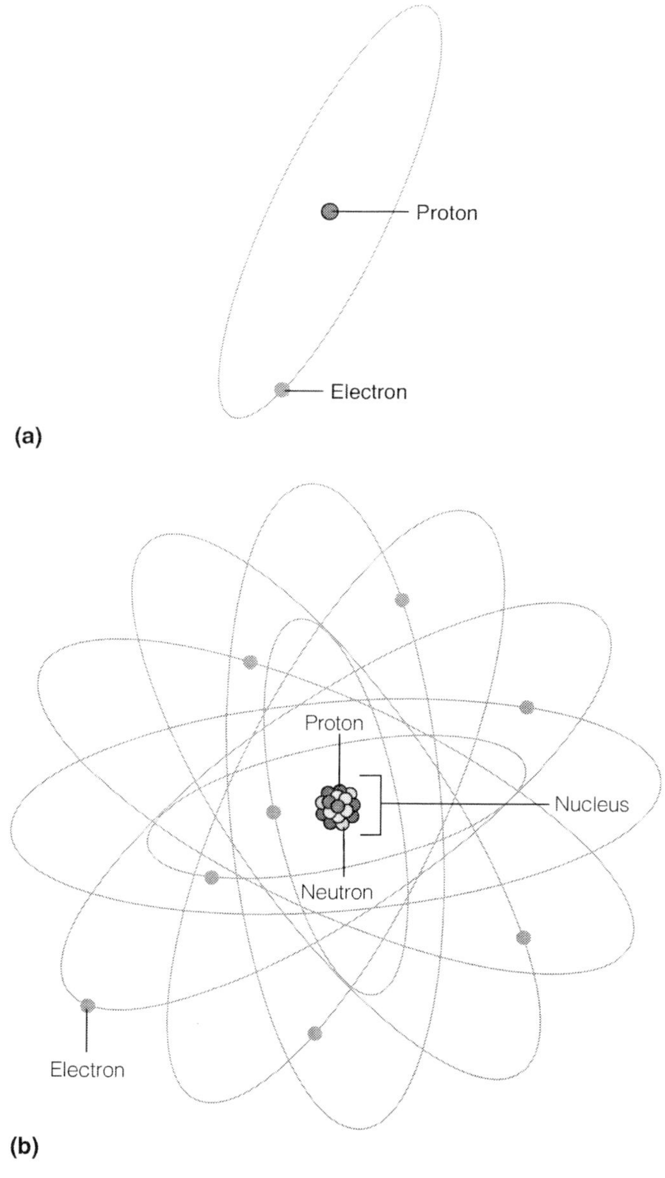

◈ FIGURE 2.3 Structures of (a) a hydrogen atom, in which an electron orbits a single proton, and (b) an oxygen atom, in which 8 electrons orbit a dense nucleus containing 8 protons and 8 neutrons.

Minerals

Earth's crust is composed of rocks, which are made up of one or more minerals. Although most people seldom think about minerals that do not have nutritional importance or great beauty and value such as gold and silver, the mineral kingdom encompasses a broad spectrum of chemical compositions. **Minerals** are defined as *naturally occurring, inorganic, crystalline* substances, each with a narrow range of *chemical compositions* and characteristic *physical properties.* Thus, neither artificial gems such as zirconia nor organic deposits such as coal and oil are minerals by definition. The crystal form of a mineral reflects an orderly atomic structure, which in turn determines the crystal shape. Solids that have random or noncrystalline atomic structures—glass and opal, for example—are described as **amorphous.** Minerals composed of a single element, such as copper and carbon (as in graphite and diamonds), are known as *native elements.* Most minerals, however, are composed of combinations of two or more elements. Although more than 3,000 minerals have been named and described, only about 20 make up the bulk of Earth's crustal rocks. The chemistry of minerals is the basis of their classification, and the most common minerals are primarily composed of the elements oxygen (O), silicon (Si), aluminum (Al), and iron (Fe). This is not surprising, because oxygen and silicon compose 75 percent of crustal rocks by weight. A memory aid for these elements in order is "OSAF."

Geologists classify minerals into groups that share similar negatively charged ions (anions) or ion groups (radicals).

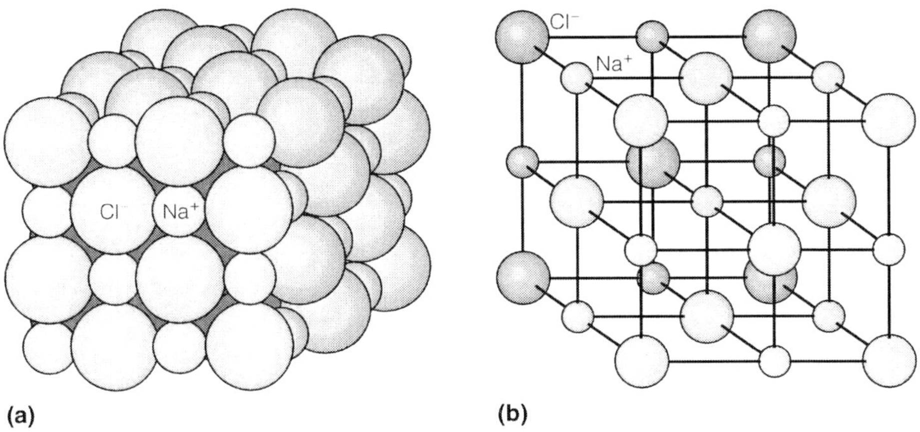

◆ FIGURE 2.4 Two models of the atomic structure of the mineral halite (NaCl), common table salt. (a) A packing model that shows the location and relative sizes of the sodium and chloride ions. (b) A ball-and-stick model that shows the mineral's cubic crystal structure.

Oxygen (O^{-2}) may combine with iron to form hematite (Fe_2O_3), a member of the *oxide group* of minerals. The *sulfide minerals* typically are combinations of a metal with sulfur. Examples include galena (lead sulfide, PbS), chalcopyrite (copper-iron sulfide, $CuFeS_2$), and pyrite (iron sulfide, FeS_2), the "fool's gold" of inexperienced prospectors. *Carbonate minerals* contain the negatively charged $(CO_3)^{-2}$ ion. Calcite ($CaCO_3$), the principal mineral of limestone and marble, is an example. Table 2.1 summarizes some of the mineral groups most important to geologists.

The most common and important minerals are the *silicates,* which are composed of combinations of oxygen and silicon with or without metallic elements. The basic building block of a silicate mineral is the silica tetrahedron (◆ Figure 2.5a). It consists of one silicon atom surrounded by four oxygen atoms at the corners of a four-faced tetrahedron. The manner in which the tetrahedra are packed or arranged in the mineral structure is the basis for classifying silicates. The tetrahedra may be arranged in layers or sheets, as in the clay minerals and mica; in long chains, as in the amphibole and pyroxene groups; and in three-dimensional networks, as in quartz and feldspar (◆ Figure 2.5b and ■ Case Study 2.1 on page 39).

Rock-Forming Silicates

Most rocks are aggregates of minerals, and feldspar is the most abundant of the silicate minerals in Earth's crust (◆ Figure 2.6). Feldspar is actually a mineral *group* that ranges in chemical composition from potassium-rich orthoclase found in granite (◆ Figure 2.7) to calcium-rich plagioclase found in basalt and gabbro. The ferromagnesian silicates are rich in iron and magnesium in addition to silicon and oxygen. Most of them are dark colored, either black, brown, or green (◆ Figure 2.8). Important among these are hornblende (amphibole), augite (pyroxene), biotite mica, and olivine, which is believed to be the most abundant mineral just below the outer crust of the planet (Table 2.2).

TABLE 2.1 Some Important Mineral Groups*

Mineral Group	Negatively Charged Ion or Ion Group	Examples	Composition
Carbonate	$(CO_3)^{-2}$	Calcite	$CaCO_3$
		Dolomite	$CaMg(CO_3)_2$
Halide	CL^{-1}, F^{-1}	Halite	NaCl
Hydroxide	$(OH)^{-1}$	Limonite	$FeO(OH) \cdot H_2O$
Native element	—	Gold	Au
		Diamond	C
Oxide	O^{-2}	Hematite	Fe_2O_3
Silicate	$(SiO_4)^{-4}$	Quartz	SiO_2
		Olivine	$(Mg,Fe)_2SiO_4$
Sulfate	$(SO_4)^{-2}$	Gypsum	$CaSO_4 \cdot 2H_2O$
Sulfide	S^{-2}	Galena	PbS

*A table of the elements and their symbols appears in Appendix 1.

Mineral Identification

Serious students of mineralogy are able to identify several hundred minerals without destructive testing. They are able to do this because minerals have distinctive physical properties, most of which are easily determined and associated with particular mineral species (Appendix 2). With practice, one can build up a mental catalog of minerals, just like building a vocabulary in a foreign language. The most useful physical properties are hardness, cleavage, crystal form, and to lesser degrees, color (which is variable) and luster. Mineralogists use the **Mohs hardness scale,** developed by Friedrich Mohs in 1812 (Table 2.3). Mohs assigned a hardness value (H) of 10 to diamond and of 1 to talc, a very soft mineral. Diamond scratches everything; quartz (H = 7) and feldspar (H = 6) can scratch glass (H = 5½–6), and calcite (H = 3) and your fingernail can scratch gypsum (H = 2).

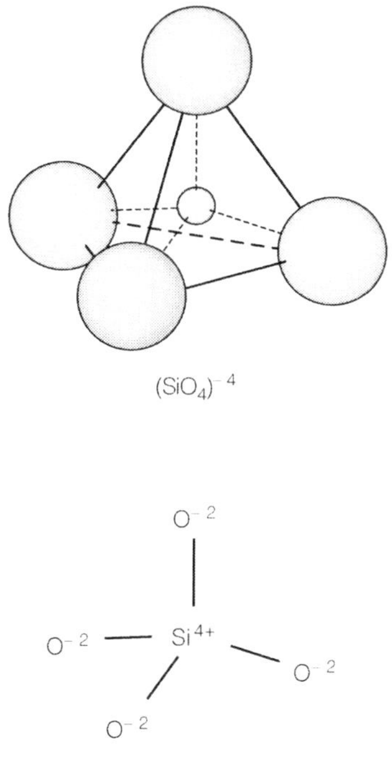

(a)

		Formula of negatively charged ion group	**Example**
Isolated tetrahedra		$(SiO_4)^{-4}$	Olivine
Continuous chains of tetrahedra	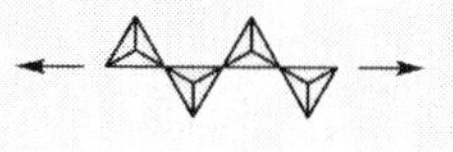	$(SiO_3)^{-2}$	Pyroxene group (augite)
	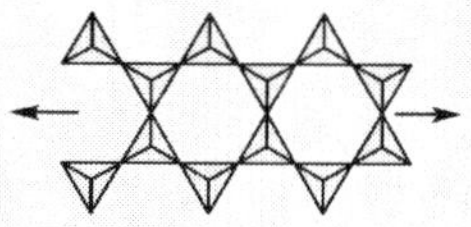	$(Si_4O_{11})^{-6}$	Amphibole group (hornblende)
Continuous sheets	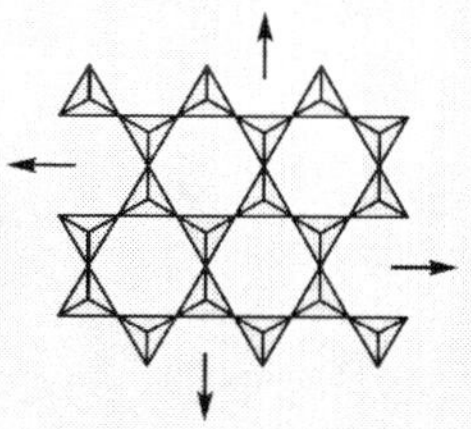	$(Si_4O_{10})^{-4}$	Mica(muscovite)
Three-dimensional networks	Too complex to be shown by a simple two-dimensional drawing	$(SiO_2)^0$ $(Si_3AlO_8)^{-1}$ $(Si_2Al_2O_8)^{-2}$	Quartz Orthoclase feldspar Plagioclase feldspar

(b)

◆ ThomsonNOW™ ACTIVE FIGURE 2.5 (a) Model of the silica tetrahedron, showing the unsatisfied negative charge at each oxygen that allows it to form (b) chains, sheets, and networks.

(a)

(b)

(c)

(d)

◆ FIGURE 2.6 Some common rock-forming silicate minerals: (a) feldspar, (b) quartz, (c) muscovite, and (d) hornblende. (a) USGS; (b)–(d) Paul Silverman/Fundamental Photographs

◆ FIGURE 2.7 This granite is an aggregate of the minerals quartz, biotite, and feldspar. Top left, Michael Dalton/Fundamental Photographs; top right and bottom left, Sue Monroe; bottom right, Paul Silverman/Fundamental Photographs

Cleavage refers to the characteristic way particular minerals split along definite planes as determined by their crystal structure. Mica has perfect cleavage in one direction; this is called *basal* cleavage, because it is parallel to the basal plane of the crystal structure. Feldspars split in two directions; halite, which has *cubic* structure, in three directions; and so on, as shown in ◆Figure 2.9. Minerals that have perfect cleavage can be split readily by a tap with a rock hammer or even peeled apart, as in the case of mica. Some minerals do not cleave but have distinctive **fracture** patterns that help one to identify them. The common crystal forms, shown in ◆Figure 2.10, can be useful in identifying a particular mineral.

◆ FIGURE 2.8 Some common ferromagnesian silicates: (a) olivine; (b) augite, a pyroxene-group mineral; (c) hornblende, an amphibole-group mineral; and (d) biotite mica. Courtesy Sue Monroe

*TABLE 2.2 Some Common Rock-Forming Minerals

Mineral	Abundance in Crust, %	Rock in Which Found
Plagioclase*	39	Igneous rocks mostly
Quartz	12	Detrital sedimentary rocks, granites
Orthoclase**	12	Granites, detrital sedimentary rocks
Pyroxenes	11	Dark-colored igneous rocks
Micas	5	All rock types as accessory minerals
Amphiboles	5	Granites and other igneous rocks
Clay minerals	5	Shales, slates, decomposed granites
Olivine	3	Iron-rich igneous rocks, basalt
Others	11	Rock salt, gypsum, limestone, etc.

*A series of six minerals within the plagioclase group from albite ($NaAlSi_3O_8$) to anorthite ($CaAl_2Si_2O_8$).
**Feldspar group of minerals.

***TABLE 2.3 Mohs Hardness Scale**

Hardness	Mineral	Common Example
1	Talc	Pencil lead 1–2
2	Gypsum	
		Fingernail 2½
3	Calcite	Copper penny 3
		Brass
4	Fluorite	Iron
5	Apatite	Tooth enamel
		Knife blade
		Glass 5½–6
6	Orthoclase	
		Steel file 6½
7	Quartz	
8	Topaz	
9	Corundum	Sapphire, ruby
10	Diamond	Synthetic diamond

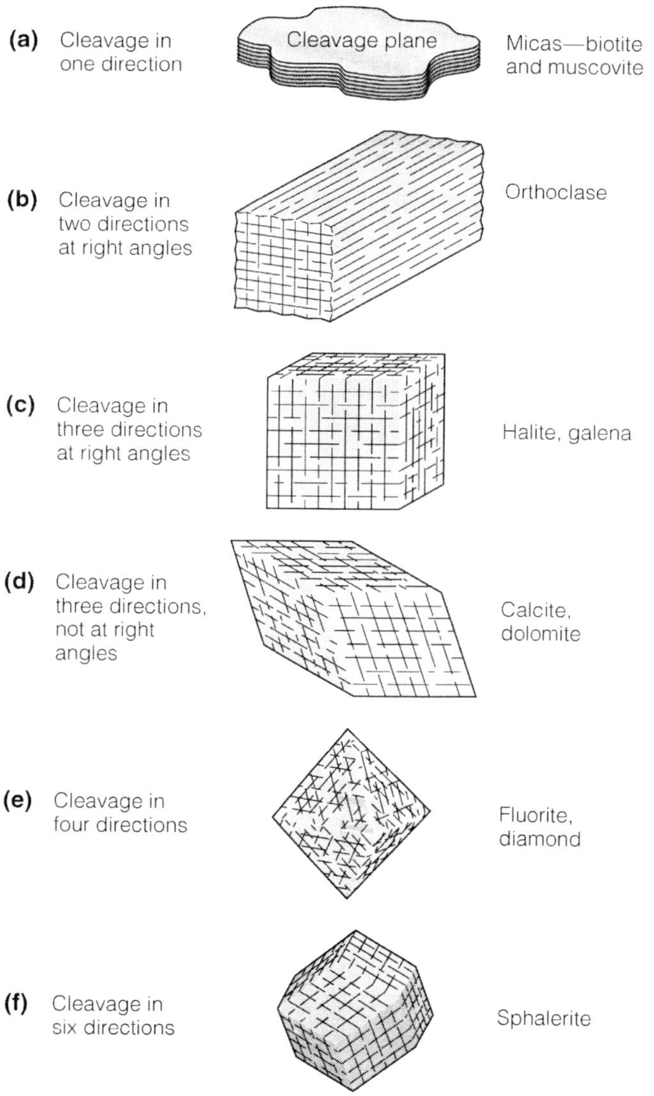

◆ FIGURE 2.9 Types of cleavage and typical minerals in which they occur.

Color tends to vary within a given mineral species, so it is not a reliable identifying property. **Luster**—how a mineral reflects light—is useful, however. We recognize metallic and nonmetallic lusters, the latter being divided into types such as glassy, oily, greasy, and earthy. (■ Case Study 2.2 on page 40 discusses some gem minerals.)

Rocks

Rocks are defined as consolidated or poorly consolidated aggregates of one or more minerals, glass, or solidified organic matter (such as coal) that cover a significant part of Earth's crust. There are three classes of rocks, based upon their origin: igneous, sedimentary, and metamorphic. **Igneous** (Latin *ignis*, "fire") rock is crystallized from molten or partly molten material. **Sedimentary rocks** include both *lithified* (turned to stone) fragments of preexisting rock and rocks that were

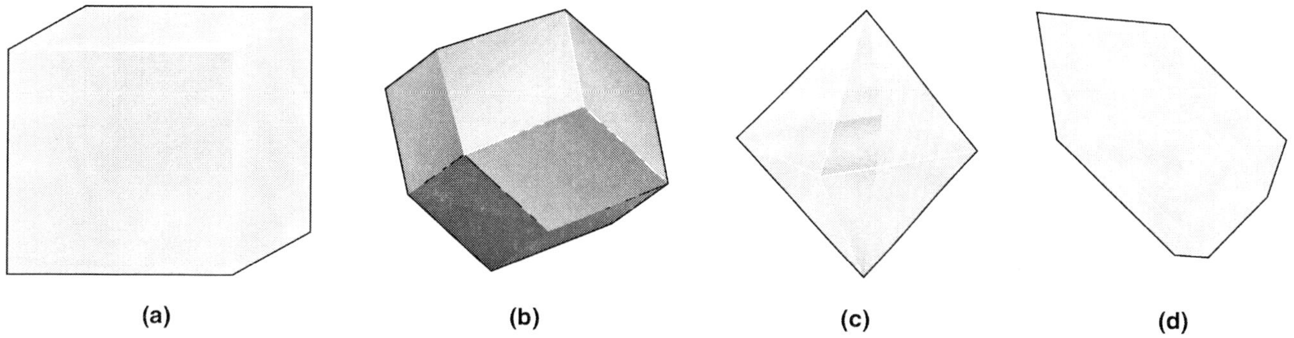

◆ FIGURE 2.10 Mineral crystal shapes: (a) cube (halite), (b) 12-sided dodecahedron (garnet), (c) 8-sided octahedron (diamond, fluorite), and (d) 6-sided hexagonal prism (quartz).

formed from chemical or biological action. **Metamorphic** rocks are those that have been changed, essentially in the solid state, by heat, fluids, or pressure within the earth.

The **rock cycle** is one of many natural cycles on Earth. The illustration of it in ◆ Figure 2.11 shows the interactions of energy, earth materials, and geologic processes that form and destroy rocks and minerals. The rock cycle is essentially a closed system; "what goes around comes around," so to speak. In a simple cycle there might be a sequence of formation, destruction, and alteration of rocks by earth processes. For example, an igneous rock may be eroded to form sediment that subsequently becomes a sedimentary rock, which may then be metamorphosed by heat and pressure to become a metamorphic rock, and then melted to become an igneous rock.

Throughout this book we emphasize those rocks that, because of their composition or structure, are involved in geologic events that endanger human life or well-being or that are important as resources.

CONSIDER THIS

Your school district is told it must remove all asbestos insulation from its older buildings at considerable cost to the taxpayers, and at the expense of teachers' salaries and educational materials. What questions should be asked of the government agency requiring this action, and what position will you take if you learn that the asbestos is the long-chain, fibrous type, rather than the sheet-structure, mica type (Case Study 2.1)?

Igneous Rocks

Igneous rocks are classified according to their texture and their mineral composition. A rock's texture is a function of the size and shape of its mineral grains, and for igneous rocks this is determined by how fast or slowly a melted mass cools. If **magma**—molten rock within the earth—cools slowly, large crystals develop, and a rock with coarse-grained, **phaneritic texture** such as granite is formed (see Figure 2.7). The resulting large rock mass formed within the planet is

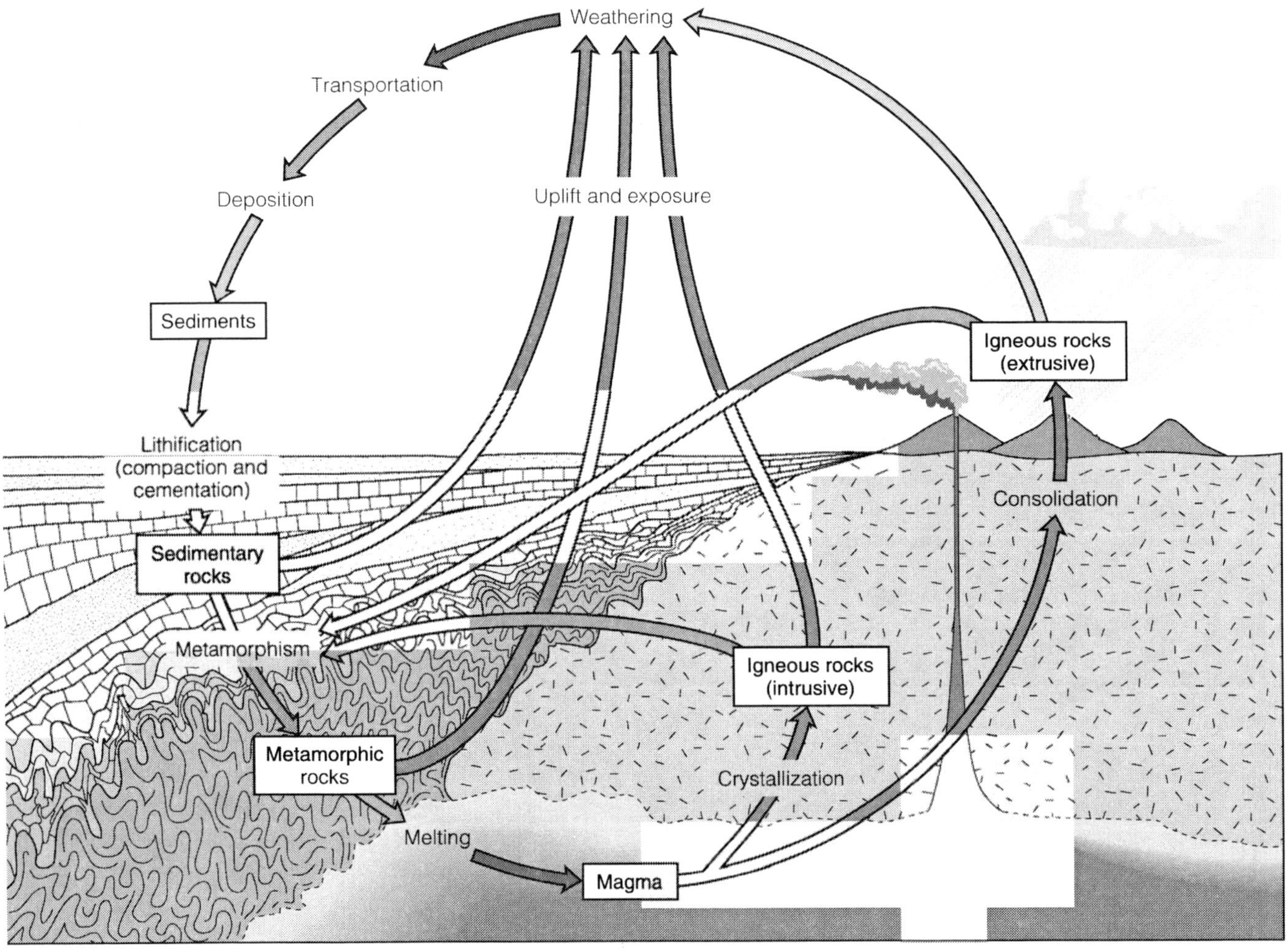

◆ FIGURE 2.11 The rock cycle. The three rock types are interrelated by internal and external processes involving the atmosphere, ocean, biosphere, crust, and upper mantle.

known as a **batholith.** Greater than 100 square kilometers (39 mi^2) in area by definition, batholiths are mostly granitic in composition and show evidence of having invaded and pushed aside the *country rock* into which they intruded. Also, many batholiths have formed in the cores of mountain ranges during mountain-building episodes. Batholiths are a type of igneous mass referred to as *plutons,* and rocks of batholiths are described as **plutonic** (after Pluto, the Greek god of the underworld), because they were formed at great depth. The Sierra Nevada of California is an example of an uplifted and eroded mountain range with an exposed core composed of many plutons (◆ Figure 2.12).

Most igneous rocks—whether on Earth, the Moon, or Mars—are composed of silicate-mineral-containing silicon, oxygen, and aluminum. Silicon combined with oxygen is the compound *silica,* SiO_2. The percentage of silica in an igneous rock is a measure of how "distilled," or fractionated, its original magma was. The higher the percentage of silica in an igneous rock, the more steps of fractionation it has gone through. Minerals that are rich in magnesium and iron and low in silica (minerals such as pyroxene, amphibole, and olivine) will crystallize and settle out of the magma first. The resulting rocks, relatively rich in magnesium and iron and poor in silica, are thus designated **mafic** (*ma-* for magnesium, *-fic* for iron). The remaining magma has less magnesium and iron and a greater percentage of silica. When the next batch of minerals settles out, perhaps with quartz crystals, rocks of *intermediate* composition are formed. During the fractionation process, the magma becomes richer and richer in silica and more impoverished in magnesium and iron. This late-stage magma is largely silica, alumina, and other elements that combine at lower temperatures. The resulting rocks are rich in feldspar and silica, and are thus described as **felsic** (*fel-* for feldspar, *-sic* for silica). "Primitive" magmas (<50% SiO_2) produce mafic gabbro and its volcanic equivalent, basalt, the rock common to Hawaii volcanoes. More "evolved" magmas (>50% SiO_2) produce rocks of intermediate composition, such as diorite and its volcanic equivalent, andesite. Highly evolved magmas produce felsic rocks such as granite and its volcanic equivalent, rhyolite. Mafic rocks are black or dark colored, and felsic rocks are whitish and pinkish. Intermediate igneous rocks vary in color between these extremes according to their percentage of silica.

Lava is molten material at Earth's surface produced by volcanic activity. Lava cools rapidly, resulting in restricted crystal growth and a fine-grained, or **aphanitic,** texture. An example of a rock with aphanitic texture is rhyolite, which has about the same composition as granite but a much finer texture. The grain size of an igneous rock can tell us whether it is an **intrusive** (cooled within Earth) or an **extrusive** (cooled at the surface of Earth) rock. The classification by texture and composition of igneous rocks is shown in ◆ Figure 2.13. Note that for each composition there are pairs of rocks that are distinguished by their texture. Granite and rhyolite, diorite and andesite, and gabbro and basalt are the most common pairs found in nature. The end member shown in Figure 2.13, peridotite, is rarely found because it originates deep within Earth. Three important and relatively common glassy (composed of amorphous or finely crystalline SiO_2) volcanic rocks are obsidian, which looks like black glass; pumice, which has the composition of glass but is not really glassy-looking; and tuff, consolidated volcanic ash or cinders (◆ Figure 2.14). Tuff has a **pyroclastic** (literally "fire-broken") texture, resulting from fragmentation during violent volcanic eruptions.

B. Pipkin

◆ FIGURE 2.12 The Sierra Nevada batholith and Mount Whitney, looking west from Owens Valley; east-central California.

Sedimentary Rocks

Sediment is particulate matter derived from the physical or chemical weathering of materials of Earth's crust and by certain organic processes. It may be transported and redeposited by streams, glaciers, wind, or waves. Sedimentary rock is sediment that has become **lithified**—turned to stone—by pressure from deep burial, by cementation, or by both of these processes. **Clastic sedimentary rocks** are those that are composed of *clasts,* fragments of preexisting rocks and minerals. Clastic sedimentary rocks are classified according to their grain size. Thus sand-size sediment lithifies to sandstone, clay to shale, and gravel to conglomerate (◆ Figures 2.15 and 2.16 a, b, and c). Other sediments result from chemical and biological activity. **Chemical sedimentary rocks,** which may be clastic or nonclastic, include chemically precipitated limestone, rock salt, and gypsum. **Biogenic sedimentary rocks** are produced directly by biological activity, such as coal (lithified plant debris), some limestone and chalk ($CaCO_3$, shell material), and chert (siliceous shells) (◆ Figure 2.16 d and e). Table 2.4 presents the classification of sedimentary rocks.

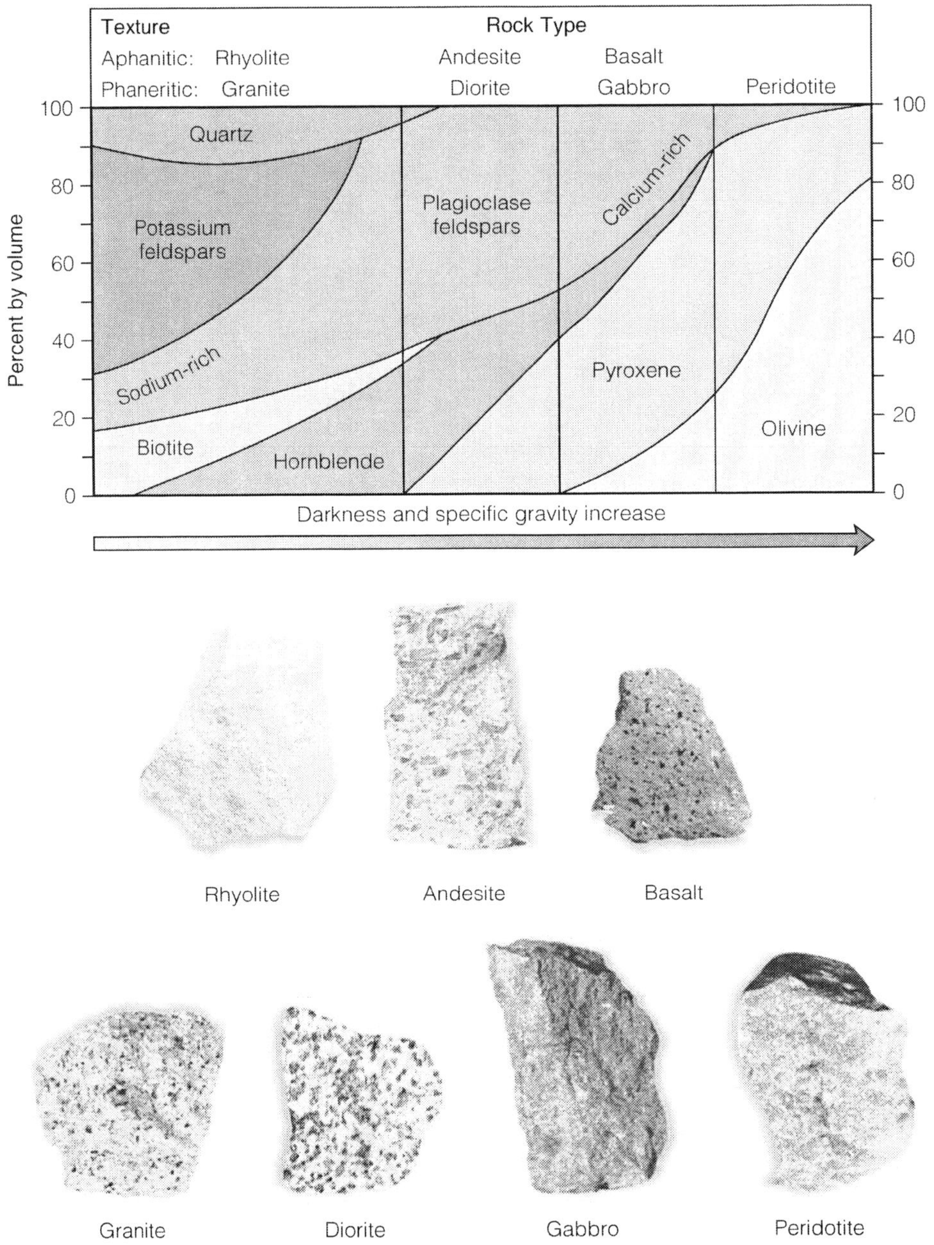

◈ FIGURE 2.13 Classification of common igneous rocks by their mineralogy and texture. Relative proportions of mineral components are shown for each rock type. Fine-grained *aphanitic* texture results from rapid cooling of molten material. Coarse-grained *phaneritic* texture results from slow cooling. Rock photos courtesy Sue Monroe

A distinguishing characteristic of most sedimentary rock is bedding, or **stratification** into layers. Because shale is very thinly stratified, or *laminated,* it splits into thin sheets. Some sedimentary rocks, such as sandstone and limestone, may occur in beds that are several feet thick. Other structures in these rocks give hints as to how the rock formed. *Cross-bedding*—stratification that is inclined at an angle to the main stratification—indicates the influence of wind or water currents. Thick cross-beds and frosted quartz grains in sandstone indicate an ancient desert sand-dune environment (◈ Figure 2.17). Some shale and claystone show polygonal *mud cracks* on bedding planes, similar to those found on the surface of modern dry lakes; these indicate desiccation in a subaerial environment (◈ Figure 2.18).

CONSIDER THIS

Forensic geologists are specialists who are sometimes called upon to help solve crimes. They can, for example, compare the mineralogy of soil on a suspect's shoes with the soil at the scene of the crime. They also use geologic clues to reconstruct events that happened millions of years ago. For instance, huge round scars (craters) on the surface of Earth can indicate ancient meteorite impacts. What are some of the "clues" in sedimentary rocks that tell us about the ancient environments in which they formed?

Ripple marks, low, parallel ridges in deposits of fine sand and silt, may be asymmetrical or symmetrical (◈ Figure 2.19). Asymmetrical ripples indicate a unidirectional current;

(a)

(b)

(c)

◆ FIGURE 2.14 Extrusive igneous rocks: (a) glassy obsidian; (b) pumice, which is formed from gas-charged magmas; and (c) tuff. The lavas that form obsidian and pumice cool so rapidly that crystals do not grow to any size. The reference scale in each photo is 5 centimeters long (5 centimeters is about 2 inches).

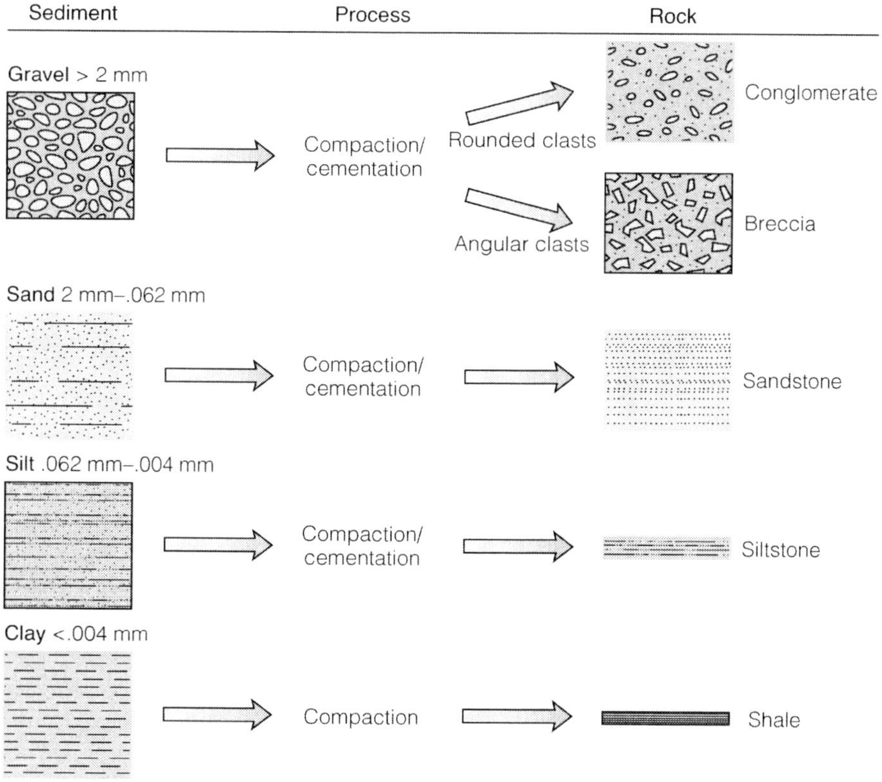

◆ FIGURE 2.15 How sediments are transformed into clastic sedimentary rocks. The sequence of lithification is *deposition* to *compaction* to *cementation* to hard rock.

(a)

(b)

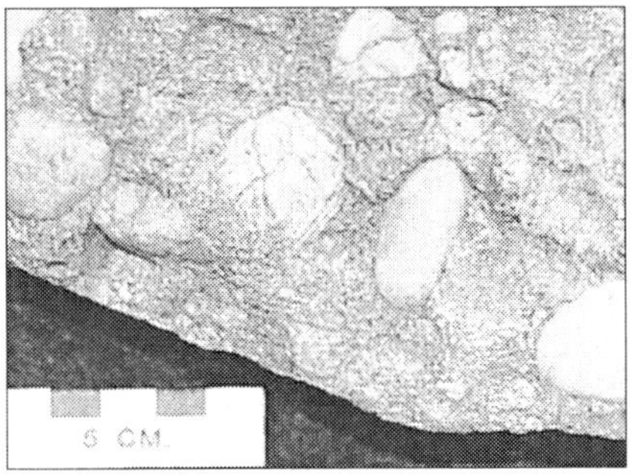

(c)

(d)

(e)

◆ FIGURE 2.16 Common *clastic* sedimentary rocks: (a) shale, (b) sandstone, (c) conglomerate. *Biogenic* sedimentary rocks: (d) limestone composed entirely of shell materials (called *coquina*), (e) coal.

★TABLE 2.4 Classification of Sedimentary Rocks

Detrital Sedimentary Rocks (clastic texture)		
Sediment	**Description**	**Rock Name**
Gravel (>2.0 mm)	Rounded rock fragments	Conglomerate
	Angular rock fragments	Breccia
Sand (0.062–2.0 mm)	Quartz predominant	Quartz sandstone
	>25% feldspars	Arkose
Silt (0.004–0.062 mm)	Quartz predominant, gritty feel	Siltstone
Clay, mud (<0.004 mm)	Laminated, splits into thin sheets	Shale
	Thick beds, blocky	Mudstone
Chemical Sedimentary Rocks		
Texture	**Composition**	**Rock Name**
Clastic	Calcite ($CaCO_3$)	Limestone
	Dolomite [$CaMg(CO_3)_2$]	Dolostone
Crystalline	Halite (NaCl)	Rock salt
	Gypsum ($CaSO_4 \cdot 2H_2O$)	Rock gypsum
Biogenic Sedimentary Rocks		
Texture	**Composition**	**Rock Name**
Clastic	Shell calcite, skeletons, broken shells	Limestone, coquina
	Microscopic shells ($CaCO_3$)	Chalk
Nonclastic (altered)	Microscopic shells (SiO_2), recrystallized silica	Chert
	Consolidated plant remains (largely carbon)	Coal

D. D. Trent

◆ FIGURE 2.17 Wind-blown sandstone exhibiting large cross-beds; Zion National Park, Utah.

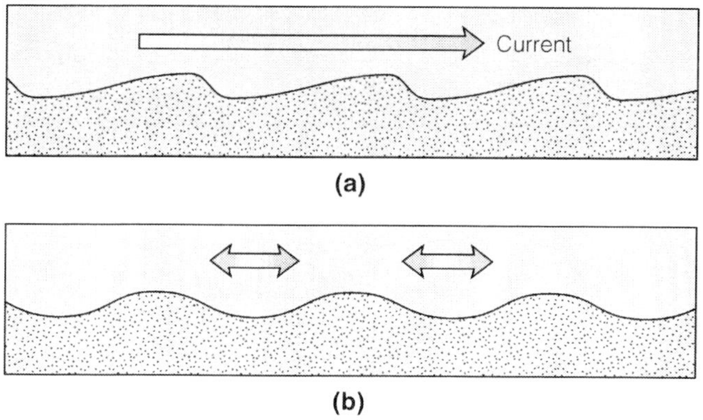

◆ FIGURE 2.19 Ripple marks found in sedimentary rocks. (a) Asymmetrical ripple marks are common in streambeds and are also found on bedding planes in sedimentary rocks. (b) Symmetrical ripple marks are due to oscillating water motion; that is, wave action.

symmetrical ones, the back-and-forth motion produced by waves in shallow water. Only in sedimentary rock do we find abundant fossils, the remains or traces of life.

Metamorphic Rocks

Rocks that have been changed from preexisting rocks by heat, pressure, or chemical processes are classified as metamorphic rocks. The process of metamorphism results in new structures, textures, and minerals. **Foliation** (Latin *folium*, "leaf") is the flattening and layering of minerals by nonuniform stresses. Foliated metamorphic rocks are classified by the development of this structure. Slate, schist, and gneiss are foliated, for example, and they are identified by the thickness or crudeness of their foliation (◆ Figure 2.20). Metamorphic rocks may also form by **recrystallization.** This occurs when a rock is heated and strained by uniform stresses so that larger, more perfect grains result or new minerals form. In this manner a limestone may recrystallize to marble or a quartz sandstone to quartzite, one of the most resistant of all rocks (◆ Figure 2.21). Table 2.5 summarizes the characteristics of common metamorphic rocks.

D. D. Trent

◆ FIGURE 2.18 Mud cracks in an old clay mine; Ione, California.

Rock Defects

Some rocks have structures that geologists view as "defects"; that is, surfaces along which landslides or rockfalls may occur. Almost any planar structure, such as a bedding plane in sedimentary rock or a foliation plane in metamorphic rock, holds potential for rock slides or falls. The orientation of a plane in space, such as a stratification plane, a fault, or a joint, may be defined by its **dip** and **strike** (Appendix 3). **Joints** are rock fractures without displacement and they occur in all rock types. They are commonly found in parallel sets spaced several feet apart (◆ Figure 2.22). **Faults** are also fractures in crustal rocks, but they differ from joints in that some movement or displacement has occurred along the fault surface. Faults also are found in all rock types and are potential surfaces of "failure." We will examine the relationships of rock defects to geologic hazards when we discuss landslides, subsidence, and earthquakes.

CONSIDER THIS

Rock climbing is a popular sport in many parts of the world. A favorite spot in the U.S. West is Yosemite Valley, a picturesque area carved by glaciers into a uniform granite body known as the Cathedral Peak Granite (see Gallery Figure 3). Why do you suppose climbers would prefer to climb near-vertical cliffs of granite, as opposed to similar cliffs composed of schist, gneiss, or shale?

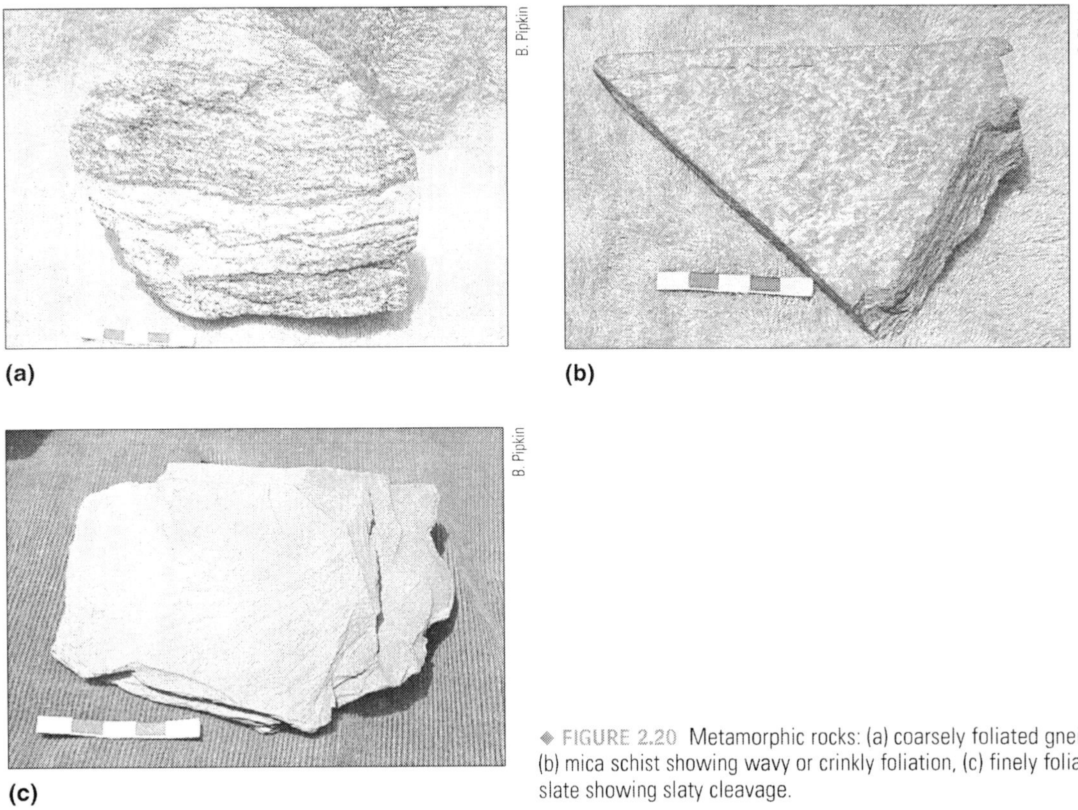

◆ FIGURE 2.20 Metamorphic rocks: (a) coarsely foliated gneiss, (b) mica schist showing wavy or crinkly foliation, (c) finely foliated slate showing slaty cleavage.

Geologic Time

Interest in extremely long periods of time sets geology and astronomy apart from other sciences. Geologists think in terms of billions of years for the age of Earth and its oldest rocks—numbers that, like the national debt, are not easily comprehended. Nevertheless, the time scales of geologic activity are important for environmental geologists because they provide a way to measure human impacts on the natural world. For example, we would like to know the rate of natural soil formation from solid rock to determine whether topsoil erosion from agriculture is too great or not. Likewise, understanding how climate has changed over millions of years is vital to properly assess current global-

◆ FIGURE 2.21 (a) Limestone recrystallizes to form white marble. (b) Quartzite, the hardest and most durable common rock, is metamorphosed sandstone. Rocks courtesy Sue Monroe

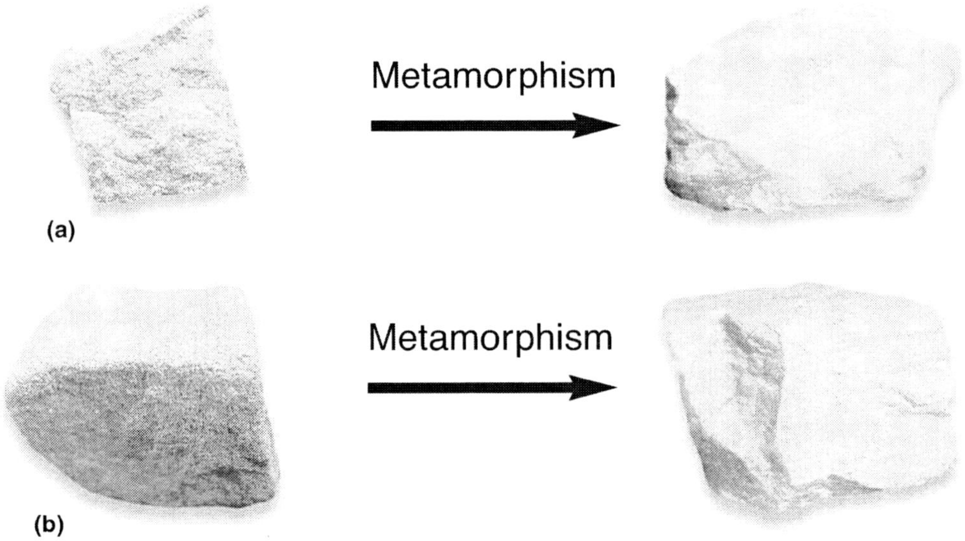

*TABLE 2.5 Common Metamorphic Rocks		
Rock	Parent Rock	Characteristics
	Foliated or Layered	
Slate	Shale and mudstone	Splits into thin sheets
Schist	Fine-grained rocks, siltstone, shale, tuff	Mica minerals often crinkled
Gneiss	Coarse-grained rocks	Dark and light layers of aligned minerals
	Nonfoliated or Recrystallized	
Marble	Limestone	Interlocking crystals
Quartzite	Sandstone	Interlocking, almost fused quartz grains

◆ FIGURE 2.22 Joints in basalt flows near Chico, California.

warming trends. Clues to past environmental change are well-preserved in many different kinds of rocks.

Geologists evaluate the age of rocks and geologic events using two different approaches. **Relative-age dating** is the technique of determining *a sequence of geological events*, based upon the structural relations of rocks. **Absolute-age dating** provides the *actual ages* for rocks in years before the present. Relative age is determined by applying geologic laws based upon the structural relations of rocks. For instance, the Law of Superposition tells us that in a stack of undeformed sedimentary rocks, the stratum (layer) at the top is the youngest. The Law of Cross-cutting Relationships tells us that a fault is younger than the youngest rocks it displaces or cuts. Similarly, we know that a pluton is younger than the rocks it intrudes (◆ Figure 2.23).

Using these laws, geologists arranged a great thickness of sedimentary rocks and their contained fossils representing an immense span of geologic time. The geologic age of a particular sequence of rock was then determined by applying the Law of Fossil Succession, the observed chronologic sequence of life forms through geologic time. This allows fossiliferous rocks from two widely separated areas to be correlated by matching key fossils or groups of fossils found in the rocks of the two areas (Figure 2.23). Using such indicator

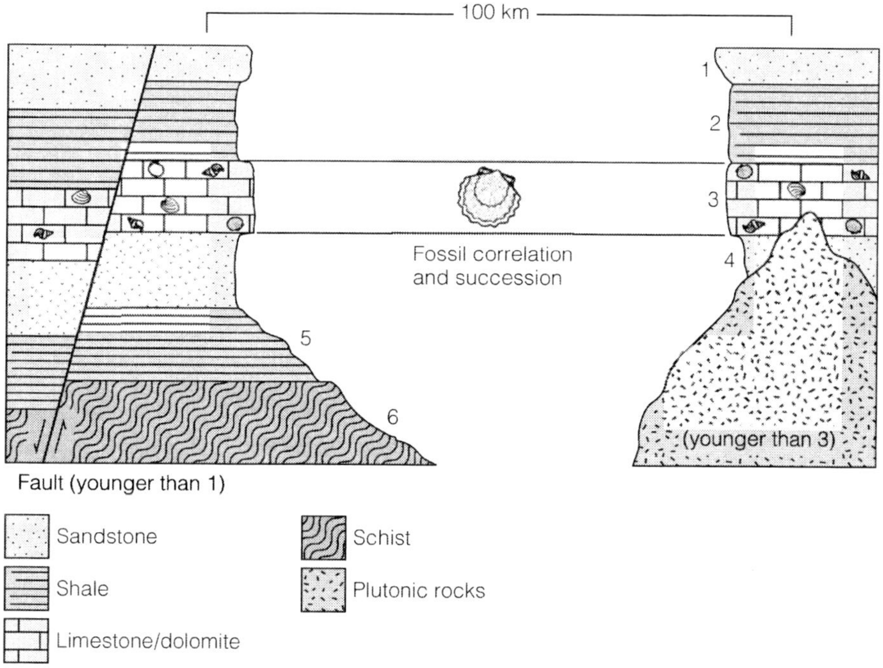

◆ FIGURE 2.23 Geologic cross sections illustrating the Laws of Superposition, Cross-cutting Relationships (intrusion and faulting), and Fossil Succession. The limestone beds are correlative because they contain the same fossils. The numerals indicate relative ages, with 1 being the youngest.

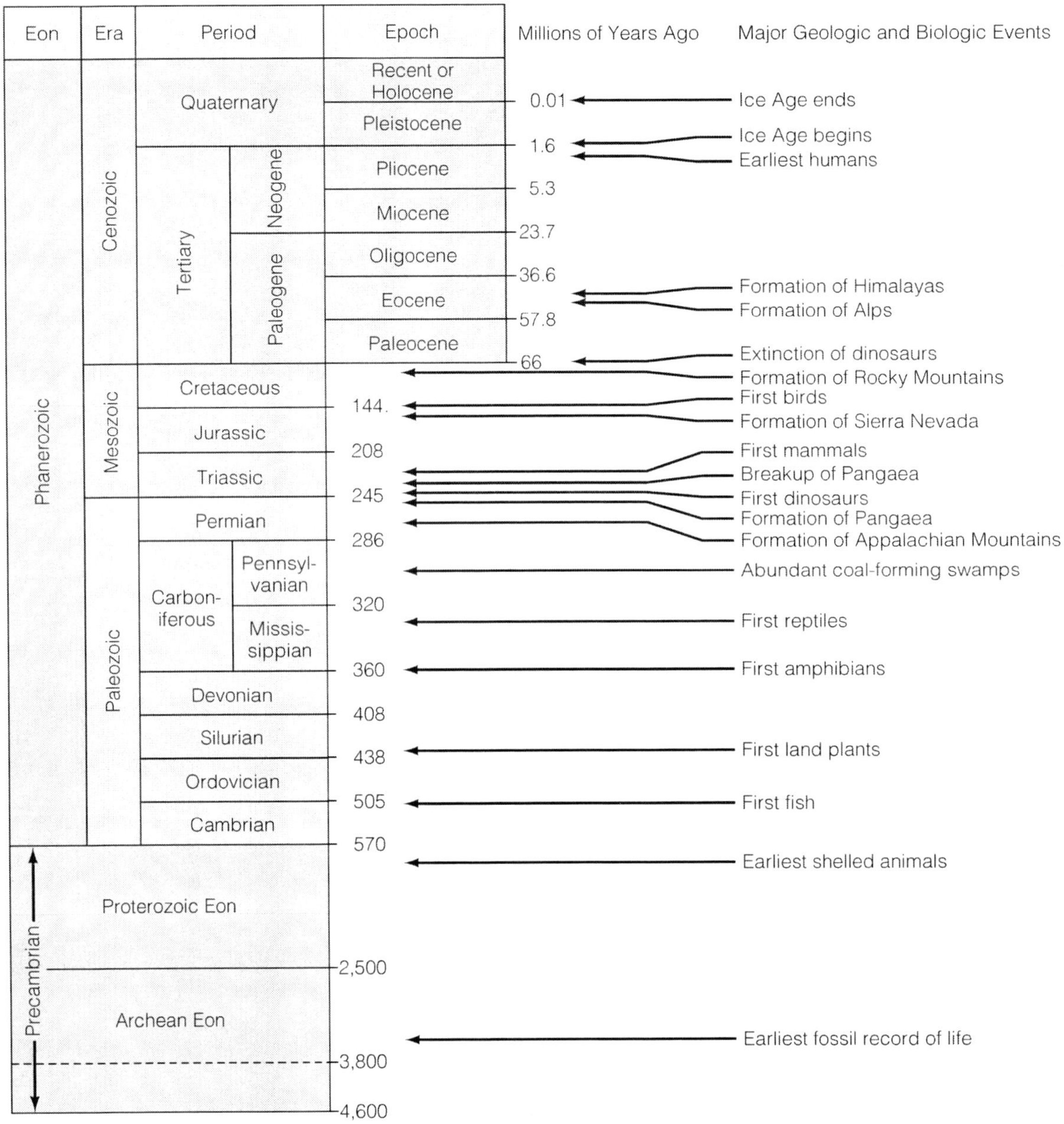

◆ FIGURE 2.24 Geologic time scale. Adapted from A. R. Palmer, Geology, 1983, 504.

fossils and radioactive dating methods, geologists have developed the geologic time scale to chronicle the documented events of Earth history (◆ Figure 2.24). Note that the scale is divided into units of time during which rocks were deposited, life evolved, and significant geologic events such as mountain building occurred. Eons are the longest time intervals, followed, respectively, by eras, periods, and epochs. The Phanerozoic ("revealed life") Eon began 570 million years ago with the Cambrian Period, the rocks of which contain the first extensive fossils of organisms with hard skeletons. Because of the significance of the Cambrian Period, the informal term *Precambrian* is widely used to denote the time before it, which extends back to the formation of Earth 4.6 billion years ago. Note that the Precambrian is divided into the Archean and Proterozoic Eons, with the Archean Eon extending back to the formation of the oldest known in-place rocks, about 3.9 billion years ago. Precambrian time accounts for 88 percent of geologic time, and the Phanerozoic for a mere 12 percent. The eras of geologic time correspond to the relative complexity of life forms: Paleozoic (oldest life), Mesozoic (middle life), and Cenozoic (most recent life). Environmental geologists are most interested in the events of the past few million years, a mere heartbeat in the history of Earth.

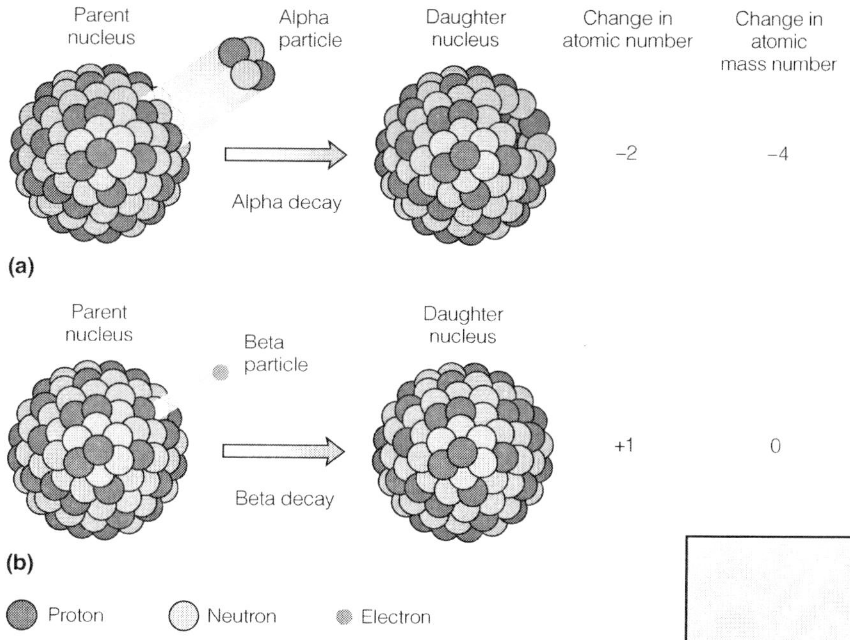

◆ FIGURE 2.25 (a) Alpha emission, whereby a heavy nucleus spontaneously emits a helium atom and is reduced 4 atomic mass units and 2 atomic numbers. (b) Emission of a nuclear electron (ß⁻ particle), which changes a neutron to a proton and thereby forms a new element without a change of mass.

Absolute age-dating requires some kind of natural clock. The ticks of the clock may be the annual growth rings of trees or established rates of disintegration of radioactive elements to form other elements. At the turn of the twentieth century, American chemist and physicist Bertram Borden Boltwood (1870–1927) discovered that the ratio of lead to uranium in uranium-bearing rocks increases as the rocks' ages increase. He developed a process for determining the age of ancient geologic events that is unaffected by heat or pressure—**radiometric dating.** The "ticks" of the radioactive clocks are radioactive decay processes—spontaneous disintegrations of the nuclei of heavier elements such as uranium and thorium to lead. A radioactive element may decay to another element, or to an isotope of the same element. This decay occurs at a precise rate that can be determined experimentally. The most common emissions are alpha particles ($^4_2\alpha$), which are helium atoms, and beta particles ($ß^-$), which are nuclear electrons. New *radiogenic* "daughter" elements or isotopes result from this **alpha decay** and **beta decay** (◆ Figure 2.25).

For example, of the three isotopes of carbon, ^{12}C, ^{13}C, and ^{14}C, only ^{14}C is radioactive, and this radioactivity can be used to date events between a few hundred and a few tens of thousands of years ago. Carbon-14 is formed continually in the upper atmosphere by neutron bombardment of nitrogen, and it exists in a fixed ratio to the common isotope, ^{12}C. All plants and animals contain radioactive ^{14}C in equilibrium with the atmospheric abundance until they die, at which time ^{14}C begins to decrease in abundance and, along with it, the object's radioactivity. Thus, by measuring the radioactivity of an ancient parchment, log, or piece of charcoal and comparing the measurement with the activity of a modern standard, the age of archaeological materials and geological events can be determined (◆ Figure 2.26).

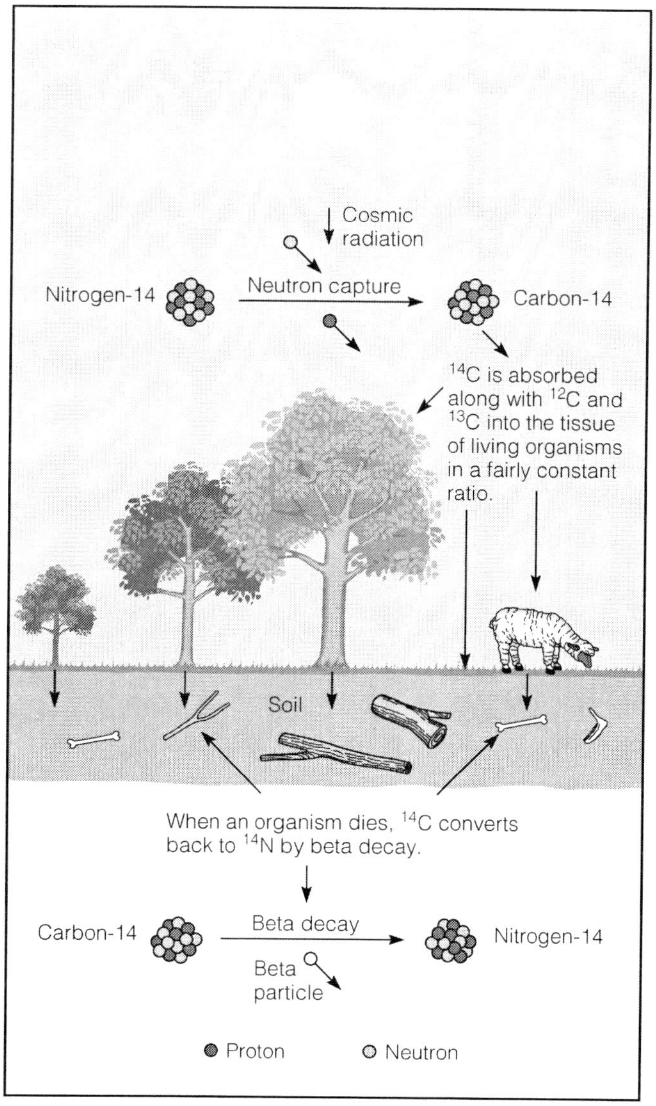

◆ FIGURE 2.26 Carbon-14 is formed from nitrogen-14 by neutron capture and subsequent proton emission. Carbon-14 decays back to nitrogen-14 by emission of a nuclear electron ($ß^-$).

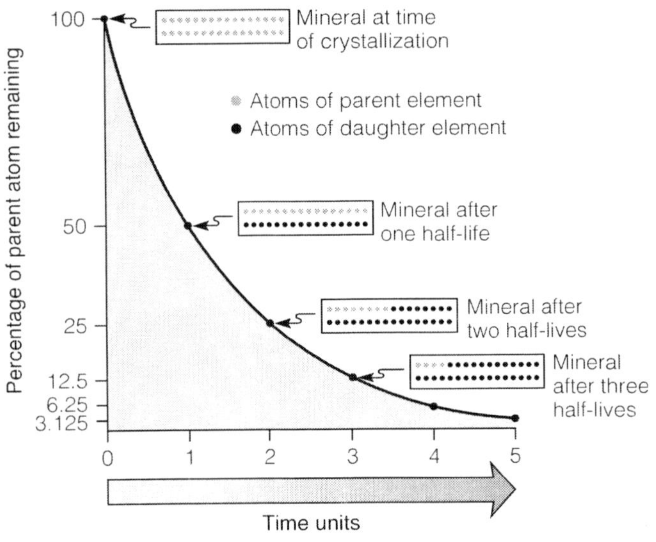

◆ FIGURE 2.27 Decay of a radioactive parent element with time. Each time unit is one half-life. Note that after two half-lives one-fourth of the parent element remains, and that after three half-lives one-eighth remains.

Carbon-14 is formed by the collision of cosmic neutrons with ^{14}N in the atmosphere, and then it decays back to ^{14}N by emitting a nuclear electron ($ß^-$).

$$^{14}_{7}N + \text{neutron} \rightarrow {}^{14}_{6}C + \text{proton}$$

$$^{14}_{6}C \rightarrow {}^{14}_{7}N + ß^-$$

Both carbon—the radioactive "parent" element—and nitrogen—the radiogenic "daughter"—have 14 atomic mass units. However, one of ^{14}C's neutrons is converted to a proton by the emission of a beta particle, and the carbon changes (or *transmutes*) to nitrogen. This process proceeds at a set rate that can be expressed as a **half-life,** the time required for half of a population of radioactive atoms to decay. For ^{14}C this is about 5,730 years.

Radioactive elements decay exponentially; that is, in two half-lives one-fourth of the original number of atoms remain, in three half-lives one-eighth remain, and so on (◆ Figure 2.27). So few parent atoms remain after seven or eight half-lives (less than 1 percent) that experimental uncertainty creates limits for the various radiometric dating methods.

Whereas the practical age limit for dating carbon-bearing materials such as wood, paper, and cloth is about 40,000 to 50,000 years, ^{238}U disintegrates to ^{206}Pb and has a half-life of 4.5×10^9 years. Uranium-238 emits alpha particles (helium atoms). Since both alpha and beta disintegrations can be measured with a Geiger counter, we have a means of determining the half-lives of a geological or archaeological sample. Table 2.6 shows radioactive parents, daughters, and half-lives commonly used in age-dating.

Age of Earth

Before the advent of radiometric dating, determining the age of Earth was a source of controversy between established religious interpretations and early scientists. Archbishop Ussher (1585–1656), the Archbishop of Armagh and a professor at Trinity College in Dublin, declared that Earth was

★TABLE 2.6 Isotopes Used in Age-Dating

Isotopes: Parent	Isotopes: Daughter	Parent's Half-Life, Years	Effective Dating Range, Years	Material That Can Be Dated
Uranium-238	Lead-206	4.5 billion	10 million to 4.6 billion	Zircon Uraninite
Uranium-235	Lead-207	704 million		
Thorium-232	Lead-208	14 billion		
Rubidium-87	Strontium-87	48.8 billion	10 million to 4.6 billion	Muscovite Biotite Orthoclase Whole metamorphic or igneous rock
Potassium-40	Argon-40	1.3 billion	100,000 to 4.6 billion	Glauconite Hornblende Muscovite Whole volcanic rock Biotite
Carbon-14	Nitrogen-14	5,730 years	less than 100,000	Shell, bones, charcoal

formed in the year 4004 B.C. Ussher provided not only the year but also the day, October 23, and the time, 9:00 A.M. Ussher has many detractors, but Stephen J. Gould of Harvard University, though not proposing acceptance of Ussher's date, was not one of them. Gould pointed out that Ussher's work was good scholarship for its time, because other religious scholars had extrapolated from Greek and Hebrew scriptures that Earth was formed in 5500 B.C. and 3761 B.C., respectively. Ussher based his age determination on the verse in the Bible that says "one day is with the Lord as a thousand years" (2 Peter 3:8). Because the Bible also says that God created heaven and Earth in 6 days, Ussher arrived at 4004 years B.C.—the extra four years because he believed Christ's birth year was wrong by that amount of time. Ussher's age for Earth was accepted by many as "gospel" for almost 200 years.

By the late 1800s geologists believed that Earth was on the order of 100 million years old. They reached their estimates by dividing the total thickness of sedimentary rocks (tens of kilometers) by an assumed annual rate of deposition (mm/year). Evolutionists such as Charles Darwin thought that geologic time must be almost limitless in order that minute changes in organisms could eventually produce the present diversity of species. Both geologists and evolutionists were embarrassed when the British physicist William Thomson (later Lord Kelvin) demonstrated with elegant mathematics how Earth could be no older than 400 million years, and maybe as young as 20 million years. Thomson based this on the rate of cooling of an initially molten Earth and the assumption that the material composing Earth was incapable of creating new heat through time. He did not know about radioactivity, which adds heat to rocks in the crust and mantle.

Bertram Boltwood postulated that older uranium-bearing minerals should carry a higher proportion of lead than younger samples. He analyzed a number of specimens of known relative age, and the absolute ages he came up with ranged from 410 million to 2.2 billion years old. These ages put Lord Kelvin's dates based on cooling rates to rest and ushered in the new radiometric dating technique. By extrapolating backward to the time when no radiogenic lead had been produced on Earth, we arrive at an age of 4.6 billion years for Earth. This corresponds to the dates obtained from meteorites and lunar rocks, which are part of our solar system. The oldest known intact terrestrial rocks are found in the Acasta Gneiss of the Slave geological province of Canada's Northwest Territories. Analyses of lead to uranium ratios on the gneiss's zircon minerals indicate that the rocks are 3.96 billion years old. However, older detrital zircons on the order of 4.0–4.3 billion years old have been found in western Australia, indicating that some stable continental crust was present as early as 4.3 billion years ago (Table 2.7). Suffice it to say that Earth is very old and that there has been abundant time to produce the features we see today (◈ Figure 2.28).

***TABLE 2.7 Earth's Oldest Known Materials**

Material	Location	Age, Billions of Years
Crust	Zircon minerals in rocks of western Australia	4.0–4.3
Rock	Zircon minerals in the Acasta Gneiss, N.W. Terr., Canada	3.96
Sedimentary rock	Isua Greenstone Belt, Greenland	3.7–3.75
Fossils	Algae and bacteria	3.5

Environmental geology deals mostly with the present and the recent past—the time interval known as the *Holocene Epoch* (10,000 years ago to the present) of the *Quaternary Period.* However, because the Great Ice Age advances of the Pleistocene Epoch of the Quaternary Period (which preceded our own Holocene Epoch) had such a tremendous impact on the landscape of today, we will investigate the probable causes of the "ice ages" in order to speculate a bit about what the future might bring. In addition, Pleistocene glacial deposits are valuable sources of underground water, and they have economic value as sources of building materials. We will see that ice ages have occurred several times throughout geologic history (see Chapter 11).

What is most impressive about geologic time is how short the period of human life on Earth has been. If we could compress the 4.6 billion years since Earth formed into one calendar year, *Homo sapiens* would appear about 30 minutes before midnight on December 31 (see Figure 2.30). The last ice-age glaciers would begin wasting away a bit more than 2 minutes before midnight, and written history would exist for only the last 30 seconds of the year. Perhaps we should keep this calendar in mind when we hear that the dinosaurs were an unsuccessful group of reptiles. After all, they endured 50 times longer than hominids have existed on the planet to date. At the present rate of population growth, there is some doubt that humankind as we know it will be able to survive anywhere near that long.

The Earth we see today is the product of earth processes acting on earth materials over the immensity of geologic time. The response of the materials to the processes has been the formation of mountain ranges, plateaus, and the multitude of other physical features found on the planet. Perhaps the greatest of these processes has been the formation and movement of platelike segments of Earth's surface, which influence the distribution of earthquakes and volcanic eruptions, global geography, mineral deposits, mountain ranges, and even climate. These *plate tectonic* processes are the subject of the next chapter.

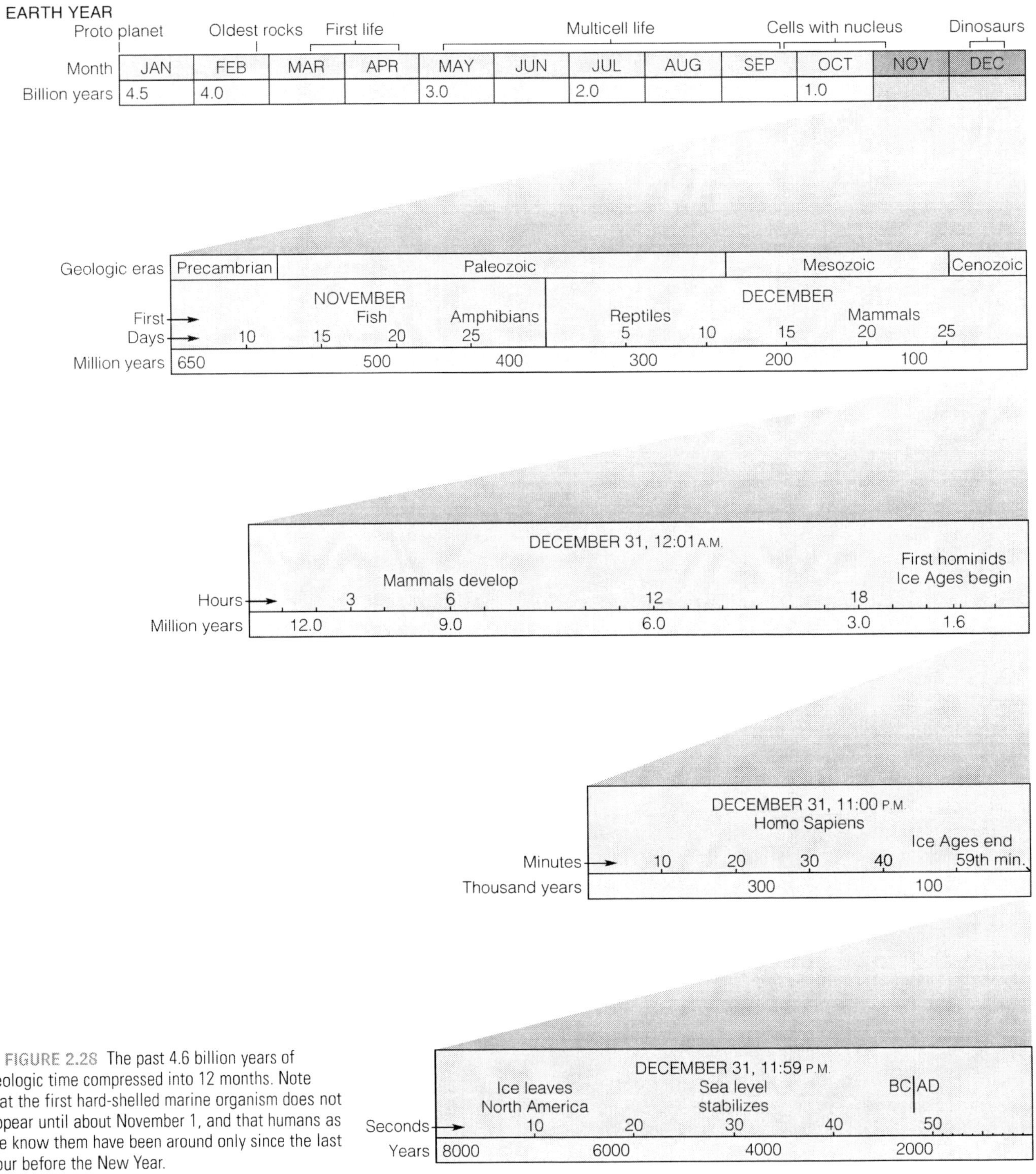

◆ FIGURE 2.28 The past 4.6 billion years of geologic time compressed into 12 months. Note that the first hard-shelled marine organism does not appear until about November 1, and that humans as we know them have been around only since the last hour before the New Year.

CASE STUDY 2.1

Minerals, Cancer, and OSHA—Fact and Fiction

When an international health organization designated quartz as "probably" carcinogenic (cancer-causing) to humans, the U.S. Occupational Safety and Health Agency (OSHA) immediately went into action. It issued a mandate requiring products containing more than one-tenth of one percent (0.1%) of quartz or "free silica" (SiO_2) to display warning signs. Because quartz is chemically inert and because it is the most common mineral species, questions arose as to whether this requirement extended to sandpaper, beach sand, and unpaved roads that may produce clouds of fine-grained quartz dust. Questions—and more—arose when some Delaware truck drivers were "ticketed" for not displaying the required "quartz on board" signs during transport of crushed rock to construction sites. Such well-meaning OSHA actions have been perceived as ridiculous. They illustrate what happens when a government authority establishes health regulations without understanding the substances involved. In this case OSHA did not understand the where, how, and chemistry of the mineral—its mineralogy. Unless one has had a long-term craving to inhale beach sand or crushed rock, there are no health hazards in the examples cited. Nonetheless, because exposure to very large quantities of any substance, even table salt or vitamins, can be harmful to health, it is certain that long-term workplace exposure to dusts from some minerals can pose health problems.

The World Health Organization's Group 1 classification of "known" carcinogenic minerals includes asbestos minerals (there are six), erionite (a zeolite), and minerals containing arsenic, chromium, and nickel. Its Group 2 of "probable" carcinogenic minerals includes all radioactive minerals and minerals containing lead, beryllium, and silica. The classification is based upon suspected relationships between specific diseases and workplace exposure to the identified substances and on laboratory experiments with animals.

A major human health concern is the relationship between lung cancer and inhaling the fine fibers of asbestos minerals. The long, fibrous crystal forms of asbestos belonging to the amphibole group (see hornblende structure, Figure 2.7) are considered the most carcinogenic. Of these, blue asbestos, crocidolite (pronounced krō-sid'-əl-īt; ◆Figure 1), is thought to be most dangerous and is believed to cause mesothelioma, a relatively rare cancer of the lining of the heart and lungs. Although crocidolite constitutes only 5 percent of all industrial asbestos, its morbidity statistics are grim indeed. For example, a 1990 survey of 33 men who, in 1953, worked in a factory where crocidolite was utilized in manufacturing cigarette filters found that 19 of them had died of asbestos-related diseases.

Common white asbestos, chrysotile (pronounced kris'-ə-tīl; ◆Figure 2), is a sheet-structure mineral similar to mica. It appears under high magnification as long fibers, which are in reality rolled sheets, much like long rolls of gift-wrapping paper. Chrysotile constitutes 95 percent of all industrial asbestos and has never been demonstrated to cause cancer at the levels found in most schools and public buildings. A study conducted in Thetford, Quebec, Canada evaluated persons who had been exposed to high concentrations of white asbestos in their workplace and found no deaths attributable to it. Many of the diagnosed chrysotile-related lung problems have been traced back to improper handling of the material and inadequate respiratory protection during asbestos removal procedures. Experimental studies at Virginia Tech have shown that respirable-sized 1-nanometer-diameter chrysotile fibers dissolve in about nine months under conditions

Janet Blabaum, Highland Geotechnical, AIPG

◆ FIGURE 1 Deadly crocidolite, "blue asbestos," magnified about 100 times. The mineral appears in its characteristic blue and green colors when viewed through a petrographic microscope. It was used in only a small percentage of commercial products.

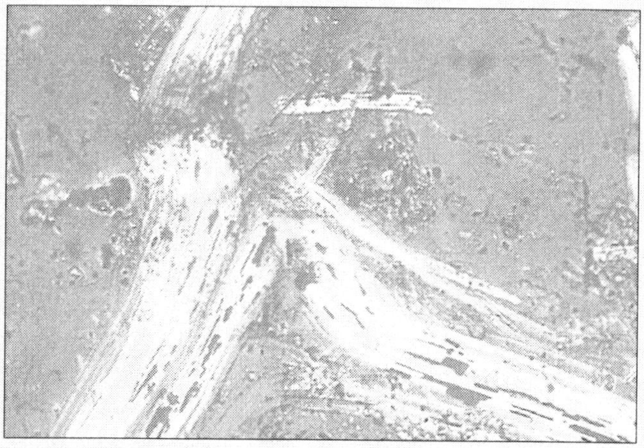

Janet Blabaum, Highland Geotechnical, AIPG

◆ FIGURE 2 Chrysotile asbestos (×100) has many uses in homes, offices, and public buildings. Although it is the target of removal in thousands of structures, at low levels it poses no documented risk. Note the difference between chrysotile and crocidolite's crystal structures.

CONTINUED

expected in lung tissue. This very short lifetime is difficult to reconcile with the general observations that asbestos-related diseases appear in asbestos workers only after many years on the job.

Inhaling finely divided airborne asbestos fibers in large quantities is definitely a health hazard. Under the Toxic Substance Control Act of 1972, the EPA mandated in 1986 that *all* asbestos minerals be treated as identical health hazards, regardless of their mineralogy, and that they be removed. By the mid-1990s the cost of removal had amounted to hundreds of millions of dollars, and the total cumulative cost for remediation of rental and commercial buildings, litigation, and enforcement was estimated at $100 billion. The EPA requirements were sweeping. They applied to buildings containing the relatively benign white asbestos and to those where airborne concentrations were so low that they could not be measured. The fear of asbestos and the EPA regulations are based on the idea that there is no safe "threshold" concentration—that even one fiber can kill. No doctor or public health official would agree with this. A New York school district's administration was besieged by irate parents when it admitted that it could not *guarantee* that *all* asbestos had been removed from its buildings. The district had spent millions of dollars to remove it.

Industry and government can ill afford to squander capital and tax dollars on poorly conceived hazard-mitigation requirements based upon the "no threshold" theory of health risk; that is, that one molecule of pesticide or one asbestos fiber can cause health problems. Many government regulations regarding "toxic substances" address risks to humans that are no greater than those incurred by routinely drinking several cups of coffee each day or eating a peanut butter sandwich for lunch. The "no threshold" criterion forces "carcinogen" classification upon otherwise benign minerals and many other useful substances and needlessly increases public anxiety. (Remember when synthetic sweeteners and cellular telephones were labeled carcinogenic?) The costs of removal and liability protection in such cases are horrendous, and a fiscal crisis in the environmental field is acknowledged. Liability extends to building owners, real estate agents who sell the buildings, banks that lend money for property purchases, and purchasers of land upon which any material classified as hazardous is found.

A person's chances of being struck by a lightning bolt are about 35 per 1 million lifetimes, and the risk of a nonsmoker's dying from asbestos exposure is about 1 in 100,000—about a third of the chance of being struck by lightning. Incidentally, the risk of dying from cigarette-related diseases is about 1 in 5 for smokers. Certainly, some substances in the environment pose dangers, and the key to risk reduction and longevity is awareness. Just as a reasonable person would not play golf during a violent thunderstorm because of the risk of being struck by lightning on the course, one should not handle hazardous substances without wearing protective gear.

CASE STUDY 2.2

A Girl's Best Friend and Other Gems

Rarity and beauty are what make precious and semiprecious gems so desirable. Those shown in ◆Figure 1 are but a sample of several hundred valued gem minerals and their varieties. The largest uncut diamond in the world, the Cullinan diamond (◆Figure 1a), was recovered in 1905 from the Premier Mine at Kimberley, Transvaal, South Africa; the huge, colorless stone weighed 3,106 carats (1.3 pounds). The smooth face of the Cullinan replica shown in the photograph replicates the stone's natural cleavage surface, which indicates that it was not the entire stone. Unfortunately, the remainder of the Cullinan lies undiscovered in a deep volcanic pipe—which is where all diamonds originate. The Transvaal government bought the diamond for $1.6 million and gave it uncut to King Edward VII of England. After months of study, an Amsterdam diamond cutter's decisive blow yielded nine large stones and about 100 smaller ones, all of them flawless. All are now in the British crown jewels collection.

Diamonds are the high-pressure form of carbon and are the hardest natural substance known. Curiously, graphite, the high-temperature, low-pressure form of carbon, is among the softest minerals known.

Diamonds are found mostly in Africa, Australia, Russia, India, and South America (Table 1). Near Yellowknife, Northwest Territories in Canada, the Ekati diamond mine opened in the late 1990s with great expectations. Diamond finds in the United States are shown on the map of ◆Figure 2. Most are solitary finds, and their sources are unknown. At Murfreesboro, Arkansas, however, thousands of diamonds have been recovered from an eroded diamond "pipe," one weighing 40 carats. For the price of admission, one can dig for diamonds here in the "blue ground" of the eroded pipe. One lucky hunter uncovered a 15-carat stone.

Precious opal (Sanskrit *upâlā*, "precious stone"; ◆Figure 1b) occurs in shallow surface diggings, and because it lacks crystalline form, it

B. Pipkin

(a)

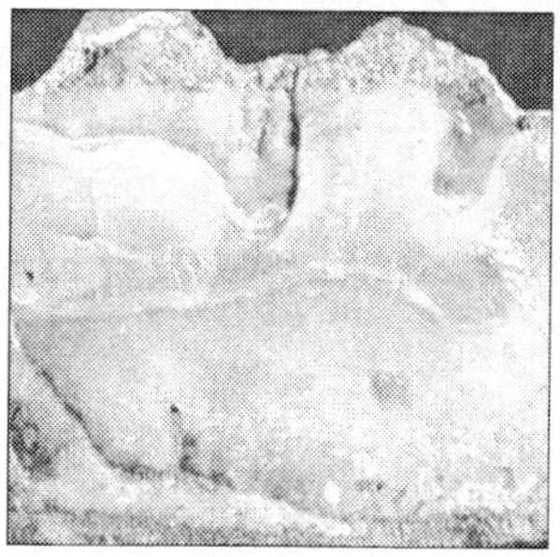

©Martin Land, Science Photo Library/Photo Researchers

(b)

©Carl Frank/Photo Researchers

(c)

◆ FIGURE 1 (a) Exact replica of the Cullinan diamond with the two largest stones cut from it, the Great Star of Africa (530.2 carats) and the Second Star of Africa (317 carats), which now reside in the British crown jewels collection. (b) Uncut precious opal. (c) Uncut emerald.

★TABLE 1 Major World Diamond Producers, 1997*

Country	Value, $ millions
Botswana	$1,461
Russia	1,210
Angola	1,035
South Africa	971
Congo	715
Namibia	410
Australia	288

*Just seven countries account for 90% of all rough-diamond production.
Source: Larry Gordon, *Los Angeles Times*, November 4, 1998.

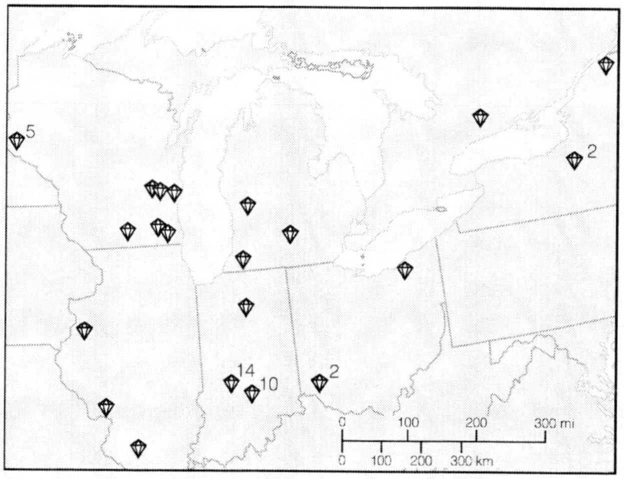

◆ FIGURE 2 Reported diamond finds in glacial drift in the Great Lakes region. Numerals indicate the number of diamonds found at the location; single finds have been reported at the other locations. After C.B. Gunn, *Gems and Gemology* 12, nos. 10 and 11 (1968), pp. 297–303, 333–334, © Gemological Inst. of America

actually is not a mineral. It is amorphous hydrated silica and gets its "fire" from an orderly arrangement of silica spheres interspersed with water molecules. Black and fire opals are the most desirable, and each stone is priced individually.

Emerald, the most valuable mineral carat-for-carat, is the precious variety of the silicate mineral beryl (◆ Figure 1c). It is found in coarse-grained granites called *pegmatites* (see Chapter 13) and in mica schist.

Quartz, plain old SiO_2, one of the most common minerals, has a host of semiprecious varieties such as purple amethyst, banded agate, yellow citrine, and smoky and rose quartz (◆ Figure 3). Searching for polishing-quality quartz varieties is the passion of many "rock hounds," and good crystal specimens also are cherished.

CONTINUED

◆ FIGURE 3 Some varieties of quartz: (a) colorless crystals, (b) smokey quartz, (c) amethyst, (d) agate, and (e) rose quartz. (a) Smithsonian Institute; (b)–(e) Sue Monroe

CASE STUDY 2.3

Minerals, Cats, and the Litter Belt

Many of Lovell, Wyoming's 2,100 inhabitants once depended upon the oil industry for their livelihood. The people there mine bentonite, which is used in oil drilling. Mixed with water, bentonite forms a "mud" that is circulated through the drill pipe and back to the ground surface outside the pipe. It forms a filter cake on the walls of the hole that prevents entry of water and loss of the drilling fluid into porous formations. It also cools the drill bit as the drill churns through hard rock and carries the drilled rock debris up to the surface. Oil prices dropped in the late 1980s, and Lovell all but dried up (Chapter 14).

Bentonite is formed by geochemical alteration of volcanic ash to form clay minerals, the most important of which is *smectite*. One of smectite's unusual properties is that it can absorb water equivalent to many times its original volume. The clay may swell as it absorbs water, forming a gooey mess, or it may crumble into fluffy pellets, depending on the amount of water and the initial size of the clay aggregates. Land areas underlain by bentonite may absorb so much water in the rainy season that they become impassable. Also, bentonitic soils can swell and cause severe damage to structures founded in it or on it (see Chapter 6). It has many uses. Native Americans once used it to remove oil and dirt from buffalo hides. It is used today in the landscaping of golf greens, to filter and clarify wine, to line landfills and ponds to prevent water loss, and in the manufacture of crayons.

Lovell was a potential ghost town without oil drilling. Bentonite to the rescue! It began when kitty litter replaced the sand box in 1948. The new product utilized a variety of clay-type minerals in the litter mix to absorb moisture. An enterprising chemist recognized the potential of bentonite, with its great moisture-absorbing properties, as cat litter. He reasoned that, with the proper grain size, the bentonite would clump when the cat did his duty, so that the soiled clumps could easily be removed from the litter pan. He was right, and the rest, as they say, is history (◆ Figure 1).

Today, almost all cat litter contains bentonite, and the Lovell mine is back in full-swing operation. The plant employs 60 people, and new houses are being built in the community to accommodate them. Bentonites in Wyoming are the result of volcanic eruptions in Cretaceous time, when dinosaurs roamed the land. The swath of mineable bentonites from Lovell to Casper has been dubbed the "Litter Belt" in Wyoming.

◆ FIGURE 1 © 1994 Margaret Miller/ The National Audubon Society Collection/Photo Researchers

A Rock Collection

Specific rocks have had historical, aesthetic, and practical significance to humans; Plymouth Rock, the Rock of Gibraltar, the biblical Rock of Ages, and the Rosetta Stone are examples. Various geologic agents of erosion have shaped the rocks of the Grand Canyon, Bryce Canyon, Yosemite, and other national parks into areas that we admire for their great beauty. Early humans lived in rock caves and threw rocks to defend themselves against their enemies and to kill animals for food. Over time humans came to build crude rock shelters, and then rock fences, and eventually beautiful castles and abbeys. Today rock and rock products are important mineral resources, used mostly as building materials. The beauty and usefulness of rocks are celebrated in these photographs.

Reed Wicander

(a)

Reed Wicander

(b)

◆ FIGURE 1 Ayers Rock, the world's largest sandstone monolith, is one of Australia's leading tourist attractions. An erosional remnant of a previous highland, it rises abruptly above the surrounding flat land. In geological terminology it is called an *inselberg* (German "island mountain") because of its resemblance to an island rising from the sea. Aborigines (native Australians) call Ayers Rock *Uluru.* In 1958 Ayers Rock and the rocks of the Olgas nearby were incorporated into Uluru National Park. (a) Ayers Rock rises 348 meters (1,141 ft) above the surrounding plain, and is scaled by 80 percent of the visitors to the park. (b) The Olgas, a circular grouping of more than 30 red domes, is a subdued but still spectacular sandstone inselberg in Uluru National Park. Its highest point, Mt. Olga, rises 460 meters (1,500 ft) above the plain. The Olgas' Aboriginal name is *katajuta,* "many heads." Ayers Rock is in the background.

From J. Monroe and R. Wicander, *The Changing Earth*, Brooks/Cole, 1997.

◆ FIGURE 2 Sixty-foot-high representations of the heads of four presidents were carved into Mount Rushmore's Harney Peak Granite in southwestern South Dakota between 1927 and 1941. The great work was possible because the Precambrian granite is relatively homogeneous (uniform) and not riddled with defects. Why would such a project not have been considered if Mount Rushmore were composed of slate, shale, or schist? Can you identify the presidents honored here?

Theodore Roosevelt Collection, Harvard Coll. Library

◆ FIGURE 3 President Theodore Roosevelt (*left*) and naturalist John Muir at Yosemite Valley in 1903. The rock on which they were photographed is Cathedral Peak Granite, one of the many granite intrusions that make up the Sierra Nevada batholith. Note the rounded exfoliation domes in the background (see Chapter 6).

B. Pipkin

◆ FIGURE 4 A skillfully assembled sandstone wall in Yorkshire, England. Yorkshire is the largest county (shire) in England, and literally thousands of kilometers of these walls are found there. This commonly occurring sandstone breaks readily into platy and blocky pieces, which makes it perfect for wall building.

Summary

Solar System

Description

Formed 5 billion years ago from an interstellar cloud (nebula).

Probable Cause

The nebula collapsed as a result of a gravitational attraction. The nebula started to spin and condense. The inner part of the cloud became the hot protosun and it was surrounded by a cooler, disc-shaped cloud, the solar nebula, which formed the large outer, Jovian planets. Rocky material closest to the sun formed the inner, terrestrial planets. Mars rocks are similar to Earth rocks.

Earth System

The Earth System has five interconnecting subsystems, or reservoirs, sometimes called "spheres." They are the solid Earth (includes lithosphere and soil), the atmosphere, the hydrosphere, the biosphere, and the extraterrestrial. Most are closed systems, such as the rock cycle and the hydrologic cycle, but energy is the exception. Continual interaction and recycling by these reservoirs is vital to the dynamic earth system.

Earth Materials

Crust

Outermost rocky layer of Earth composed of rocks that are aggregates of minerals.

Minerals

Naturally occurring inorganic substances with a definite set of physical properties and a narrow range of chemical compositions.

Classification By

Chemistry of anions (negative ions or radicals), such as oxides, sulfides, carbonates, etc. Silicates are the most common and important mineral group and are composed of silica tetrahedra $(SiO_4)^{-4}$ units that are joined to form chains, sheets, and networks.

Identification

Physical properties such as hardness, cleavage, crystal structure, fracture pattern, and luster.

Rocks

Consolidated or poorly consolidated aggregates of one or more minerals or organic matter.

Classes

Igneous—rock formed by crystallization of molten or partially molten material.

Sedimentary—layered rock resulting from consolidation and lithification of sediment.

Metamorphic—preexisting rock that has been changed by heat, pressure, or chemically active fluids.

Classification by

Texture and composition.

Structures

Many planar rock structures may be viewed as defects along which landslides, rockfalls, or other potentially hazardous events may occur. Stratification in sedimentary rocks and foliation in metamorphic rocks are common planar features. Faults and joints occur in all classes of rock.

Rock Cycle

A sequence of events by which rocks are formed, altered, destroyed, and reformed as a result of internal and external earth processes.

Geologic Time

Relative Time

The sequential order of geologic events established by using basic geologic principles or laws.

Geologic Time Scale

The division of geologic history into eons, eras, periods, and epochs. The earliest 88 percent of geologic time is known informally as Precambrian time. *Precambrian* rocks are generally not fossiliferous. Rocks of the Cambrian Period and younger contain good fossil records of shelled and skeletonized organisms. Environmental geology is concerned mainly with geologic events of the present epoch, the Holocene, and the one that preceded it, the Pleistocene.

Absolute Age

Absolute dating methods use some natural "clock." For long periods of time, such clocks are known rates of radioactive decay. A parent radioactive isotope decays to a daughter isotope at a rate that can be determined experimentally. This established rate, expressed as a half-life, enables us to determine the age of a sample of material.

Age of Earth

Comparing present U/Pb ratios with those in iron meteorites, we can extrapolate backward and obtain an Earth age of 4.6 billion years.

Rocks and Crust

The oldest known in-place rocks are 3.96 billion years old. Dating of minerals derived from older crust yields an age of 4.0–4.3 billion years for stable continental crust.

Key Terms

absolute-age dating
alpha decay
amorphous
aphanitic texture
atom
atomic mass
atomic number
batholith
beta decay
biogenic sedimentary rocks
chemical sedimentary rocks
clastic

clastic sedimentary rocks
cleavage
dip
element
extrusive
fault
felsic
foliation
fracture
half-life
igneous rocks
intrusive
ion
isotope
joint
lava
lithified
luster
mafic
magma
metamorphic rocks
mineral
Mohs hardness scale
nucleus
phaneritic texture
plutonic
pyroclastic
radiometric dating
recrystallization
relative-age dating
rock cycle
rocks
sedimentary rock
stratification
strike

Study Questions

1. What physical and chemical factors are the bases of the rock-classification system? What are the three classes of rocks, and how does each form? Identify one characteristic of each class that usually makes it readily distinguishable from the others.
2. How and where do batholiths form? What type of rock most commonly forms in batholiths?
3. What are some of the planar surfaces in rocks that may be weak and thus lead to various types of slope failure?
4. How may structures in sedimentary rocks be used to reconstruct past environments?
5. How may absolute-age-dating techniques be used to the betterment of human existence?
6. Earth is 4.6 billion years old, and humans have been on Earth for only a few hundred thousand years. What changes of a global nature have humans invoked on Earth's natural systems (water, air, ice, solid Earth, and biology) in this short length of time? Which impacts are reversible, and which ones cannot be reversed or mitigated?
7. What is the most common mineral species? Name several rocks in which it is a prominent constituent.
8. What is cleavage? How can it serve as an aid in identifying minerals?
9. The most common intrusive igneous rock is composed mostly of (1) the most common mineral and (2) a mineral of the most common mineral group. Name the rock and its constituent minerals.
10. What mineral is found in both limestone and marble?

For Further Information

Books and Periodicals

Albritton, C. C. 1984. Geologic time. *Journal of geological education* 32, no. 1: 29–47.

Bowring, S. A., I. S. Williams, and W. Compston. 1989. 3.96 Ga gneisses from the Slave province, Northwest Territories, Canada. *Geology* 17: 971–975.

Brown, V. M. and J. A. Harrell. 1991. Megascopic classification of rocks. *Journal of geological education* 39: 379.

Dietrich, R. V. and Brian Skinner. 1979. *Rocks and rock minerals.* New York: John Wiley & Sons.

Eicher, D. L. 1976. *Geologic time,* 2nd ed. Englewood Cliffs, N.J.: Prentice-Hall, Inc.

Gunter, Mickey Eugene. 1994. Asbestos as a metaphor for teaching risk perception. *Journal of geological education* 42: 17.

Harvey, Carolyn, and Mark Rollinson. 1987. *Asbestos in the schools.* New York: Praeger.

Hurley, Patrick. 1959. *How old is the earth?* New York: Anchor Books.

Libby, W. F. 1955. *Radiocarbon dating.* Chicago: University of Chicago Press.

Mackenzie, Fred T. G. and Judith A. Mackenzie. 1995. *Our changing planet: An introduction to Earth system science and global environmental change.* Englewood Cliffs, N.J.: Prentice-Hall, Inc.

National Research Council, National Academy of Sciences. 1993. *Solid-earth sciences and society.* Washington, D.C.: National Research Council Commission on Geosciences, Environment, and Resources, National Academy of Sciences Press.

Newcott, William R. 1998. Return to Mars. *National geographic* 194, no. 2: 2–29.

Nuhfer, E. B., R. J. Proctor, and Paul H. Moser. 1993. *The citizen's guide to geologic hazards.* Arvada, Colo.: American Institute of Professional Geologists.

Parker-Pope, T. 1997. Cat litter breathes new life into region of bentonite mines. *The Wall Street Journal,* April 1.

Skinner, H. Catherine W. and Malcom Ross. 1994. Geology and health. *Geotimes* (Washington, D.C.: American Geological Institute), January: 11–12.

________. 1994. Minerals and cancer. *Geotimes,* January: 13–15.

Snow, T. P. 1993. *Essentials of the dynamic universe.* St. Paul, Minn.: West Publishing Company.

Monroe, James, Reed Wicander, and Richard Hazlett. 2006. *Physical geology: exploring the Earth.* Belmont, California: Brooks/Cole.

ThomsonNOW™

Assess your understanding of this chapter's topics with additional comprehensive interactivities at http://www.thomson.com/login, which also has up-to-date web links, additional readings, and exercises.

Northern California

Geology of Northern California

Frank DeCourten

Department of Earth Science Sierra College

Standing more than 10,000 feet (3,000 m) above the surrounding terrain, Mt. Shasta is the largest volcano in northern California and symbolizes the dynamic geologic processes that have shaped a spectacular landscape.

U.S. Department of the Interior
U.S. Geological Survey

National Elevation Data Set Shaded Relief of California

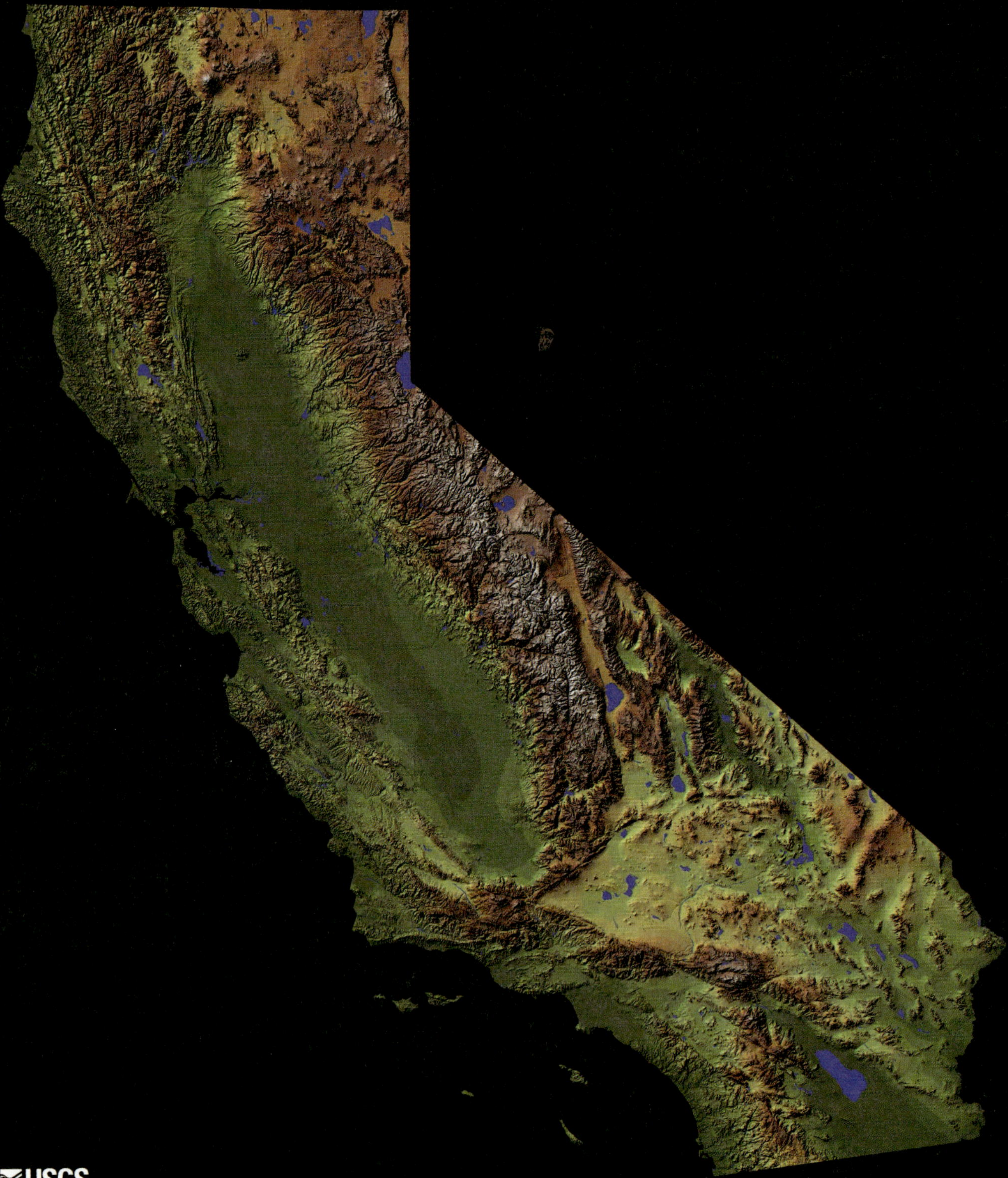

USGS
science for a changing world

Geography · *The National Map*

ESSENTIAL QUESTIONS TO ASK

Northern California.1 Introduction

What are northern California's physiographic provinces?
What is the Farallon subduction zone?
What two types of plate boundaries exist in northern California today?
What are terranes, how do they originate, and why are they important in northern California?

Northern California.2 The Sierra Nevada: California's Geologic Backbone

What is the Sierra Nevada batholith?
What kinds of rocks surround the Sierra Nevada batholith?
When and how was the modern Sierra Nevada uplifted?
What types of gold deposits occur in the Sierra Nevada?
What is the Mother Lode?

Northern California.3 The Klamath Mountains

In what ways are the Klamath Mountains and the Sierra Nevada similar?
What kinds of rocks comprise the ophiolites in the Klamath Mountains and what tectonic events do they signify?
What mineral resources occur in the Klamath Mountains?

Northern California.4 The Great Valley

What factors have led to the formation of fertile soils in the Great Valley?
What was the origin of the sedimentary rocks in the Great Valley Sequence?
What is the origin of the natural gas produced in the Great Valley?
Why is the Great Valley so prone to chronic flooding?

Northern California.5 The Northern Coast Ranges

What is the Franciscan complex and how did it form?
How did the San Andreas fault system originate?
What is the Salinian block?
How old are the northern Coast Ranges and what tectonic forces elevated them?

Northern California.6 Volcanoes of the Cascade Range and Modoc Plateau

What California volcanoes are part of the Cascade Range?
What is the Cascadia subduction zone?
What kind of volcanic activity typifies the Cascade Range?
Why is Mount Lassen an especially interesting volcano in the Cascade Range?

Northern California.7 The Basin and Range of Northeast California

What tectonic forces are responsible for the pattern of alternating mountains and valleys in the Basin and Range province?
What mountains and basins in northern California belong to the Basin and Range province?

Northern California.8 The Northern California Ice Ages

During what time periods did northern California experience Ice Age conditions?
What landscape features resulted from the Pleistocene Epoch glaciations in northern California?
How did the Pleistocene landscape of northern California differ from the modern setting?

Northern California.9 Northern California Earthquakes

What plate tectonic settings are associated with northern California earthquakes?
What is the likelihood of another major earthquake in northern California?
What might be the effects of a large northern California earthquake?
Given the severity of the potential hazards, how can the effects of earthquakes be minimized in northern California?

Northern California.10 Living on the Edge: Coastal Hazards in Northern California

In comparison to the coast of southern California, why is the northern California shoreline so rugged and scenic?
What coastal hazards exist in northern California?
How do human activities affect coastal hazards?

Printed in Canada

Northern California.1
Introduction

California's Varied Landscape: California is arguably the best place in the world to study geology. Few areas of comparable size are as geologically varied, physiographically diverse, or so spectacularly scenic as the Golden State. California's abundant natural resources reflect in large measure its rich geological history, and earthquakes, floods, and mass wasting events underscore the importance of ongoing geologic processes for residents and visitors. With the boundary between two of the largest lithospheric plates on the planet running for more than 1,300 kilometers along the western side of the state, California is an outstanding natural laboratory for studying Earth processes and plate interactions.

During the 1960s and 1970s, our understanding of earth dynamics shifted dramatically as the modern theory of plate tectonics was developed and refined. Many of the new concepts of that era either were developed in California or were specifically formulated to explain and/or reinterpret its incredibly varied geological features. Today, with a new theoretical framework and vastly improved tools for exploration and analysis, scientists continue to consider California a geologic paradise.

The rocks in California tell an amazing story of the evolution of land and life at the edge of North America. Although it would require many years to explore all the geologic wonders in California, a first course in physical geology is an excellent start to a lifetime of fascinating adventure in the Golden State. The purpose of this chapter is threefold: (1) to introduce the broader aspects of California's geologic setting, (2) to outline the major tectonic events that have shaped its landscape over geologic time, and (3) to examine the impacts on humans of the state's ongoing geologic evolution. I also hope you will consider this chapter an invitation to begin your own lifelong exploration of one of the most magnificent landscapes in the world.

California's Physiographic Provinces: The California landscape can be divided into a dozen regions of distinctive and characteristic geology, landforms, climate, geomorphic trends, soils and vegetation, and drainage. These natural areas are known as **physiographic provinces**. After the geologic alignment of the state's major mountain systems, most of the geomorphic provinces are oriented in a northwest-to-southeast trend (Figure NC.1). Hence some of the provinces, such as the Sierra Nevada and the Coast Ranges, extend from the northern part of the state to the southern portion. For the purposes of the present discussion, we will arbitrarily define northern California as the region between Monterey Bay (latitude approximately 36.5° N) and the California–Oregon border. In this portion of California, seven natural physiographic provinces comprise the landscape: the northern Sierra Nevada, the northern Coast Ranges, the northern Central Valley, the Klamath Mountains, the Cascade Range, the Modoc Plateau, and portions of the Basin and Range. Though there are consistent geologic patterns within each of these regions that make them distinctive and identifiable, none can be regarded as simple or monotonous. Each of the physiographic provinces in northern California is a varied and fascinating geological realm with endless opportunities for applying the knowledge you have gained in your physical geology course. Collectively, they represent a region of such complex origin that scientists have yet to develop completely satisfactory explanations of all of the geologic features in this varied terrain. The common attribute of all the physiographic provinces of northern California is that, one way or another, each reflects the consequences of plate tectonic interactions along the western margin of North America over the past 500 million years.

Geologic Map of California: Geologic maps are indispensable assets in exploring the natural history and geologic setting of any region. Such maps show the distribution of rock types and ages on the surface, along with information about the orientation of the various rock bodies, the nature of their contact with adjacent rock masses, and the trends and extent of geologic structures such as faults and folds. This information is essential in unraveling the geologic history of a region because it reveals spatial patterns in the distribution of rocks of various types, ages, and degree of deformation that reflect the tectonic and geologic events of the past. In addition, geologic maps are of critical importance in locating surface and subsurface deposits of earth resources and in identifying the areas most susceptible to various geologic hazards.

Figure NC.1 The Physiographic Provinces of California

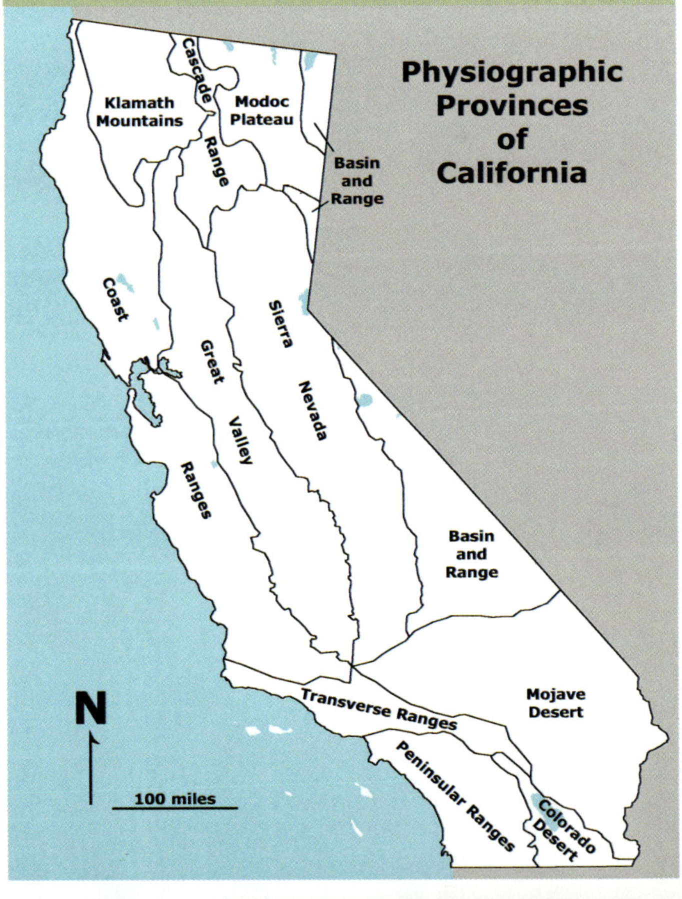

Map courtesy of the U.S. Geological Survey

▶ **Figure NC.2** Geological Map of California

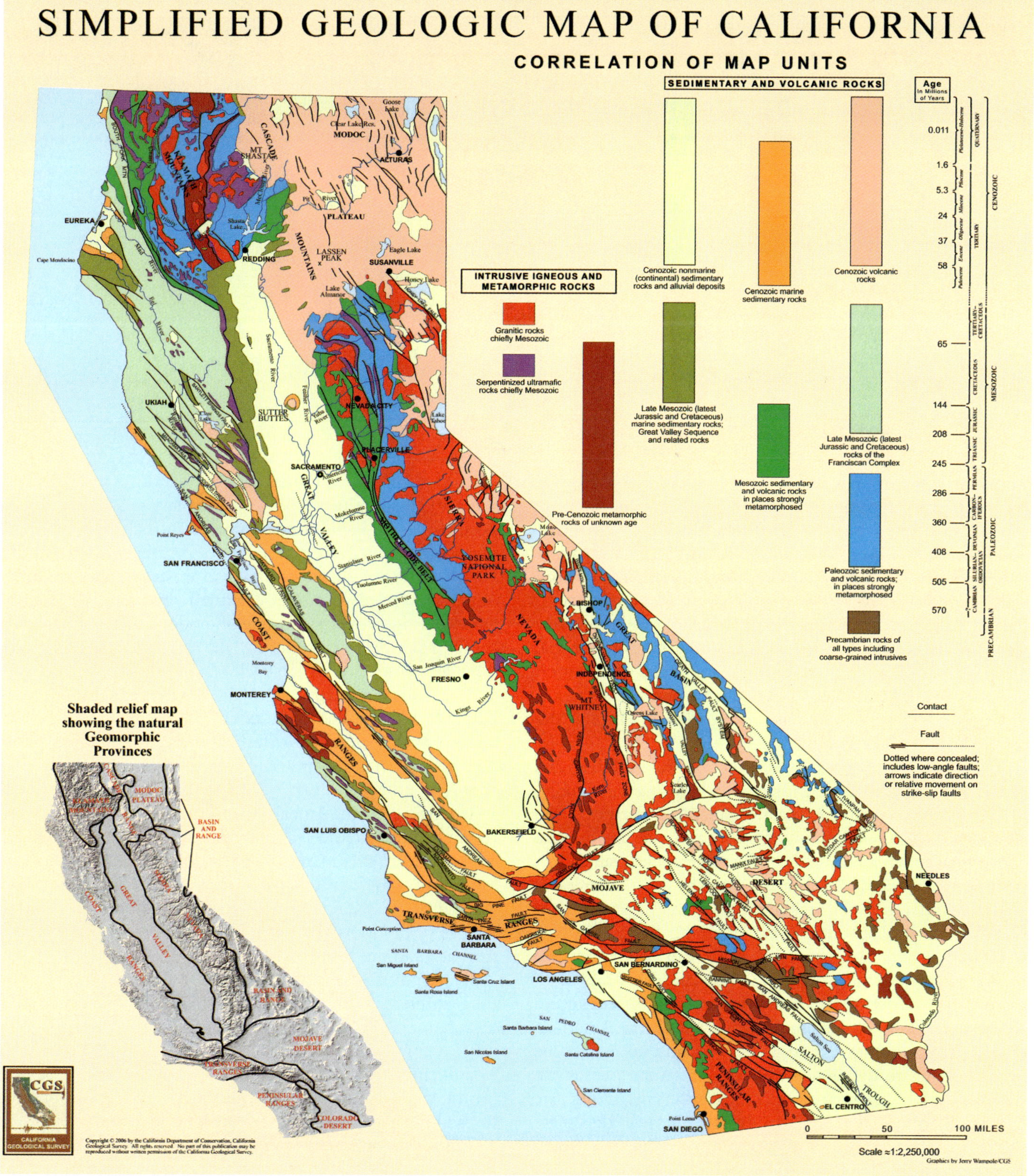

The geologic map of California (Figure NC.2) has been compiled by the California Geological Survey over many decades of geologic mapping. This map is scaled to show the entire state and, as such, it portrays only the broader distribution of various California rocks and structures. Nonetheless, you will notice at a glance the strong similarity between the physiographic provinces in California and the distribution of various rock types. For example, notice the similarity in location, trend, and extent of the Sierra Nevada province and the large area of red, blue, and green

▶ **Figure NC.3** Interactions between The Farallon, Pacific, and North American plates over the past 40 million years.

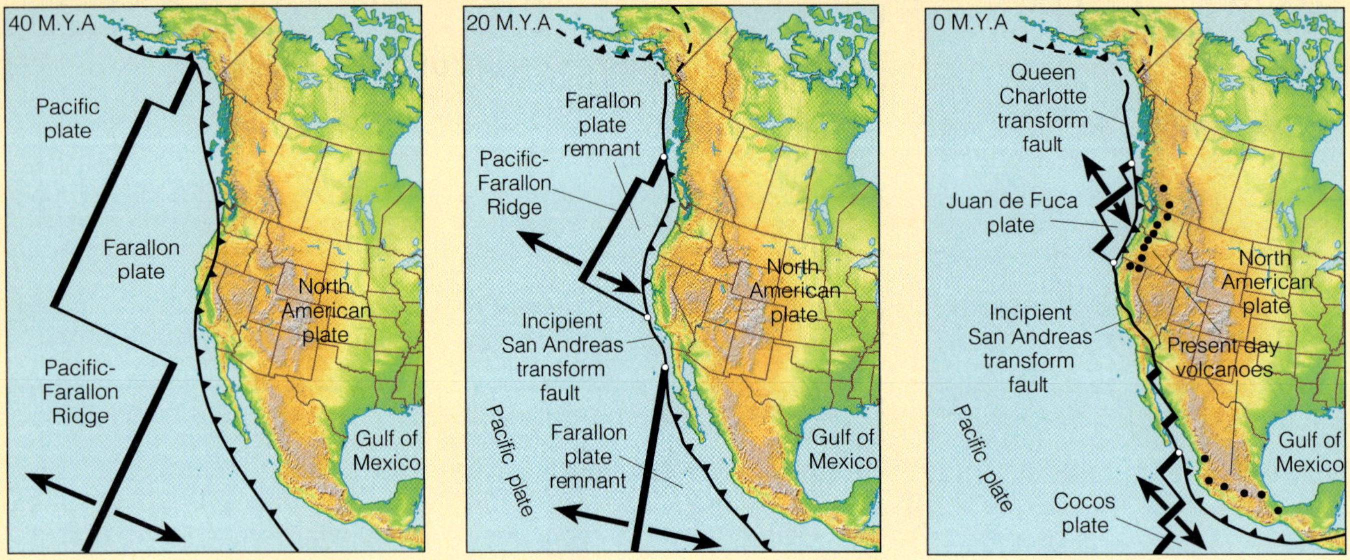

colors on the geologic map. These colors represent the Mesozoic-age plutonic igneous rocks that comprise the core of the Sierra Nevada (red) and the older rocks that were metamorphosed by the emplacement of the magma (blue and green). A very similar pattern is observed in the Klamath Mountains physiographic province northwest of the Sierra Nevada, suggesting that the two regions share some common geologic traits. The similarity between the physiographic provinces of California and the distribution of various types and ages of rock in the state underscores the importance of the geologic foundation in shaping the character of the landscape.

The Westward Migration: For the past 500 million years convergent plate tectonic interactions have prevailed along the western margin of North America. Early in the Mesozoic Era, about 200 million years ago, the rate of plate convergence increased significantly as the ancient supercontinent of **Pangaea** began to break up. The North American plate separated from the northern part of the supercontinent via the opening of the Atlantic Oceanic basin. Seafloor spreading in this basin propelled the North American plate to the west while the Eurasia and African plates moved in the opposite direction. Several different oceanic plates were subducted under the leading edge of the North American plate as it slowly moved west. The last and largest of the oceanic plates to descend beneath North America was the **Farallon plate**, remnants of which still exist along the western margin of North and Central America. This plate tectonic interaction along the western edge of North America produced the **Farallon subduction zone**, which was established in mid-Mesozoic time and persisted from some 160 million years ago until about 30 million years ago. Many of the major geologic trends in California are the result of this long history of plate convergence.

Origin of the Modern Transform Plate Boundary: The Farallon plate originated at an oceanic ridge to the west and southwest of North America. On the other side of that ancient spreading center, the Pacific plate was sliding to the west, while the Farallon plate moved in the opposite direction toward the advancing margin of North America. A little less than 30 million years ago, the western edge of North America collided with, and eventually overran, the Pacific-Farallon ridge near the latitude of modern Los Angeles. As a consequence of this collision, the North American plate came into contact with the Pacific plate (Figure NC.3).

The collision between North America and the Pacific-Farallon ridge ended plate convergence in the California region and established the modern **transform plate boundary**. After the initial collision of the ridge and the continent, the transform plate boundary expanded north and south as North America continued moving west, overrunning more of the ridge in the process (Figure NC.4). The relatively small Cocos, Rivera, and Juan de Fuca plates represent modern remnants of the ancient Farallon plate. The famous San Andreas fault system developed as a consequence of the transform plate boundary between the North America and the Pacific plates.

The geologic setting of modern northern California is thus influenced by two different kinds of boundaries between the North American and oceanic plates to the west: a transform boundary from Monterey Bay to Cape Mendocino and a remnant convergent boundary to the north. We will explore the consequences of these interactions in more detail in the sections that follow.

California and Accreted Terranes: One of the consequences of the long history of plate convergence along the western margin of North America was the accretion of numerous blocks of rock to stable crust of the continent. Such blocks added to the edge of a continent by plate convergence are known as **accretionary terranes**, or simply **terranes**. Five hundred million years ago, there was no land where the Pacific Coast now stands. The crust of California

▶ Figure NC.4 Major terranes of western North America. In addition to the Franciscan Complex and the Salinian block illustrated here, dozens of smaller terranes have been identified in northern California.

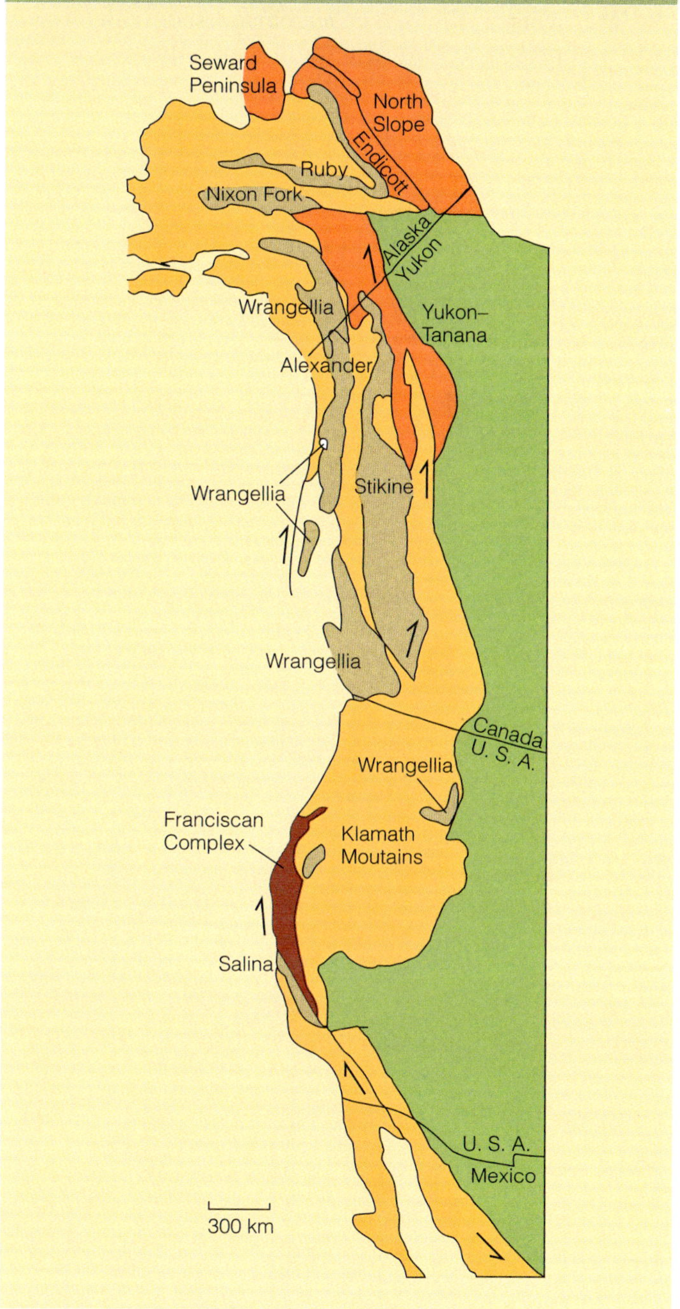

has since been assembled in piecemeal fashion as seamounts, island arcs, coral reefs, and small continental blocks that were carried on subducting oceanic plates collided with the western edge of the continent and were embedded into the existing margin. The rocks in the various terranes were metamorphosed and deformed as each was sutured into North America like pieces of a mosaic. In this manner, North America grew incrementally westward with the addition of each fragment. Geologists are in general agreement that about 100 such terranes were accreted to the western margin of North America since the breakup of Pangaea about 200 million years ago. Several others were added in earlier accretionary events in the Paleozoic Era, before North America separated from Pangaea. It was not until Cenozoic time that all of the terranes were in place along the Pacific Coast and the modern California landscape began to emerge.

California, as a whole, therefore represents a geologic collage, an amalgam of pieces assembled through the convergence of plates along the west edge of North America over the past 500 million years. Northern California is especially intriguing because here both a remnant of an ancient convergent boundary and the modern transform boundary to the south continue to shape the landscape. It is not surprising that geologists find northern California such a fascinating region. It is a place where the geologic past meets the dynamic present, and there is no place in the world better suited for geologic exploration. Let's look a little closer.

Section Northern California.1 Summary

- The geology and landscape of northern California is extremely varied, with seven different physiographic provinces, each with distinctive rock assemblages and geologic histories.
- Convergent plate boundaries have existed for 500 million years in the northern California region. After North America separated from Pangaea early in the Mesozoic Era, the rate of convergence increased as several oceanic plates were subducted under the west-moving continent.
- The subduction of the Farallon plate in the Mesozoic Era resulted in many of the geologic trends that can be observed in modern California. About 30 million years ago, plate convergence ended as the transform boundary between the Pacific and North American plates began to develop. The unique geologic setting of modern northern California is shaped by both a remnant of the Mesozoic subduction zone and the continuing evolution of the more recent transform plate boundary.

Northern California.2 The Sierra Nevada: California's Geologic Backbone

The Sierra Nevada is California's best known mountain system. Stretching for more than 700 kilometers from Lake Almanor in the north to Tehachapi in the south, this northwest-trending mountain system is home to three national parks and the highest peak in the coterminous United States at Mount Whitney (14,495 feet/4,418 meters above sea level). Winter storms passing east from the Pacific Ocean over the high Sierra Nevada produce heavy snowfall that is an essential supply of water to the entire state. Materials released from the weathering of Sierra Nevada bedrock help sustain the fertility of California's rich agricultural soils. It was in the Sierra Nevada foothills that gold was discovered in the 1800s, and the course of California history was forever changed. Without the Sierra Nevada, California would simply not be California.

The bedrock of the Sierra Nevada is dominated by the Mesozoic Sierra Nevada **batholith**, one of the largest and most complex masses of granitic rock in the world. Adjacent to the batholith, and sometimes as **xenoliths** and **roof pendants** within it, are older Mesozoic and Paleozoic metamorphic rocks that were invaded by the subterranean magma. More recent episodes in the evolution of the Sierra Nevada are documented by Cenozoic **volcanic** and sedimentary rocks that rest on the granite-metamorphic basement.

The Sierra Nevada Batholith: The geologic map of California clearly indicates that the Mesozoic granite of the Sierra Nevada batholith comprises the core of the Sierra Nevada. Vast exposures of such light-colored **plutonic** rock in the high country of Yosemite, Kings Canyon, and Sequoia National Parks, is in part what led John Muir to refer to the Sierra as "the range of light" (Figure NC.5). The **Sierra Nevada batholith** is a complex assemblage of perhaps as many as 200 individual **plutons** representing magma bodies emplaced mostly between 140 million and 80 million years ago at depths of 10 to 30 kilometers beneath the surface. Most of the plutons comprising the Sierra Nevada batholith consist of **felsic** rock such as **granite**, rich in light-colored quartz, potassium feldspar, and sodium-rich plagioclase (Figure NC.6). Some of the plutons are richer in the darker ferromagnesian minerals and consist of rock more similar to **diorite** or **gabbro**.

The Sierra Nevada batholith represents the deep roots of a Mesozoic **volcanic arc** that developed **along** the western margin of North America above the Farallon subduction zone. In this ancient subduction zone, magma was produced by partial melting and migrated upward through older crustal material to build a chain of andesitic volcanoes that rivaled the modern Andes Mountains of South America. Recent geologic studies have revealed evidence of large **calderas** formed by explosive eruptions in this ancient volcanic chain. Late in the Cenozoic Era, the uplift of the Sierra Nevada resulted in the nearly complete erosion of the volcanic edifice that originally concealed the batholitic rocks. However, isolated remnants of the ancient volcanoes can still be seen in places along the crest of the range (Figure NC.7).

Paleozoic-Mesozoic Accretionary Terranes of the Northern Sierra Nevada: The magma that formed the Sierra Nevada batholith was emplaced into older rocks that either accumulated on the seafloor west of the margin of

▶ **Figure NC.5** Vast exposures of granitic rock characterize the Sierra Nevada, John Muir's "Range of Light".

Frank DeCourten

▶ Figure NC.6 Close up of granite from the Sierra Nevada Batholith. Dark crystals are biotite and hornblende, white crytals are plagioclase and potassium feldspar, and the gray translucent crystals are quartz.

Frank DeCourten

▶ Figure NC.7 Dark-colored rocks of the Ritter Range in the eastern Sierra Nevada are remnants of the Mesozoic Sierra volcanic arc.

Dick Hilton

Paleozoic North America or were accreted to the edge of the continent in convergent plate interactions that preceded the Farallon subduction zone. These old rocks were deformed and metamorphosed during several different accretion events and altered by the heat and fluids associated with magma rising from the Farallon subduction zone in later Mesozoic time. Metamorphic rocks such as marble, slate, schist, **serpentinite**, and **greenstone** are common in the pre-batholithic terranes of the Sierra Nevada. Some of

▶ Figure NC.8 Folds in these Paleozoic sedimentary rocks along the Yuba River were developed during the collision between a block of oceanic rocks and western edge of North America in the Mesozoic Era.

Frank DeCourten

these rocks have been affected by multiple periods of metamorphism. In addition, these rocks are commonly folded and faulted by the compressional stress generated along ancient convergent plate boundaries (Figure NC.8).

It has taken geologists many decades to unscramble the complicated history recorded in the metamorphic rocks surrounding the Sierra Nevada batholith. Within the granitic core of the Sierra Nevada, exposures of these rocks are found as roof pendants and xenoliths (Figure NC.9) preserved in the plutonic rocks. In the foothills north and west of the batholithic core of the Sierra Nevada, extensive exposures of metamorphic rocks have revealed a complex history of metamorphism and accretion. Because there has been much greater Cenozoic uplift in the southern Sierra Nevada than in the north, erosion there has progressed to a deeper level in the crust and much of the older rock that formerly surrounded the granitic plutons has been removed.

Within the western foothills metamorphic belt, geologists have identified three primary units of metamorphosed rocks that are separated from each other by major **faults**. These faults probably were produced when seamounts, undersea volcanic arcs, oceanic crust, or other fragments of lithosphere collided with the edge of North America. The metamorphosed and deformed rocks comprising these

▶ **Figure NC.9** Dark-colored xenoliths in Sierra granite represent fragments of older rock assimilated into the magma prior to crystallization.

Frank DeCourten

terranes are almost entirely of oceanic origin, and generally older than the tectonic events that sutured them to the North American continent. Some represent mafic magma erupted by undersea volcanoes, and the characteristic **pillow structures** can still be observed (Figure NC.10). Other terranes include oceanic sedimentary rocks such as **chert**, **mudstone**, and **limestone** that have been metamorphosed to varying degrees. There is strong evidence that some of the accreted fragments traveled thousands of kilometers before the now-vanished plates carrying them sank into ancient subduction zones. **Ophiolites**, sequences of rocks representing oceanic lithosphere, are present in some terranes, and probably originated as fragments splintered from the descending oceanic plates sutured into the various terranes. Adding complexity to some of the terranes is the presence of **mélanges**, chaotic mixtures of deformed and metamorphosed rocks that are formed in subduction zone settings. We will learn more about ophiolites and mélanges when we explore the northern Coast Range, where they formed within the much younger Farallon subduction zone.

▶ **Figure NC.10** The rounded pillow structures in this outcrop of mildly metmorposed basalt in the western metmorphic belt indicate eruption of lava on the seafloor.

Frank DeCourten

Cenozoic Rocks of the Sierra Nevada: At the end of the Mesozoic Era about 65 million years ago, the Sierra Nevada was an elevated volcanic terrain perched on a complex basement comprised of older rocks intruded by the concealed Sierra Nevada batholith. As Cenozoic time (the Tertiary Period) began, igneous activity appears to have temporarily subsided in the Sierra region, and erosion was beginning to attack the summits of the dormant volcanoes. The volcanic highland extended into what is now western Nevada and the ancient shore of the ancestral Pacific Ocean was located along the foothills of the dormant volcanic arc in central California. Rivers draining the volcanic highland ran west across the site of the modern Sierra Nevada. These ancient rivers steadily wore away the volcanic bedrock, and transported great quantities of sediment to the ocean basin to the west. This long period of erosion in the Early Cenozoic Era produced a pronounced plain, which has since been elevated hundreds of feet above the floors of modern stream valleys in the Sierra Nevada region (Figure NC.11).

During the Eocene Epoch, about 50 million years ago, river-deposited sediments began to accumulate in the ancient river systems of the Sierra Nevada. In the stream channels, coarse gravel deposits formed. On the floodplains, fine sand, silt, and mud accumulated. These river-deposited sediments are still preserved in many places in the foothills of the Sierra Nevada as colorful exposures of conglomerate, sandstone, and mudstone. Geologic mapping in the early 1900s demonstrated that exposures of the ancient river gravels were aligned as elongated ribbons, indicating the location and drainage pattern of the 50-million-year-old river channels. Soon after the California Gold Rush began, prospectors discovered that some of the Eocene-age river gravels contained rich concentrations of gold in the form of nuggets, flakes, and fine particles. The richest gravel deposits became known as the "auriferous gravels." Such gold-bearing sediments were intensely mined and processed in the mid-1800s, primarily by washing away the weakly-cemented material with powerful steams of water and separating the small gold particles from the loosened sand, pebbles, and cobbles (Figure NC.12). Known as hydraulic mining, this process was phased out after 1884 because so much sediment had been washed from the Sierra slopes that agriculture and river navigation downstream in the Central Valley was being adversely affected. The best exposures of the auriferous gravels today are in the scars and excavations left from the hydraulic mining activities more than century ago.

During the Oligocene and Miocene Epochs, 35 million to 5 million years ago, volcanic activity resumed along what is today the crest of the Sierra Nevada and areas to the east. The initial eruptions were violent caldera-forming explosions that sent great flows of ash and fragments of volcanic

▶ **Figure NC.11** Beyond the deep canyon of the Yuba River on the Sierra Nevada west slope, the flat surface reflects a lengthy interval of erosion in the early Tertiary Period.

Frank DeCourten

▶ **Figure NC.12** Hydraulic mining in the Sierra Nevada involved washing Eocene sediments with powerful jets of water. Photo taken in 1905 at Junction City, California.

Bettmann/Corbis

▶ **Figure NC.13** Oligocene tuff near Soda Springs, California consists of ash partcles and white pumice fragments welded into a coherent rock.

Frank DeCourten

rock surging west through the canyons carved by the Eocene-age rivers. These **pyroclastic** deposits hardened into **tuffs** and **welded tuffs** that partially filled the ancient canyons (Figure NC.13). The later eruptions produced **basalt** flows, volcanic **breccias**, and volcanic mudflow deposits known as **lahars**. The Tertiary-age volcanic rocks completely filled some of the valleys of the western slope of the ancestral Sierra Nevada, forcing the ancient streams from their channels. In the past few million years, as the modern Sierra Nevada rose, vigorous down-cutting has left the erosion-resistant valley-filling Tertiary volcanic rocks perched high above modern river beds. Sinuous ridges capped by Tertiary volcanic rocks demonstrate the phenomena of **inverted topography** in many places in the Sierra foothills (Figure NC.14).

Recent Uplift of the Sierra Nevada: Though the rocks of the Sierra Nevada document a long history of geologic unrest, good evidence suggests that the modern range reflects a relatively recent pulse of uplift. Although the Sierra Nevada was probably an elevated tract of land since the mid-Mesozoic Era, recent geologic studies suggest that 5 million to 10 million years ago, the mountain system rose to its current elevation, primarily by westward tilting along **normal faults** located along the eastern escarpment. These faults, many of which remain active today, have uplifted and tilted the range to the west, producing a spectacularly rugged eastern escarpment and a gently inclined western slope. The steep eastern escarpment of the Sierra Nevada was a formidable barrier to the migration of people during the Gold Rush. Even today, only a few highways cross the crest of the northern Sierra Nevada through passes that range in elevation from more than 3,000 meters (9,945 feet) to just over 2,200 meters (7,259 feet) above sea level.

Geologists have not yet determined the precise cause for the recent uplift of the Sierra Nevada. The forces affecting the Sierra region are complicated by its proximity to regions of differing plate tectonic interactions. In the Basin and Range province to the east, tensional stresses are stretching the crust and producing the numerous normal faults, such as those along the eastern escarpment of the Sierra Nevada. To the west, the modern transform boundary between the Pacific and North American plate produces shear stresses that tend to move the Sierra Nevada block to the northwest. Recent studies also suggest that the dense lower crust of the southern Sierra Nevada may have been removed in late Cenozoic time by heat associated with material upwelling from the mantle. The removal of the dense root beneath the Sierra Nevada would have increased the buoyancy of the Sierra Nevada, causing it to rise. All of these factors may have been involved in the recent ascent of the Sierra Nevada. Whatever the precise cause of uplift may be, there is no doubt that it is still continuing. The northern portion of the Sierra Nevada is currently rising at a rate of 2 to 3 millimeters per year, estimated on the basis of stream incision rates and the increased tilt of ancient streambeds.

The Mother Lode: Gold and Geology of the Sierra Foothills: California is known as the "Golden State" for a reason: since the initial discovery in the mid-1800s, more than 115 million ounces of gold have been produced in the state, an amount equivalent to a volume of more than 190 cubic meters! At current prices, the cumulative value of California gold is more than $110 billion. Though gold has been found in many places in California, more than 75% of the amount recovered historically has come from the western foothills of the Sierra Nevada. It was also in this region that James Marshall first noticed the glittering nuggets in the American River in 1849 and ignited one of the most dramatic human migrations in history.

Figure NC.14 Inverted topography in the Sierra Nevada foothills is a common consequence of Miocene volcanic activity.

James S. Monroe

Gold was initially discovered in California as flakes and nuggets resting in modern river gravels. Such deposits are known as **placers**, and early miners invented several techniques to wash the gold particles from the loose river sediments. Soon, miners began to exhaust the placer deposits, and looked upstream from the gravel bars to find the source of the gold particles. They eventually found two sources for the placer gold. First, it was discovered that the Eocene-age river deposits were also rich in gold, particularly the lower channel-filling conglomerates. The "auriferous gravels" were more difficult to work than modern placers because they were more consolidated and generally exposed high above the modern streams. The invention of the hydraulic mining methods mentioned earlier helped to solve these difficulties, but the disastrous environmental consequences ended this practice in the 1880s. At the same time, prospectors discovered a second source of the placer gold: quartz veins in the bedrock in the western foothills metamorphic belt were found to contain large masses of native gold. Such gold-bearing quartz veins are called **lodes**, and the largest concentration of such ores in the Sierra foothills became known as the Mother Lode (Figure NC.15).

Geologists are still not certain about the source of all the gold in California, but the variety of ore types suggests multiple origins. Much of the Mother Lode gold may have been originally disseminated in the oceanic metamorphic rocks or in the younger igneous bodies. Some of the gold may have originated in the volatile gases associated with the granitic magma that intruded the Sierra Nevada basement during the Mesozoic era. During metamorphism related to either burial, tectonic accretion, or the emplacement of magma, hot fluids were introduced into the rocks of the Mother Lode belt and migrated through them along fractures, faults, or shear zones. Interactions between the circulating hot (hydrothermal) fluids and the metamorphic rock concentrated gold in the quartz-rich fluids. The hydrothermal fluids eventually cooled as they circulated through the fractured metamorphic rock, leaving veins of quartz laced with pure gold. When the gold-rich metamorphic rocks were eroded by Cenozoic-age streams, the heavy gold particles accumulated in nearby stream sediments and the less dense components of the bedrock were washed farther downstream.

In the Mother Lode belt, all three types of gold deposits—placers, auriferous gravels, and lodes—were present. It is not surprising that this region was the primary target of the Gold Rush prospectors and miners. Though production of gold in California fell dramatically after the 1860s, people occasionally still find gold in the streams of the Sierra Nevada.

▶ **Figure NC.15** Specimen of gold from Grass Valley, California. Inset map shows location of the Mother Lode region in the western foothills of the Sierra Nevada.

National Museum of Natural History, Specimen #R12197, Photo by D. Penland.

Section Northern California.2 Summary

- The Sierra Nevada is California's geological backbone and the largest mountain system in the state. It is a region of majestic scenery with three national parks, and the highest peaks in the coterminous United States.

- The core of the Sierra Nevada is comprised of the Sierra Nevada batholith, a large mass of granite emplaced into older rocks during the Mesozoic Era. Magma rising from the Farallon subduction zone constructed the batholith beneath a volcanic arc similar in origin to the modern Andes Mountains of South America. Surrounding the Sierra Nevada batholith are older metamorphic rocks in which numerous accreted terranes can be recognized. The metamorphic rocks form roof pendants and xenoliths in the batholithic rocks and extensive bedrock exposures in the western foothills. Cenozoic sedimentary and volcanic rocks record more recent periods of erosion and pyroclastic volcanic eruptions in the Sierra Nevada region.

- Though the Sierra Nevada was an elevated region during the Mesozoic Era, the uplift of the modern mountains appears to have occurred primarily during the past 5 million to 10 million years, when normal faulting along the eastern side was initiated. The Sierra Nevada was lifted and tilted to west during this time, creating the gentle western slope and the abrupt eastern escarpment.

- Gold in the Sierra Nevada was concentrated in placers, Eocene stream gravels, and in lode deposits of the western foothills. In the famous Mother Lode belt of the Sierra Nevada foothills, all three types of gold deposits are present.

Northern California.3
The Klamath Mountains

The Klamath Mountain region is a rugged highland with a general elevation of 1,500 to 2,150 meters (5,000 to 7,000 feet) above sea level. This elevated terrain is deeply incised by the Klamath River and its major tributaries, the Trinity and Salmon Rivers. These rivers carry water to the Pacific Coast through winding and spectacularly rugged canyons (Figure NC.16). The deep river gorges separate the Klamath highland into several distinct mountain ranges, including the Trinity Alps and the Siskiyou, Marble, and Salmon Mountains, some of which extend northward into southern Oregon. These rugged mountains are generally lower than the Sierra Nevada to the south, with the maximum elevations reaching only about 2,750 meters (9,000 feet). The Klamath Mountains separate the northern Coast Range on the west from the volcanoes of the Cascade Range to the east. Unlike most other geologic provinces in northern California, the Klamath Mountains lack a prominent northwest orientation. Instead, the individual mountains in this province have a crude north-to-south alignment or a weakly curved trend.

Figure NC.16 The rugged canyon of the Salmon River in the Klamath Mountains privince.

Dick Hilton

Even a quick glance at the geologic map of California reveals a strong similarity in outcrop colors and patterns of the Klamath Mountains region and the Sierra Nevada (see Figure NC.2). These similarities arise from comparable, though not identical, geologic histories in the two regions. In a general sense, the Klamath Mountains can be considered a northwest extension of the geologic trends of the Sierra Nevada. However, the continuity between the two regions is broken by the young volcanic rocks of the southern Cascade Range and by the sediments of the northern part of the Central Valley. The Klamath Mountains shares with the Sierra Nevada a long history of subduction-related accretion, beginning in the Paleozoic Era during which numerous oceanic terranes collided with the western edge of North America. The dozen or so terranes recognized by geologists in the Klamath region are separated from each other by major east-dipping fault zones. Granitic plutons of Mesozoic age were emplaced into the accreted terranes in several areas of the Klamath region, though such rocks are not as widespread in the Klamath region as they are in the Sierra Nevada. Both the Sierra Nevada and the Klamath Mountains experienced glaciation during the Pleistocene ice ages, though the lower elevations in the Klamath region resulted in smaller and less extensive glaciers. The Klamath Mountains region also experienced gold mineralization during the late Mesozoic Era, though none of the Klamath gold districts were as rich as the Mother Lode of the Sierra Nevada. Despite these similarities, the geologic story of the Klamath Mountains is not exactly the same as the Sierra Nevada. In the following sections, we focus on the unique attributes of the Klamath Mountains with respect to the similar Sierra Nevada.

Klamath Mountain Terranes and Ophiolites: In general, the Klamath Mountains region consists of numerous oceanic terranes representing fragments of crustal material that were embedded into the western margin of North America since Early Paleozoic time. The fragments include metamorphosed volcanic and sedimentary rocks that represent volcanic island arcs, submarine plateaus, reeflike bodies of limestone, and deep ocean sediments that were intensely deformed during accretion. It has taken geologists many years to unravel the complicated tectonic pattern in the Klamath Mountains, but it now appears that as many as a dozen different accreted terranes exist in the region. Some of these terranes no doubt represent the same fragments recognized in the western metamorphic belt of the Sierra Nevada. In general, the Klamath terranes become younger from east to west, a pattern that probably reflects the westward growth of North America through a series of plate collisions along the western edge of the continent. Some of the Klamath terranes may have originated thousands of kilometers from North America, whereas others were probably of local origin. Fossils discovered recently in the eastern Klamath terranes suggest that rocks as old as Late Precambrian, nearly 600 million years old, are included in the oldest accreted fragments.

► **Figure NC.17** Ophiolites are sequences of rock that represent oceanic lithosphere emplaced on land at convergent plate boundaries.

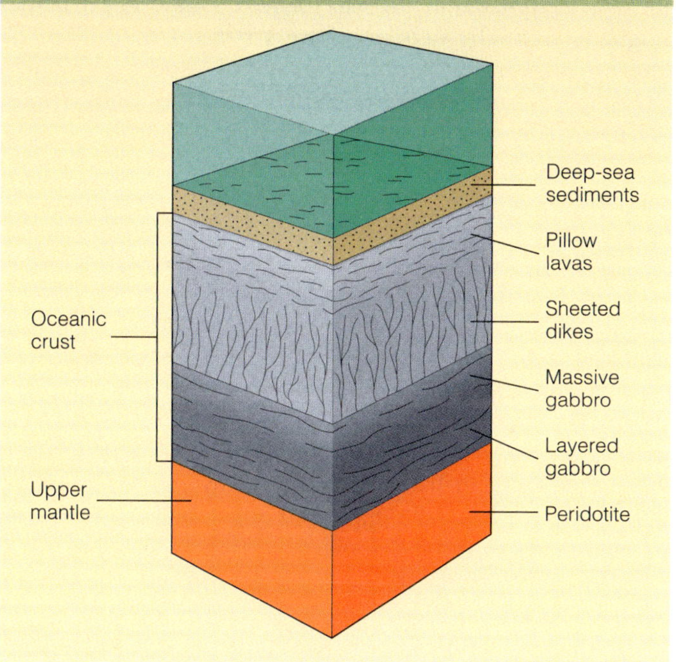

Of particular interest in the Klamath region are extensive exposures of **mafic** and **ultramafic** rocks known as ophiolites, sequences of igneous rocks thought by geologists to represent disrupted oceanic lithosphere. Ophiolites consist, in ascending order, of upper mantle **peridotite** overlain by layered and massive gabbro, sheeted basalt dikes, and basaltic pillow lavas (Figure NC.17). Ophiolite sequences can be emplaced on land during plate convergence simultaneously with the accretion process. As relatively thin and dense oceanic plates sink into subduction zones, fragments of the descending lithosphere are sometimes incorporated into the accretionary mass. The heat, pressure, and chemically active fluids that accompany plate convergence usually result in significant deformation, disruption, and metamorphism of the ophiolite sequences. In many exposures of the Klamath Mountains ophiolites, the mafic igneous rocks have altered into the metamorphic rock serpentinite, but careful studies can still reveal their origin as components of oceanic lithosphere. We will learn more about sepentinite in our examination of the northern Coast Ranges, where they are also abundant.

In the Klamath Mountains region, the Trinity and Josephine ophiolites are two of the largest and best known rock sequences of their type in the world. These two rock assemblages represent slivers of oceanic lithosphere that were incorporated into the accreted terranes at different times and different places in the Klamath region. The Trinity ophiolite is part of the Paleozoic–Early Mesozoic Eastern Klamath-Yreka **superterrane**, which is very similar to the Northern Sierra terrane of roughly the same age to the south. Most geologists feel that the Trinity and the Northern Sierra terranes were originally part of the same accreted oceanic plate. The mid-Mesozoic Josephine ophiolite is younger, about 162 million years old, and is exposed in the northwestern part of the Klamath region. The Josephine ophiolite may be in part equivalent to similar rocks in the western metamorphic belt of the Sierra Nevada. Neither the Josephine nor the Trinity ophiolite sequences are preserved in their entirety in the Klamath Mountains; they were disrupted and metamorphosed during accretion. Nonetheless, there is little doubt that ophiolites of the Klamath Mountains represent fragments of accreted oceanic lithosphere.

Mesozoic Plutonic Rocks of the Klamath Mountains: The Mesozoic Farallon subduction zone that generated great volumes of felsic magma in the Sierra Nevada region extends north to the Klamath region. However, the magma that intruded the Klamath basement was emplaced mostly as isolated plutons of relatively limited size and did not coalesce into larger masses similar to the Sierra Nevada batholith. The relatively small exposures of light-colored granitic rocks in the Klamath Mountains contrast with the darker metamorphic rock of the terranes to create spectacular landforms such as Castle Crags (Figure NC.18). The Mesozoic-age plutons of the Klamath region range in age from about 136 million to 174 million years, and tend to be somewhat more mafic (containing more ferromagnesian silicate minerals) than the granitic rocks comprising the Sierra Nevada.

Gold and Chromite Mineralization: The geologic similarity between the Klamath Mountains and the Mother Lode region of the Sierra Nevada did not escape the notice of prospectors and miners in the mid-1800s. Gold was discovered in placer gravels along the Klamath and Trinity rivers soon after the Gold Rush began. Additional gold deposits were eventually located in ancient river sediments and in the metamorphic basement rocks. Ultimately, the Klamath Mountains became the second largest gold producing region in California.

During and just after World War II, deposits of chromite were extensively mined in the Klamath region. Chromite is an iron chromium oxide mineral ($FeCr_2O_4$) that is the only source of chromium, a valuable metal used in the manufacture

► **Figure NC.18** Castle Crags is an exposure of 167-million-year old granite in the Klamath Mountains.

Frank DeCourten

of specialized steel alloys and in corrosion-resistant steel plating. Chromite is very rare in crustal rocks, but comprises a significant fraction of the upper mantle material. In the Klamath Mountains region, the mantle slices preserved in the Trinity and Josephine ophiolites contained numerous pods of chromite-bearing rock that were rich enough to be mined. Much of the ore was depleted after the 1960s, and production of chromite from the Klamath Mountains has since declined dramatically. For a time, however, the Klamath Mountains was the primary source for this valuable strategic metal.

Section Northern California.3 Summary

- The Klamath Mountains province is a mountainous upland deeply incised by the Klamath River and its tributaries. Several distinct mountain ranges comprise the region, with elevations up to 2,750 meters (9,000 feet). In contrast to the strong northwest orientation of California's major landforms, the ranges in the Klamath Mountains province have a weak north-to-south alignment or a curved trend.
- The bedrock of the Klamath Mountains province has strong similarities with the Sierra Nevada to the south, and can be considered as a northwest continuation of the geologic patterns of the later. In both areas, metamorphic bedrock comprising multiple terranes is intruded by Mesozoic granite plutons and overlain by Cenozoic-age sedimentary and volcanic rocks. The rocks comprising some of the Klamath Mountains terranes are almost identical in type, age, and origin to those of the northern Sierra Nevada. At least a dozen terranes have been identified in the Klamath Mountains, some of which can be assembled into larger superterranes.
- Ophiolites, representing fragments of oceanic lithosphere, are prominent in the terranes of the Klamath Mountains region. The Josephine and Trinity ophiolites are dominated by mafic igneous rocks commonly altered to greenstone and serpentinite.
- Gold and chromite have been mined from the Klamath Mountains region. The gold probably originated in a manner similar to the Mother Lode deposits of the Sierra Nevada. The chromite deposits are concentrated in the mantle rocks present within the Klamath Mountains ophiolite sequences.

Northern California.4
The Great Valley

A Great Depression: Nestled between the Sierra Nevada on the east and coastal mountains to the west, the Great Valley of California is a vast elongated basin extending nearly 700 kilometers (430 miles) and averaging about 80 kilometers (50 miles) wide. The northern portion of the Great Valley is known as the Sacramento Valley, and the southern two thirds is designated as the San Joaquin Valley. These names originate from the two large rivers systems that drain the interior basin from the north and south, respectively. These rivers meet southwest of Sacramento in the Delta region and eventually drain into the San Francisco Bay.

The Sacramento Valley is a remarkably flat interior basin surrounded by elevated terrain to the west (Coastal Ranges), north (Klamath Mountains and Cascade Range), and east (the Sierra Nevada). The Sacramento River system receives runoff from these adjacent highlands via several major tributaries including the Feather, Yuba, American, Pit, and Bear Rivers (Figure NC.19). Because northern California receives significantly more rain than southern California, the Sacramento River is California's largest river, carrying about 18 million acre-feet of water, six times more than the San Joaquin River, toward the sea annually. This water is a critically important resource for agriculture, industry, commerce, and domestic use statewide.

Soils and Agriculture in the Great Valley: The surface of the Sacramento Valley is covered by recent and Pleistocene-age **alluvium** washed into the bottomlands by streams draining the adjacent highlands. These stream sediments consist of a heterogeneous assemblage of channel gravels, river bank sands, silt, and clay deposited on the broad floodplain, and in some places, peat deposits representing plant litter that accumulated in lakes and wetlands. The rivers draining the Sacramento Valley typically follow meandering courses that shift continuously across the

Figure NC.19 The drainage basin of the Sacramento River.

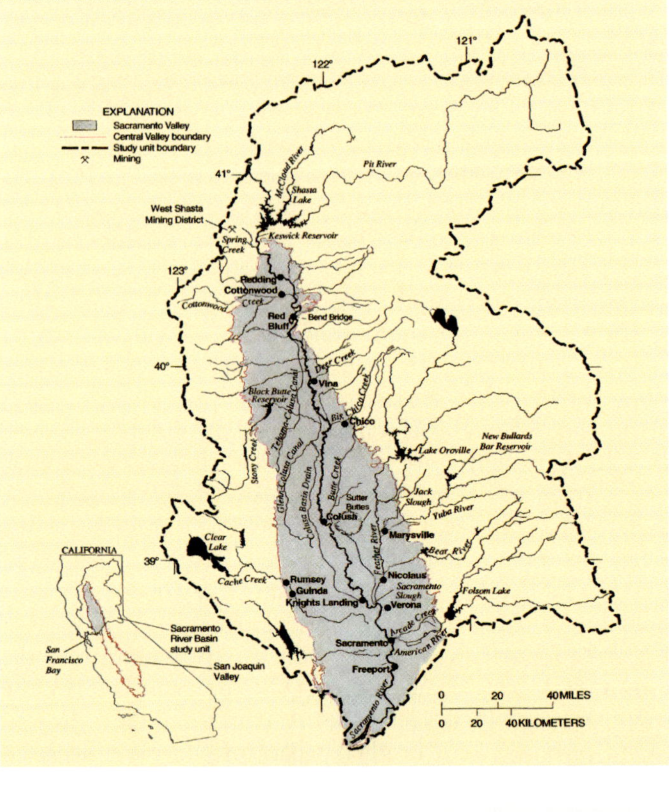

Map courtesy of the U.S. Geological Survey.

nearly flat floodplain. Over time, the migrating rivers have deposited thick sequences of interlayered river sediments. These alluvial deposits are generally several hundred meters thick, but can be much thicker in the lower areas on the west side of the valley.

The soils developed on the alluvium of the Great Valley are among the most fertile in the world. These soils, coupled with abundant water, mild climate, and a lengthy growing season have made the Great Valley one of the most agriculturally productive regions in North America. The annual value of California agricultural products exceeds $32 billion and most of the products are grown in the Great Valley. In the Sacramento Valley, the principal crops are rice, almonds, walnuts, orchard fruits, grapes, and feed grains. Even before cultivation, the fertility of the Great Valley was noted by early explorers who reported that the Sacramento River flowed through an incredibly lush grassland, with large stands of oaks trees and numerous wetlands.

Soil fertility is the consequence of several inter-related factors, including the texture, chemistry, micro flora and fauna, moisture, and organic content of the soil. However, some vital nutrients can be obtained by plants only from chemicals released during the weathering of mineral grains in the soil. These mineral nutrients include potassium, magnesium, and calcium. The alluvium of the Sacramento Valley is naturally rich in minerals such as feldspar and mica that release these nutrients as they undergo chemical weathering. These mineral grains, in turn, originated primarily in the plutonic rocks and associated metamorphic terranes of the Sierra Nevada and the Klamath Mountains. The agricultural productivity of the Sacramento Valley soils reflects, in part, the vast exposures of these rocks in the upper reaches of the Sacramento River watershed.

Deeper Structure of the Great Valley: Beneath the alluvium and soils at the surface, the bedrock of the Great Valley is comprised of a thick sequence of mostly Mesozoic and Cenozoic sedimentary rocks that are downfolded in a great asymmetrical **syncline** **(Figure NC.20)**. These strata are called the Great Valley Sequence, an enormous mass of oceanic sediments exceeding 6,000 meters in maximum thickness. Because they are folded downward to depths of 15,000 feet or more, the strata of the Great Valley Sequence are only exposed at the surface in the foothills of the highlands on either side of the valley. To the east, along the Sierra Nevada foothills, the rock layers dip gently toward the center of the valley, although they are steeply upturned to the west, where they are exposed in the foothills of the Coast Ranges (Figure NC.21).

Figure NC.20 Generalized diagram of the subsurface structure of the Great Valley, Coast Ranges, and Sierra Nevada.

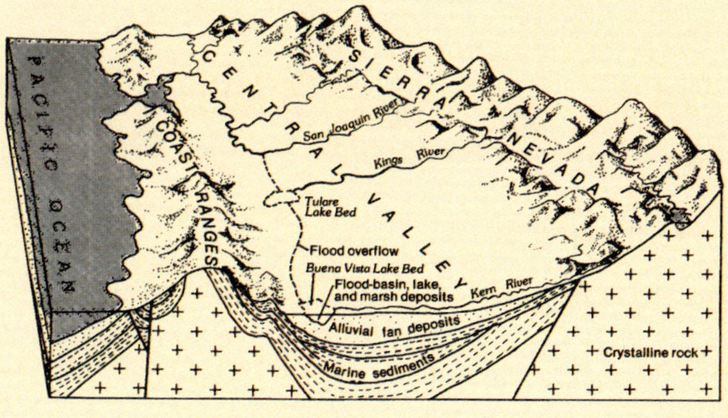

Diagram courtesy of the U.S. Geological Survey.

The Great Valley Sequence is comprised mostly of sandstone and shale representing sediment that accumulated in a deep **forearc basin** that developed in the Mesozoic Era. This basin was located between the offshore trench and the ancestral Sierra volcanic arc above the Farallon subduction zone (Figure NC. 22). Much of the sediment eroded from the volcanic and metamorphic rocks in the Klamath and Sierra Nevada regions was transported to the deep seafloor by **turbidity currents**. These great undersea landslides probably were triggered by volcanic eruptions, earthquakes, or storm events along the ancient continental margin. Repeated deposition of sediment by turbidity currents leads to the development of **submarine fans** on the seafloor, and geologists recognize sediment of this type in many portions of the Great Valley Sequence. Some of the sediments in the Great Valley Sequence are rich in organic matter, primarily the remains of microorganisms that continually rained down to the seafloor from the shallow waters above. Over time, this organic matter has become transformed in the subsurface to the combustible hydrocarbons that comprise natural gas. Where these gases are trapped in the porous subsurface sandstones, productive gas wells can be developed. From 1977 to 1998, the annual production of natural gas in the Sacramento Valley region averaged about 100 trillion cubic feet. The majority of this gas originates in the Great Valley Sequence.

Above and Below the Great Valley Sequence: The sediments of the Great Valley Sequence were deposited on top of oceanic lithosphere that can still be recognized by geophysical techniques. The deeply buried oceanic crust in the Great Valley is 7 to 8 kilometers thick and the **moho** beneath is 11 to 16 kilometers below the surface. Surprisingly, this relatively dense oceanic basement appears to be perched above the western extension of the Sierra Nevada batholith. The presence of a thick slab of dense oceanic rock deep beneath the Great Valley may act like a gravitational anchor, helping to explain why this region has remained so low with respect to the surrounding terrain.

The forearc basin in which the Great Valley Sequence accumulated persisted into the early part of the Cenozoic Era. By about 25 million years ago, the rate of subsidence decreased and the basin began to fill in the north. Starting in Late Cenozoic time, the oceans gradually drained away from the Great Valley, and all of the sediments washed into the lowland accumulated in terrestrial environments such as river beds, lakes, and swamps. As we will soon see, volcanoes in the Sierra Nevada and Cascade Range became active at roughly this same time, and some of the erupted material flowed or fell into the Great Valley. Consequently, along the modern periphery of the Great Valley, the Late Cenozoic

▶ **Figure NC.21** Submarine fan deposits of the Great Valley Sequence exposed in the foothills of the Coast Ranges dipping steeply to the east (left).

Frank DeCourten

▶ **Figure NC.22** The Mesozoic forearc basin in which sediments of the Great Valley Sequence accumulated was located between the ancestral Sierra Nevada volcanic arc and oceanic trench related to the Farallon subduction xone.

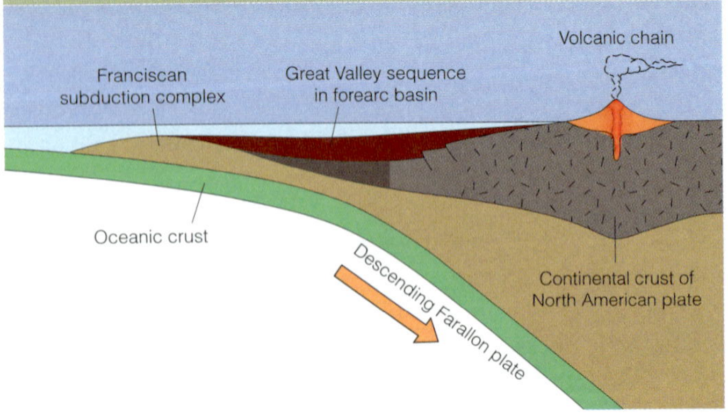

Diagram courtesy of the U.S. Geological Survey.

▶ **Figure NC.23** The Sutter Buttes are the only prominent mountains in the Great Valley of northern California.

Frank DeCourten

rocks are characterized by interlayered volcanic and terrestrial sedimentary rocks.

The Sutter Buttes: About 12 kilometers northwest of Marysville stand the Sutter Buttes, a circular cluster of ragged volcanic peaks, about 17 kilometers (10 miles) in diameter, that rise to an elevation of more than 650 meters (2,100 feet). These peaks are such a striking feature in the monotonously flat interior of the Great Valley that they have attracted attention ever since humans first arrived in northern California (Figure NC.23). The core of the Sutter Buttes consists of several rhyolite domes emplaced as shallow intrusions from 1.6 million to 1.3 million years ago. Surrounding this core is a circular zone of slightly older pyroclastic material of andesitic composition. Geologists

▶ **Figure NC.24** Flowing on a low gradient with an extensive flood plain, the Sacramento River is subject to seasonal flooding that places millions of people at risk.

James S. Monroe

generally consider the Sutter Buttes to be the southernmost expression of volcanic activity related to the Cascade Range to the north. However, development of the Sutter Buttes may also have been influenced by the migration of the Mendocino triple junction. We will review both of these aspects of northern California geology in later sections.

Flooding in the Sacramento Valley: Because it is a nearly flat surface drained by California's largest river (Figure NC.24), the Sacramento Valley is prone to recurring floods (Figure NC.25). The annual flood cycle generally peaks in late winter and early spring as heavy winter rains give way to the melting of the snow pack in the bordering mountains. Before people settled there, the floodplain adjacent to the Sacramento River and its tributaries was repeatedly inundated in a cyclic pattern. The annual floods renewed the floodplain soils and scoured the channel of accumulated sediment and debris.

Today, the intense agricultural and residential development in the Sacramento Valley has resulted in a serious flood hazard for millions of people. Billions of dollars have been invested in flood control measures that include large dams in the foothills, miles of levees, and river bypass channels. Even with these measures, the threat of seasonal floods has not been eliminated from the Sacramento Valley. Attempts to reduce flooding along the Sacramento River have also diminished the beneficial effects of this natural process.

▶ **Figure NC.25** Flooding in the Sacramento Valley, 2005–2006. Photo on the left was taken from space on Dec. 10, 2005 immediately prior to heavy late winter rains. Photo on the left is from Jan. 4, 2006. Note that the Sutter Buttes were entirely circled by flood waters from the Sacramento River.

Photos from NASA Earth Observatory.

Section Northern California.4 Summary

- The Great Valley is an elongated lowland situated between the Sierra Nevada on the east and the Coast Ranges on the west. The northern portion of the Great Valley is referred to as the Sacramento Valley, named for the large south-flowing river system that drains it. The watershed for the Sacramento River includes the highlands of the northern Sierra Nevada, Cascade Range, Klamath Mountains, and the northern Coast Ranges. Because these mountainous regions receive more precipitation than other regions of the state, the Sacramento is California's largest river.

- Great quantities of alluvium have been deposited across the floor of the Great Valley by rivers draining the adjacent highlands. The fertility of the soils of the Great Valley reflects, in part, the concentration of vital mineral nutrients derived from the weathering of rocks in the adjacent watersheds. The natural fertility of the Great Valley soils, coupled with the prevailing climate and abundance of water, has made the Great Valley one of the most productive agricultural regions in the world.

- In the subsurface of the Great Valley, Mesozoic and Cenozoic sedimentary rocks of the Great Valley Sequence are downfolded in the form of a large syncline. The Great Valley Sequence sediments accumulated in a deep forearc basin associated with the Farallon subduction zone. Most of the sandstone and shale in these rocks was transported to the deep ocean floor by turbidity currents and accumulated as submarine fans. Organic matter in parts of the Great Valley Sequence has produced significant quantities of natural gas.

- The Great Valley Sequence accumulated on dense oceanic lithosphere that lies deep under the Great Valley. This dense slab of rock is thought to be related to the low elevation of the region in relation to the surrounding highlands.

- The Sutter Buttes are a distinctive cluster of volcanic domes that erupted in the Sacramento Valley 1.6 million to 1.3 million years ago. The magma that surfaced in this location may be related to the Cascade Range volcanoes to the north or to the migration of the Mendocino triple junction along the coast.

- Because it is a nearly flat valley drained by California's largest river, the Sacramento Valley is naturally prone to cyclic flooding. Over geologic time, the recurring floods that inundated the floodplain have added alluvium to the valley surface and renewed soil components. Flood control measures, such as levees and upstream dams, are necessary to protect floodplain communities and industries, but they also limit the beneficial effects of the annual flood cycles.

Northern California.5
The Northern Coast Ranges

The rugged north coast of California is one of the most scenic seashores in the world. In most places along this rocky coast, beaches are narrow and sea cliffs rise from the water's edge to elevations of several thousand meters in just a few kilometers. Spectacular examples of coastal landforms such as **sea stacks, sea arches**, and **marine terraces** (Figure NC.26) provide evidence of recent emergence of the northern California Coast. The vigorous erosion that accompanies such an actively rising coast is responsible for the spectacular ruggedness of the Northern California seaside.

The relatively young coastal mountains of California define the Coast Ranges physiographic province that extends more than 600 kilometers (960 miles) from the Transverse Ranges in the south to the Oregon border, and beyond. Within this region, coastal mountains such as the Santa Cruz Mountains, the Mendocino Range, the Gabilan Range, and the Diablo Ranges of northern California are all aligned in a consistent northwest trend. The Coast Ranges are generally oriented parallel to the Great Valley and Sierra Nevada provinces to the east, and have a similar extent. This suggests that the geologic evolution of these regions may be linked to

▶ **Figure NC.26** Steep cliffs, sea stacks, and sea arches characterize northern California's emergent coast.

Sue Monroe

some of the same tectonic events. However, the recent emergence of the Coast Ranges was strongly affected by the evolution of the San Andreas fault system and the development of the modern transform boundary between the Pacific and North American plates. Thus, though the older bedrock of the Coast Ranges can be related to broader tectonic patterns, the story of their more recent emergence involves plate tectonic events that affected only the coastal region.

The California Coast Ranges have traditionally been divided into northern and southern portions, with San Francisco Bay arbitrarily dividing the two. However, for sound geologic reasons, it is more meaningful to divide the Coast Ranges into northern and southern portions separated by the San Andreas fault system. This great fault zone slices diagonally across the Coast Range, and the areas on opposite sides have unique and distinctive geologic histories. By this definition, the northern Coast Ranges extend northward from approximately Monterey Bay and the Gabilan Range to the Oregon border. It is this part of the larger Coast Ranges province that we will refer to as the North Coast Ranges, and on which we will focus in the following sections.

The Chaotic Franciscan Complex: In most of the Northern Coast Ranges, the basement rock consist of an incredibly complex assemblage of intensely deformed and metamorphosed sandstone, shale, chert, basalt, and plutonic rocks. In many exposures of these rocks, layering and other internal structures have been obliterated by internal movements such that some geologists have described it as a "churned" rock assemblage. In the early 1900s, the name Franciscan was applied to these mangled rocks for their prevalence in the San Francisco Bay region. Since then, the Franciscan rocks have become one of the most famous and well studied rock units in the world. For decades the complexities of the Franciscan assemblage defied geologic explanation. With the advent of the plate tectonic paradigm in the 1970s, geologists began to unscramble the origin of this chaotic rock sequence. Today, there is general agreement that the Franciscan rocks represent a subduction complex, a mixture of rocks that form in association with subducting oceanic lithosphere.

The age of rocks comprising the Franciscan complex ranges from about 200 million to 80 million years. Recall that about 140 million years ago, the Farallon plate began to subduct beneath western North America, generating the magma that ultimately formed the granite batholith of the Sierra Nevada (see Figure NC.22).Offshore of that ancient continental margin, a deep trench developed on the oceanic floor. In this trench, sand and rock fragments were washed from the continent and accumulated in great thickness. Such sediments comprise **graywacke**, an impure sandstone that is the most common rock type in the Franciscan assemblage. The Franciscan graywacke deposits commonly exhibit **graded bedding**, suggesting that turbidity currents transported them into the deep basin. Other deep ocean sedimentary rocks in the Franciscan complex also include fine-grained shale and chert (Figure NC.27). In addition to the sedimentary components of the Franciscan rocks, mafic igneous

▶ Figure NC.27 Chert of the Franciscan complex exposed in Marin County. Most of the layers are 5 cm thick.

Sue Monroe

rocks such as pillow basalt, gabbro, and peridotite are present. The igneous materials probably comprised portions of the Farallon plate that originated at ancient spreading ridges, as submarine lava flows, or in the upper part of the mantle. As the Farallon plate collided with North America, and was forced downward, the igneous rocks were subjected to high temperatures, extreme pressures, and chemically active fluids including hot seawater. Alteration of the iron and magnesium-bearing silicates in the igneous components of the Farallon plate produced such minerals as chlorite, pumpellyite, antigorite, chrysotile (asbestos), and lizardite. These greenish-colored minerals are abundant in greenstone and serpentinite (Figure NC.28), both of which are common in the Franciscan complex. Serpentinite plays an important role in creating the "churned" appearance of the Franciscan complex. It is significantly less dense than the igneous rocks from which it develops. Under the temperature and pressure conditions that prevail in subduction zones, serpentinite can become mobile and plastic, squeezing upward into the overlying rocks to disrupt the original layering.

The internal structure of the Franciscan rocks is so complex that it bewildered geologists for decades. The original layers of the sedimentary rocks are highly deformed (Figure NC.29) or even completely obliterated by serpentinite intrusion and tectonic deformation. The igneous rocks are usually metamorphosed, fragmented, and broken into isolated blocks. The entire rock assemblage is further complicated my numerous faults and shear zones that separate the sequences into discontinuous blocks. Such a mixture of deformed, jumbled, and altered rocks is known as a mélange, a French term for a mixture of different components. The origin of these rocks was a mystery to geologists until the plate tectonic theory provided a new model for interpreting the complexity of the Franciscan assemblage.

The Franciscan rocks, as a whole, are now thought to be a subduction complex, a heterogeneous mass of deformed and altered rocks that forms in and near subduction zones. The sedimentary components of the Franciscan complex

▶ **Figure NC.28** Serpentinite, like this sample from the northern Coast Ranges, is California's official State Rock.

Frank DeCourten

represent sand, silt, and ooze that accumulated in the trench or on the Farallon plate as it approached North America during the Mesozoic Era. Some of these sediments were scraped off the plate as it descended beneath the leading edge of North America and deformed by the immense pressure generated in the zone of plate convergence. Some sedimentary rocks were also pulled downward by the descending Farallon plate to depths of 15 kilometers or so beneath the surface. In this high-pressure and low-temperature environment, **blueschist** was produced, a metamorphic rock rich in the bluish mineral glaucophane. The igneous components of the Farallon plate descended to even greater depths and encountered higher temperature conditions where the serpentintite and greesnstone formed. Slices of rock were sometimes broken from the descending slab and thrust over or crushed against the edge of the continent. In addition, the Farallon plate was evidently carrying isolated seamounts, lava plateaus, and coral reefs that became detached and embedded into the Franciscan mélange. Geologists currently recognize no fewer than nine such micro-terranes in the Franciscan Complex in the San Francisco Bay region. The chaotic nature of the Franciscan complex thus reflects the diversity of rocks types, the varying degrees of metamorphism, the accretion of exotic terranes, and the intensity of deformation that result from plate subduction.

The subduction of the Farallon plate persisted under the northern Coast Ranges for more than 130 million years, extending well into the Cenozoic Era. During this time, a thick wedge of Franciscan rocks developed along the leading edge of North America, but the coastal mountain ranges had not been elevated from the seafloor. About 30 million years ago, the plate boundary along the west coast began to shift from a convergent type to a **transform** boundary. The emergence of the modern Coast Ranges is linked to development of the modern transform boundary between the Pacific and North American plates.

Evolution of the San Andreas System: Recall that the Farallon plate originated at an offshore spreading ridge that separated it from the Pacific plate moving to the west (see Figure NC.3). Segments of this spreading ridge were offset by transform faults to create an irregular divergent boundary. Through the Mesozoic and Early Cenozoic Eras, the North American plate was moving faster to the west than was the Farallon plate to the east. Consequently, the west edge of North America eventually collided with the spreading ridge as the intervening Farallon plate was consumed by subduction. This collision first occurred at the approximate latitude of modern Los Angeles about 30 million years ago (Figure NC.30). One consequence of this collision was the creation of a point where three plates—the Farallon, Pacific,

Figure NC.29 Deformed chert in the Marin Headlands terrane of the Franciscan complex.

James S. Monroe

Figure NC.30 Evolution of the San Andreas fault system as a transform boundary between the Pacific and North America Plates. The Juan de Fuca Plate north of the Mendocine Triple Junction is a remnant of the Farallon plate.

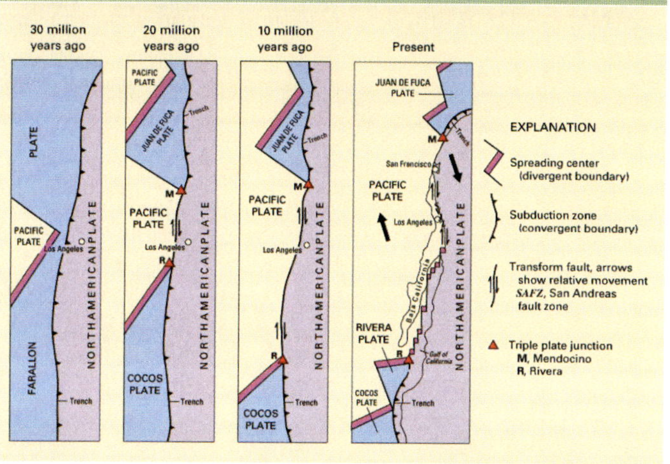

Diagram courtesy of the U.S. Geological Survey.

Figure NC.31 Major Fault of California, including the San Andreas and related stike-slip faults that represent the boundary between the Pacific and North American plates.

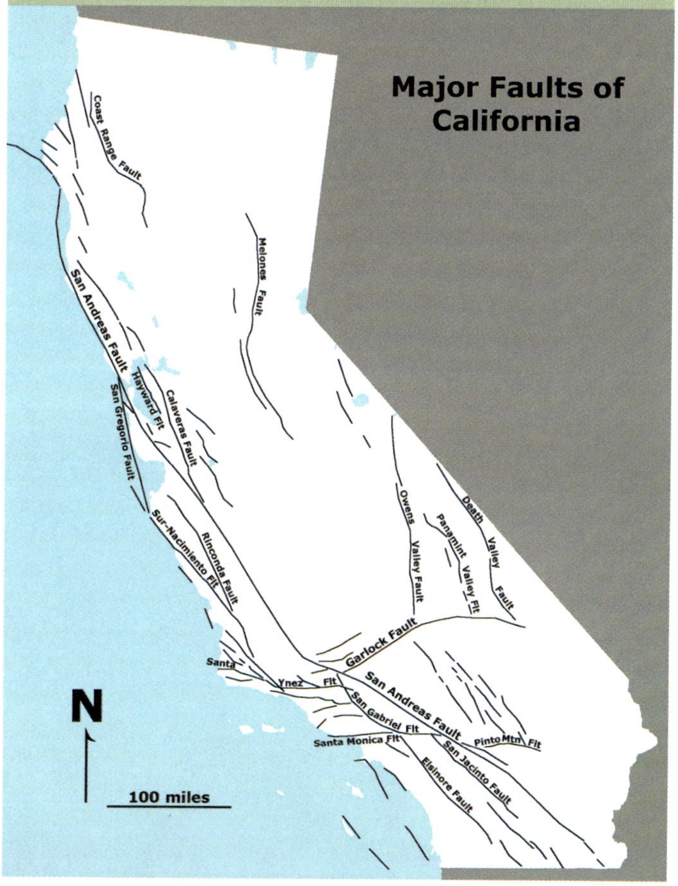

Map courtesy of the U.S. Geological Survey.

and North American—met. Such points are called **triple junctions**, and they signify places where different kinds of plate boundaries exist in proximity to each other.

After the initial triple junction developed, the North American plate continued to overrun more of the spreading ridge. This caused the original triple junction between the Farallon, Pacific, and North American plates to separate into two (see Figure NC.30). The continued westward motion of the North American plate caused the two triple junctions to migrate in opposite directions along the continental margin. The northern triple junction migrated through the San Francisco Bay area about 10 million years ago. Today, the Mendocino triple junction is located near Cape Mendocino, more than 200 kilometers (130 miles) north of San Francisco. The corresponding point to the south is the Rivera triple junction, located near the southern tip of Baja California.

The collision between the spreading ridge and the North American continent brought the Pacific plate into contact with the North America plate. Because both plates had westerly motions, this new boundary ended the long era of plate convergence and subduction. Instead, differences in the rates and directions of plate motion gave rise to a **transform** plate boundary, in which the two plates slide laterally past each other. The development of the transform boundary between the North America and Pacific plate initiated the famous San Andreas fault system (Figure NC.31). The San Andreas fault is perhaps the best known fault in the world, but it is only one of hundreds of faults in the zone that bears the same name. Collectively, these faults accommodate the relative motion between the North American and Pacific plates through **right-lateral strike-slip** displacement (Figure NC.32). It is important to understand that the San

Figure NC.32 The right-lateral displacement on the San Andreas Fault is illustrated by the offet of this stream channel in the southern Coast Ranges. View is to the west from the North American plate (foreground) to the Pacific plate (distance).

John S. Shelton

Figure NC.33 Major faults in the San Francisco Bay area. The blue arrows represent their average displacement rates.

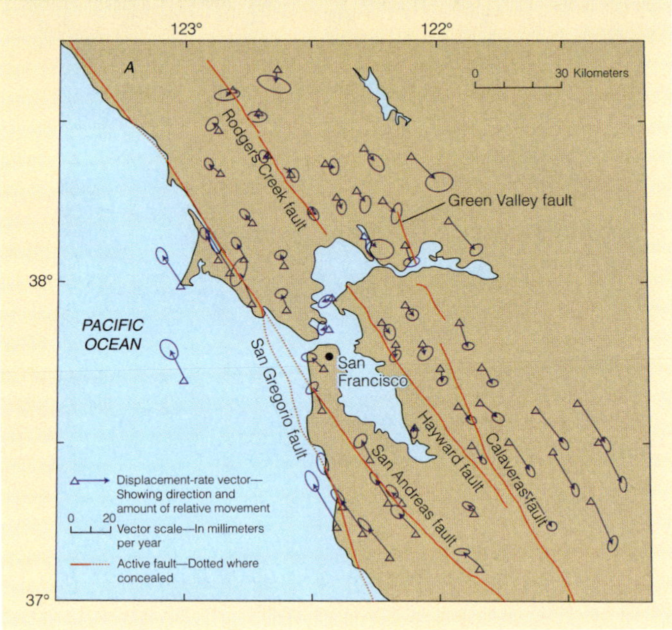

Map courtesy of the U.S. Geological Survey

Figure NC.34 Granite of the Salinian Block with dark-colored xenoliths at Point Reyes National Seashore. This granite is very similar to the plutonic core of the Sierra Nevada,

Dick Hilton

Andreas fault system is actually a zone of dominantly strike-slip faults that ranges from 100 kilometers to nearly 300 kilometers wide. The collective motion of these faults allows the Pacific plate to slip to the northwest relative to the North American plate at an average rate of about 5 centimeters (2 inches) per year. This may sound very small, but it is enough to displace older geologic features by more than 300 kilometers (185 miles) since the transform boundary originated.

In the San Francisco Bay area, the San Andreas fault is one of several large right-lateral faults that comprise the plate boundary (Figure NC.33). Other major faults in this region include the Calaveras, Hayward, and San Gregorio faults that are nearly, but not exactly, parallel to the plate boundary. The San Andreas fault extends offshore from Point Arena to the Mendocino triple junction. The San Andreas fault zone produces continuous earthquakes, some of which pose serious threats to California residents. We will examine the pattern of seismic activity along the San Andreas in a later section.

The Salinian Block—An Immigrant from the South: In the northern Coast Ranges, rocks of the Franciscan complex are absent west of the San Andreas fault. In places such as Point Reyes National Seashore, the basement rocks consist of Mesozoic-age plutonic and metamorphic rocks overlain by Cenozoic-age sediments (Figure NC.34). These rocks comprise a block that can be traced south from Point Arena to the southern Coast Ranges. Because such materials are especially prominent around the Salinas Valley, geologists refer to this mass as the Salinian block (Figure NC.35).

▶ Figure NC.35 Right lateral displacement along the San Andreas Fault has moved the Salinian Block more than 500 kilometers to the northwest.

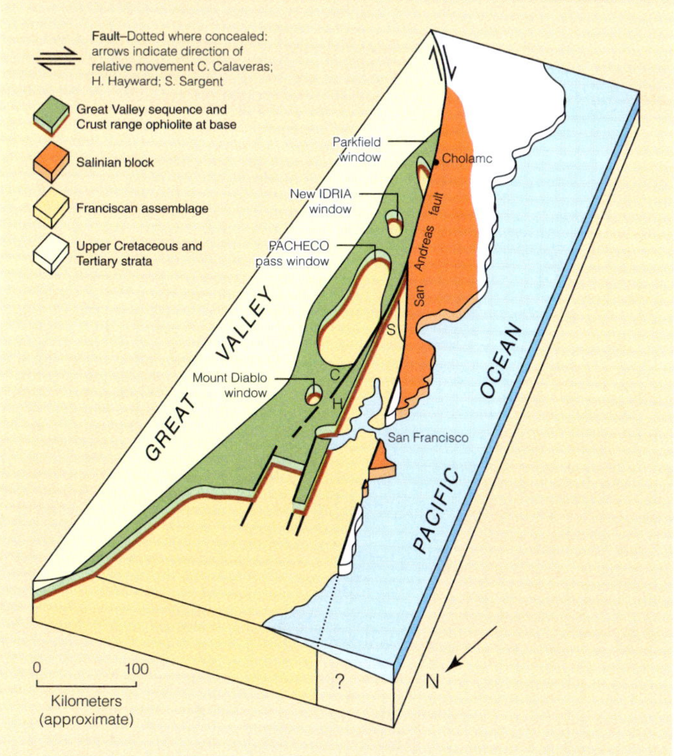

Diagram courtesy of the U.S. Geological Survey.

The granites of the Salinian block are 78 million to 110 million years old, and are remarkably similar in age and composition to the granites of the southern Sierra Nevada. In addition, metamorphic rocks associated with the granite basement of the Salinian block are almost identical to the roof pendants of the Sierra Nevada. These similarities suggest that the Salinian block was part of the Sierra Nevada batholith before the development of the San Andreas fault system. The Salinian block has since been transported a minimum of 550 kilometers (340 miles) to the northwest by right-lateral displacement along this fault.

The Rise of the Northern Coast Ranges: Overlying the basement rocks of both the Salinian block and the Franciscan complex are as much as 6,000 meters (nearly 20,000 feet) of Cenozoic-age sedimentary rock that records the emergence of the modern Coast Ranges. This sedimentary record is complicated by motion along the San Andreas system, but it is clear that there were several different basins where oceanic and terrestrial sediments accumulated. The older deposits in these basins are mostly mud, ooze, and sand that settled out on the continental shelf. Some sequences, such as the Monterey Formation (Miocene Epoch; 11 million to 22 million years old), accumulated in deep water, where organic material became concentrated in the mud and ooze (Figure NC.36). The Monterey Formation is believed to be the source rock of much of California's petroleum.

By Late Pliocene time, 3 million to 4 million years ago, the coastal basins were receiving mostly terrestrial sediments, indicating that the northern Coast Ranges were rising from the seafloor during this time. However, the complex association of Late Cenozoic river, estuary, and beach deposits in the northern Coast Ranges indicates that the precise timing and rate of uplift was somewhat variable. Accompanying the uplift of the modern northern Coast Ranges was the eruption of volcanic ash and lava in the areas around Sonoma (3 million to 8 million years ago) and Clear Lake (2 million to 10,000 years ago). Ash from these eruptions is commonly found within the layers of Late Cenozoic sedimentary rocks. This relatively recent volcanic activity in the northern Coast Range is probably related to the northward migration of the Mendocino triple junction. In the Clear Lake volcanic region, shallow residual magma is probably responsible for the steam that has produced as much as 2,000 megawatts of electricity annually at The Geysers geothermal field.

The forces that raised the north Coast Ranges in Late Cenozoic time are continuing to elevate the region. The general northwest alignment of mountain ranges and drainages is attributable to the orientation of the strike-slip faults that influenced the Late Cenozoic uplift. Several of the mountain ranges are still being squeezed upward by compression generated by small differences in the alignment of active faults. The coastal mountains of northern California have only recently emerged as land, and they will grow even higher as the Pacific plate continues to slide past the North American plate.

▶ **Figure NC.36** Folded layers of the Monterey Formation in the Monterey Peninsula area of the northern Coast Ranges.

Dick Hilton

Section Northern California.5 Summary

- The northern Coast Ranges include mountains such as the Mendocino Range, the Santa Cruz Mountains, and the Diablo Range that are aligned parallel to the Great Valley and Sierra Nevada. These are relatively young mountain ranges, and their continuing uplift is related to the emergence of the northern coast.
- The bedrock of the northern Coast Ranges is dominated by the Mesozoic-age Franciscan complex, an incredibly varied mixture of sedimentary and metamorphic rocks known as a mélange. The Franciscan complex accumulated in or adjacent to the Farallon subduction, and includes several exotic terranes that originated far from northern California.
- The San Andreas fault zone cuts diagonally across the northern Coast Ranges and is part of the modern transform boundary between the Pacific and North American plates. This boundary evolved when North America collided with and overran the Farallon-Pacific plate spreading center, beginning about 30 million years ago. Right-lateral strike slip motion on the San Andreas and other faults helps to accommodate the northwest motion of the Pacific plate relative to the North American plate.The Salinian block is a large fragment of crust in the northern Coast Ranges that has been transported at least 550 kilometers to the northwest by right-lateral displacement along the San Andreas fault zone.
- The mountains of the northern Coast Ranges are all relatively young, with uplift beginning only 3 million to 4 million years ago. The rise of the Coast Ranges is documented by the shift from marine to terrestrial sediments in Late Pliocene time. The forces that lifted the Coast Ranges appear to be related to the interaction of blocks separated by various faults within the San Andreas fault system and to the northward migration of the Mendocino triple junction.

Northern California.6
Volcanoes of the Cascade Range and the Modoc Plateau

The landscape of northeastern California is dominated by the Cascade Range and the adjacent Modoc Plateau (see Figure NC.1), which together create a scenic volcanic wonderland in this part of the state. The Cascade Range includes majestic Mount Shasta (4,319 meters/14,161 feet high; Figure NC.37), Mount Lassen (3,188 meters/10,457 feet high), and scores of other smaller volcanoes of all types. The Modoc Plateau is a high lava plain east of the bordering Cascades averaging about 1,350 meters (4,500 feet) in elevation (Figure NC.38). Many volcanoes are perched on this arid plateau, but the largest of them is the Medicine Lake volcano (Figure NC.39) near the eastern margin of Cascade Range. This immense shield volcano rises to an elevation of 2,375 meters (7,795 feet) and covers more than 2,000 square kilometers (770 square miles).

Both the Modoc Plateau and the Cascade Range in northern California represent only a portion of larger geologic provinces that extend into adjacent states. The Cascade Range extends northward from the southern terminus at Mount Lassen for more than 800 kilometers (500 miles) into southern British Columbia. Other prominent Cascade volcanoes include Oregon's Crater Lake, Mount Baker, and Mount Hood, along with Mount Rainier and Mount St. Helens of Washington. The Modoc Plateau is likewise continuous with the Columbia Plateau of Oregon, Washington, and Idaho. The Modoc Plateau encompasses nearly 26,000 square kilometers; the Columbia Plateau is roughly 20 times larger.

The principle volcanic features of the Cascade Range and Modoc Plateau are all relatively young, generally resulting from eruptions that occurred during the past 3 million years (Figure NC.40). Earlier Cenozoic-age volcanic activity did occur in the region, but the structures built during this time have either been eroded or obscured beneath the younger lava flows. The most recent activity

► **Figure NC.37** Mt. Shasta is the second highest composite in the Cascade Range and measures about 20 km in diameter at the base.

Diagram courtesy of the U.S. Geological Survey.

▶ Figure NC.38 Crater Mountain in Lassen County is one of several young shield volcanoes that erupted the basaltic lava in the Modoc Plateau region.

James S. Monroe

▶ Figure NC.39 Map showing location of Mt. Shasta and the Medicine Lake shield volcano, 60 kilometers to the east.

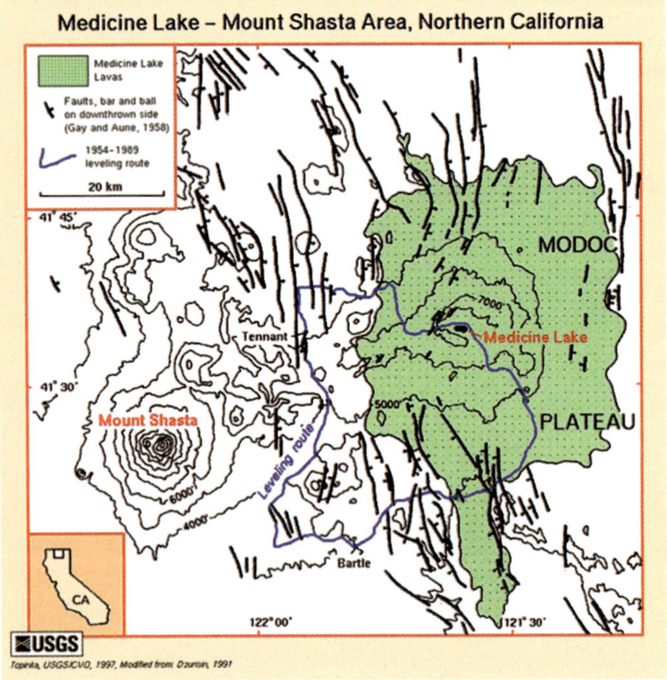

Map courtesy of the U.S. Geological Survey.

▶ Figure NC.40 Recent eruptions of the volcanoes of the Cascades Range. Note that Mt. Shasta and Mt. Lassen in California have both been active in the past several centuries.

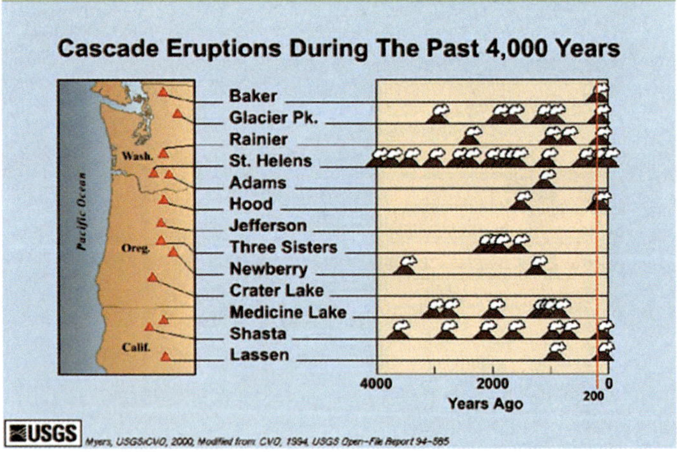

Map courtesy of the U.S. Geological Survey Cascade Volcano Observatory.

occurred in the years 1914 to 1917, when Mount Lassen experienced a series of small summit eruptions (Figure NC.41). The recent volcanic activity in the Cascade Range and Modoc Plateau is related to the plate convergence that exists along the margin of North America, north of the Mendocino triple junction. Here, oceanic plates are descending beneath the continent in the Cascadia subduction zone, in the process generating magma that continues to rise into the Cascade region.

The Cascadia Subduction Zone: The Cascadia subduction zone extends for 1,200 kilometers (750 miles) along the west edge of North America from southern British Columbia to northern California (Figure NC.42). Along this convergent plate boundary, the Juan de Fuca plate (and

▶ **Figure NC.41** A summit eruption from Mt. Lassen in 1915.

▶ **Figure NC.42** Plate tectonic setting of the Pacific Northwest. Subduction of Juan De Fuca plate beneath North America accounts for the volcanism of the Cascade Range.

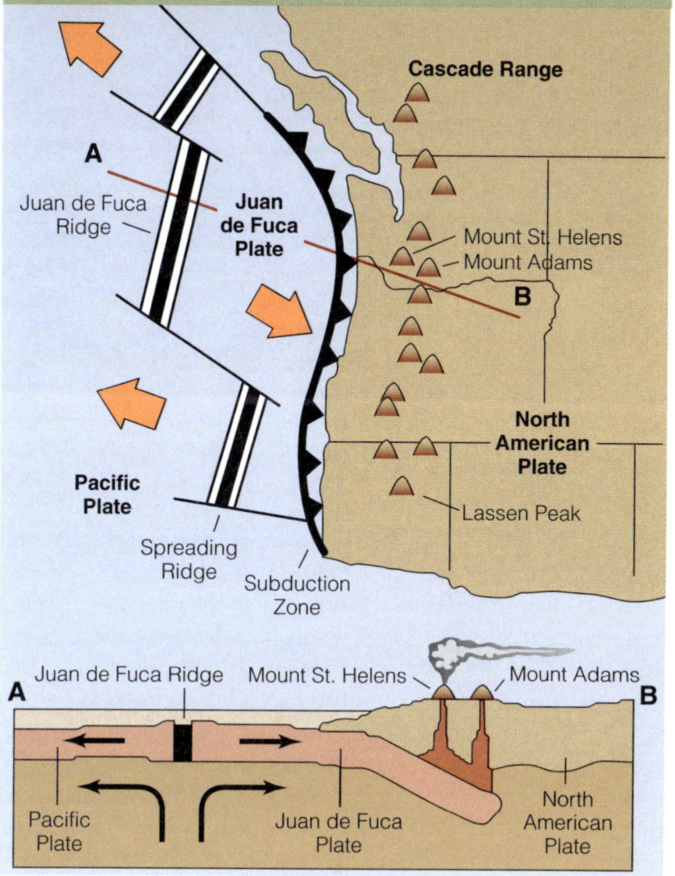

Diagram courtesy of the U.S. Geological Survey.

its southernmost segment, the Gorda plate) is moving under the advancing continent at a rate of 3 to 4 centimeters per year. Water released from the sinking slab causes subsurface rock to melt, forming magma bodies beneath the Cascade Range. Magma produced in this way is dominantly andesitic in composition and commonly contains large amounts of dissolved gases such as water vapor, carbon dioxide, and sulfurous vapors. The larger volcanoes in the Cascade chain are **composite** cones built through a series of explosive eruptions, such as the 1980 eruption of Mount St. Helens, alternating with less violent effusions of viscous magma.

The Cascadia subduction zone is terminated to the south by the Mendocino fracture zone, a transform fault that separates the Juan de Fuca (Gorda) plate from the Pacific plate (see Figure NC.42). Because there is no plate convergence south of the Mendocino fracture zone, no subduction-related volcanic activity occurs in that part of California. Mount Lassen, at the southern end of the Cascade Range, is located just slightly north of the landward projection of the Mendocino fracture zone.

California's Cascade Volcanoes: Among the hundreds of volcanoes in the California portion of the Cascade Range, three merit special attention because of their size and potential for future activity. Though future eruptions could occur almost anywhere in the region, Mount Shasta, Medicine Lake, and Mount Lassen volcanoes appear to have been the most active over the past several centuries. Each of these volcanic centers has a unique history, and poses distinctive threats to residents and visitors.

Mount Shasta (see Figure NC.37) is one of the largest **composite volcanoes** in the Cascade chain. This volcano was constructed over the past 100,000 years by hundreds of eruptions, including at least 13 in the past 10,000 years. These eruptions produced andesitic lava flows, pyroclastic flows, and debris flows, some of which extended well beyond the base of the volcano. Shastina, a small volcanic cone on the western flank of Mount Shasta (Figure NC.43), resulted from eruptions that occurred between about 9,300 and

▶ **Figure NC.43** Mt. Shasta and Shastina, seen from the north

Lyn Topinka/USGS

▶ **Figure NC.45** Mt. Lassen, the southern most peak in the Cascade Range.

Sue Monroe

9,700 years ago. The most recent activity at Mount Shasta was about 200 years ago, when several small debris flows, possibly triggered by steam eruptions, originated from the summit region.

At Medicine Lake, eruptions began about 700,000 years ago, when great volumes of basalt lava flowed onto the surface, building the base of this large shield volcano (Figure NC.44). The eruption of voluminous basalt flows is generally associated with divergent plate boundaries or **hot spots**, and is unusual in subduction-related volcanic areas. The basalt eruption at Medicine Lake is probably related to the proximity of the region to the Basin and Range province immediately to the east. In this area, the early stages of continental rifting may be in progress as a new divergent boundary develops in Nevada. Perched atop or along the gently sloping flanks of the Medicine Lake volcano are numerous **cinder cones** that provide sources for decorative rock, concrete aggregate, and roadbed material. The more recent eruptions at Medicine Lake, some probably less than 100 years old, have produced silica-rich magma that resulted in obsidian flows and rhyolite domes on the higher slopes.

▶ **Figure NC.44** Basalt Flow and Cinder Cone in the Medicine Lake Highland of Northern California.

Rick Hazlett

Mount Lassen (Figure NC.45) is unusual with respect to the other volcanoes of the Cascade Range in that it is not a composite volcano. Rather, it is a **lava dome** volcano, a bulbous steep-sided structure formed by viscous magma above a volcanic conduit. The rock comprising the lava dome at Mount Lassen is mostly **dacite**, representing sticky silica-rich magma that oozed up the conduit a little more than 11,000 years ago. Before the development of the Mount Lassen lava dome, a much larger composite volcano existed to the west. Geologists have named this ancient volcano Mount Tehama (Figure NC.46), and it was the site of numerous violent volcanic blasts over the past 600,000 years. Mount Lassen is the only California volcano to be observed erupting in historic time. Between 1914 and 1921, a series of small eruptions, mostly producing ash and steam, occurred near the summit of the lava dome. In May of 1915, however, somewhat larger eruptions sent glowing avalanches of debris down the flanks of Mount Lassen, and produced a small dacite lava flow on the. Today, visitors to Lassen Volcanic National Park can explore hot springs, boiling mud pots, and gas-discharging **fumeroles** that provide hints that the volcanic system is still active.

Volcanic Hazards in Northern California: All of the volcanoes of the Cascade Range and Modoc Plateau region are geologically young, and it is almost certain that there will be future eruptions in the region. In addition, the presence of magma at shallow depths may be responsible for the geothermal activity around Clear Lake in the northern Coast Ranges. Though these areas are distant from large population centers, future eruptions could produce widespread impacts, adversely affecting distant communities as well as local residents and visitors. A major volcanic eruption in the Cascade Range or elsewhere in northern California could also affect agriculture, transportation, communications, water quality, timber resources, and recreation over a broad area.

Volcanic eruptions can threaten human welfare by generating destructive flows of lava or debris, by filling the skies with clouds of ash and other fine rock particles, and by emitting noxious gases such as carbon dioxide, chlorine, or acidic

▶ **Figure NC.46** The original profile of Mt. Tehama near the lava dome of Lassen Peak.

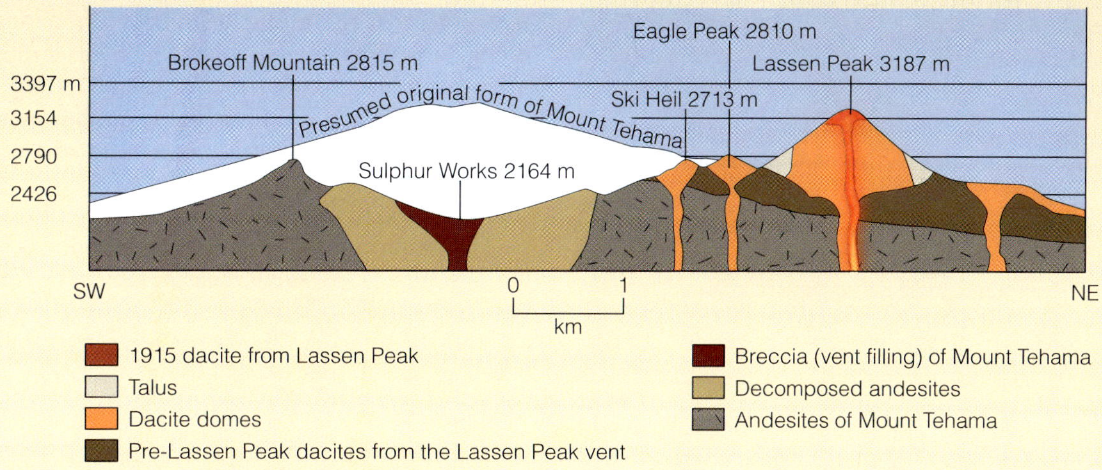

Diagram courtesy of the U.S. Geological Survey.

compounds. Of these multiple potential hazards, large flows of pyroclastic material from Cascade volcanoes would be the greatest threat in northern California. The geologic evidence indicates that between 300,000 and 360,000 years ago, an explosive eruption at Mount Shasta produced a volcanic debris flow that traveled 45 kilometers from the volcano and buried an area exceeding 450 square kilometers in rubble (Figure NC.47). Even larger volcanic mudflows, known as lahars, resulted from prehistoric eruptions at both Mount Lassen and Mount Shasta. And during the relatively small eruptions of 1915, ash particles from Mount Lassen fell as far away as Elko, Nevada, 500 kilometers (300 miles) to the east.

Mindful of the serious implications a large eruption in northern California would have, geologists have prepared volcanic hazard maps of the most sensitive areas and have helped develop evacuation plans for disaster response agencies. In addition, both Mount Shasta and Mount Lassen (along with other areas in central and southern California) are continuously monitored by scientists from the U.S. Geological Survey for signs of renewed activity. In the absence of any means to accurately predict future volcanic eruptions, the combination of hazard assessments beforehand, strategic preparedness, and monitoring will help limit the harmful consequences of future eruptions.

▶ **Figure NC.47** The hummocky terrain in the foreground was produced by an avalanche of volcanic debris from Mt. Shasta more than 300,000 years ago.

Dick Hilton

Section Northern California.6 Summary

- The Cascade Range of northern California is part of a chain of mostly composite volcanoes that extends north into British Columbia. This chain includes the famous volcanoes of the Pacific Northwest including Mount Rainier, Mount St. Helens, Mount Hood, and Crater Lake. In northern California, Mount Lassen and Mount Shasta are the most prominent volcanoes of the Cascade Range. The Modoc Plateau to the east is a high volcanic table land with numerous shield volcanoes, such as that in the Medicine Lake highland, and cinder cones.

- The magma that sustains the volcanic activity in the Cascade Range arises from the Cascadia subduction zone, in which the Juan de Fuca plate descends beneath North America. Water released from the subducting plate induces the formation of magma bodies more than 100 kilometers beneath the surface. This magma is generally intermediate to felsic in composition, relatively viscous, and laden with magmatic gases. These factors commonly lead to explosive eruptions and the construction of composite volcanoes in the Cascade Range.

- Mount Shasta is a large composite volcano built in a series of eruptions over the past 100,000 years. Mount Lassen is a lava dome volcano that was last active in the early 1900s. The Medicine Lake highland is a large shield volcano, with smaller cinder cones and obsidian flows on its flanks and summit. The most recent eruptions in the Medicine Lake region may have occurred only 100 years ago. All of these volcanoes are potential sites of future eruptions in northern California.

- Because they all have the potential for future activity, the Cascade Range volcanic centers pose serious volcanic hazards for people in northern California. The geologic record of previous eruptions demonstrates that the Cascade volcanoes can produce pyroclatic flows, volcanic mudflows known as lahars, and ash falls that could affect communities hundreds of kilometers away. These threats can be mitigated by continuous monitoring of the volcanic areas, assessment of the potential hazards beforehand, and preparations for responses to the future eruptions.

Northern California.7
The Basin and Range of Northeast California

East of the Modoc Plateau in extreme northeastern California, the volcanic table land is broken by numerous large normal faults into a series of elevated blocks and lower basins. Principal among these mountains and valleys are the Warner Range, rising to 3,017 meters (9,892 feet), and Surprise Valley, lying 1,600 meters (5,200 feet) lower immediately to the east. The displacement between the Warner Range and Surprise Valley is a result of the motion along normal faults along the base of the steep eastern escarpment of the mountain block. These faults comprise the Surprise Valley fault zone, which has a total cumulative of more than 14,000 feet. Normal faulting in this area began as early as 14 million years ago, but much of the displacement has occurred in more recent times. The Honey Lake basin to the south of Surprise Valley is also a depression bounded by normal faults.

The pattern of alternating high mountains and intervening basins in these small areas of northern California continues eastward for hundreds of kilometers into Nevada and portions of the adjacent states. This corrugated landscape, covering more than 770,000 square kilometers (300,000 square miles) of western North America is known as the Basin and Range province. In this immense region, tensional forces are stretching the Earth's crust and fracturing it into blocks bounded by normal faults. The hanging walls of the normal faults move downward to form the basins or **grabens**, while the mountain blocks remain high to form **horsts**. Surprise Valley and the Honey Lake basin are only two of hundreds of depressions formed as the crust in the Basin and Range province yields to tensional stresses.

Geology and Geothermal Resources of the Warner Range region: The bedrock of the Warner Range consists mostly of volcanic rocks erupted since about 36 million years ago. Included in this thick accumulation are lahar deposits, tuffs of various types, and felsic to intermediate lava flows. These older volcanic rocks appear to be related to subduction of the Farallon plate. Younger basaltic (more mafic) lava flows in the region are about 3 million to 8 million years old and probably signify a shift from subduction-related volcanic activity to the rifting-related eruptions. The prominent normal faults are relatively young, displacing the entire succession of volcanic rocks.

Numerous hot springs and "mud volcanoes" occur along a north-to-south zone near the base of the Warner Range, within the Surprise Valley fault zone. Exploratory drilling along the fault zone has encountered water as hot as 170° F at depths from 500 to 1,500 meters (1,600 to 4,900 feet). This geothermal activity probably reflects the interaction between water moving down from the surface and hot young volcanic rock in the subsurface. Since 1984, the hot water from the Surprise Valley springs has been used to heat schools and hospitals in the small communities of the valley. In 1993 alone, this saved roughly $50,000 in fossil fuel costs.

The Lake Tahoe Graben: Though it rests along the eastern edge of the Sierra Nevada physiographic province, the geological structure of the Lake Tahoe basin is similar to the grabens of the Basin and Range province. This magnificent lake fills a deep depression created by normal faulting to a depth of nearly 500 meters (more than1,600 feet). Volcanic activity 3 million to 4 million years ago created a natural dam of basalt and andesite at the north end of the graben that blocked the drainage of the upper Truckee River. The basin-forming normal faults are still active in

▶ **Figure NC.48** Several scarps mark the trace of active faults on the floor of the Lake Tahoe Basin. Note the large blocks of rock on the lake bottom that document an ancient catastophic landslide.

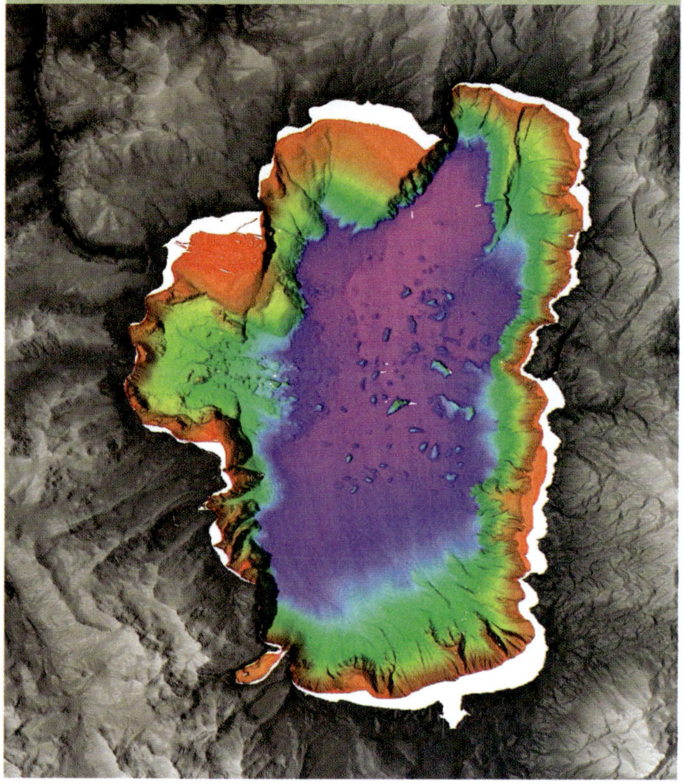

Image courtesy of the U.S. Geological Survey

the Lake Tahoe region, producing many small to moderate earthquakes. Ancient landslide rubble on the floor of Lake Tahoe is thought to have resulted from a major landslide triggered by a powerful Ice Age earthquake. Recent imaging of the lake (Figure NC.48) reveals numerous fault scarps resulting from recent displacement, a further indication of active faulting.

Section Northern California.7 Summary

- The Basin and Range province of extreme northeast California is a landscape of alternating mountains and valleys that extends eastward into Nevada and adjacent states. The corrugated terrain is the result of intense normal faulting caused by the tensional stresses that affect the entire Basin and Range region of western North America. The mountain blocks such as the Warner Range represent the footwall (or horsts) of the normal faults, whereas basins such as Surprise Valley represent the downthrown grabens. The Honey Lake and the Lake Tahoe basins also developed as grabens displaced downward by normal faults along the eastern edge of the Sierra Nevada province.

- The Warner Range is the most prominent mountain range of northeast California, and is composed primarily of Cenozoic-age volcanic rocks. Hot springs and mud volcanoes along the fault zone at the base of the eastern escarpment indicate the presence of young and hot volcanic rock in the shallow subsurface. These hot springs have been used locally as a source of energy for space heating.

- The Lake Tahoe basin lies in a graben that developed through the downward displacement of rocks along normal faults to the east and west. The elongated graben slopes gently to the north, and was blocked by lava flows that erupted 3 million to 4 million years ago. The natural volcanic dam impounded the waters of the ancestral Truckee River, creating the beautiful alpine lake. Ongoing seismic activity in the Lake Tahoe region suggests that the normal faults are still active and the basin is still evolving.

Northern California.8
The Northern California Ice Ages

During the Pleistocene Epoch, 1.8 million to about 10,000 years ago, all of the regions of northern California were affected by multiple climatic oscillations. In the higher elevations of the Sierra Nevada, Klamath Mountains, and the Cascade Range, numerous valley glaciers formed, and in the basins of the Great Valley and Basin and Range, lakes grew as water accumulated in the lower terrain (Figure NC.49). Even the coastal regions were affected, because sea level fell when the glaciers were expanding and rose when they were in retreat. Based on evidence from the Sierra Nevada, where the effects of the Pleistocene glaciers are most striking, geologists have identified four major glaciations corresponding to periods of cool climatic conditions. Some of these glaciations appear to be comprised of multiple cycles of glacial growth and decline. The glacial periods were separated by warm interglacial intervals when glaciers melted back and the lowland lakes diminished. The multiple ice ages of the Pleistocene Epoch in northern California signify a period of erratic oscillation in climate, not a long interval of enduring cold.

Because of their oscillatory nature, establishing a detailed and accurate history of glacial events from the geological evidence is difficult in most places. Each glacial advance tends to obliterate or obscure the evidence for the previous one. Geologists are in general agreement, however, that the two most recent glacial epochs in northern California peaked at about 160,000 years ago (the Tahoe glaciation) and 20,000 years ago (the Tioga glacation). Of these two glaciations, the earlier Tahoe appears to be longer and more severe than the later Tioga. There is also geologic evidence for earlier glaciations, but it is more fragmentary and localized than for either the Tahoe or the Tioga event.

▶ **Figure NC.49** Donner Lake in the northern Sierra Nevada, rests in a U-shaped glacial trough carved by Pleistocene glaciers.

Frank DeCourten

Effects of Valley Glaciers: The overall effect of glacial erosion in mountainous regions was to increase the relief and sharpness of the landscape through the development of U-shaped glacial troughs, **horns, arêtes**, and **cirques**. Although these landforms are displayed in spectacular fashion around Yosemite National Park in the central Sierra Nevada, they are also present in many other mountain regions of northern California. Wherever the Pleistocene glaciers existed, glacial **polish** and **striations** on bedrock surfaces are common (Figure NC.50). Such erosional features are widespread in the Trinity Alps, in the higher peaks of the Cascade Range, and in the alpine terrain west and south of Lake Tahoe. The Pleistocene-age valley glaciers also left extensive deposits of **moriane** of several types in these same mountain areas, along with **glacial erratics** (Figure NC.51).

▶ **Figure NC.50** Glacial striations on bedrock the northern Sierra Nevada.

Frank DeCourten

Lowland Lakes and Wetlands: During the cool climatic intervals of the Pleistocene Ice Ages, increased precipitation and reduced evaporation led to the development of large lakes and wetlands in the lower areas of northern California. Ice Age lakes that form in response to cool climate cycles are known as **pluvial** lakes and there were several of them at various times in the Honey Lake basin, in Surprise Valley, and in the Great Valley. In the Great Valley, large lakes formed repeatedly during the Pleistocene Epoch. One of the largest Ice Age lakes in the Great Valley existed between 800,000 and 600,000 years ago and covered the entire valley from its north end to the southern extremity. At other times, there were several smaller lakes with

Figure NC.51 Erratics of light-colored granite perched on dark-colored volcanic bedrock in the northern Sierra Nevada.

Dick Hilton

extensive wetland habitats between them. The lush native vegetation provided forage for a remarkable menagerie of Ice Age creatures, including giant bison, elk, ground sloths, mammoths, large cats, and bears. The remains of these creatures have been found in excavation sites throughout the Great Valley.

The Ice Age Landscape of San Francisco Bay: Earlier than about 600,000 years ago, the Sacramento and San Joaquin Rivers did not flow to the Pacific Ocean via San Francisco Bay as they now do. Instead the drainage from the surrounding highlands accumulated in one or more lakes in the Central Valley. Tectonic activity in the northern Coast Ranges had already produced a valley where the bay now is, but only local streams flowed into and through it. Sometime after 600,000 years ago, water from central California began to flow into the San Francisco Bay, where it mingled with ocean water to create the large estuary that exists today. However, cycles of climate change during the Late Pleistocene caused sea level to fluctuate several times. During cool intervals leading to glacial advances, sea level fell by as much as 120 meters (390 feet), causing water to withdraw from San Francisco Bay to the west. About 20,000 years ago, for example, the shoreline of the Pacific Ocean was located beyond the Farallon Islands, more than 30 kilometers (18 miles) west of the modern Golden Gate Bridge. At that time, San Francisco Bay was a broad river valley where streams from the Great Valley met with local tributaries before turning west through a valley that then existed in the Golden Gate area. Since about 20,000 years ago, the climate has become warmer, the Late Pleistocene glaciers melted, and sea level rose. By about 8,000 years ago, sea water had once again invaded San Francisco Bay to re-establish the estuary of modern times (Figure NC.52).

Figure NC.52 Sea level rise over the past 20,000 years has converted the San Francisco Bay area from a river valley to an estuary.

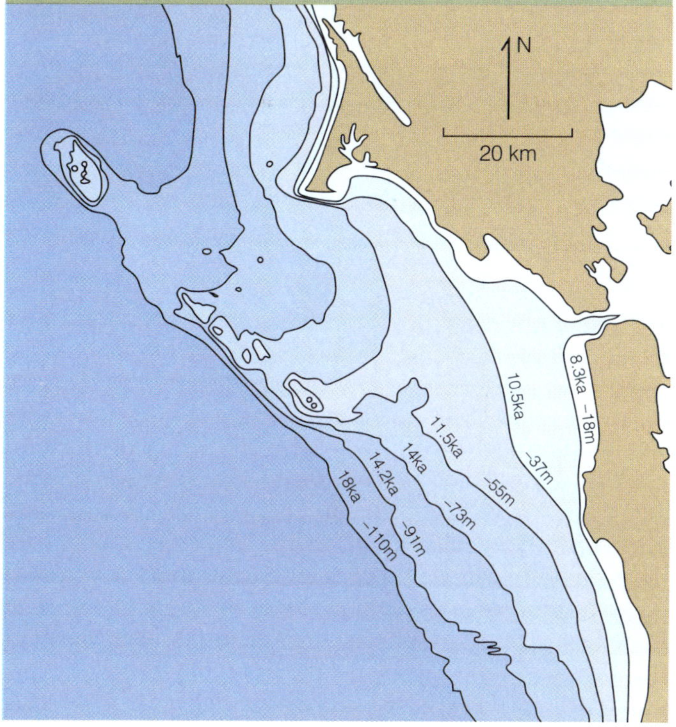

Map courtesy of the U.S. Geological Survey.

Section Northern California.8 Summary

- During the Pleistocene Epoch, California was affected by multiple cycles of climatic oscillations. Cool intervals resulted in the growth of valley glaciers in the higher elevations of the Sierra Nevada, Klamath Mountains, and the Cascade Range, and pluvial lakes developed in the lowlands.

- The two best documented glacial intervals are the Tioga glaciation about 20,000 years ago and the preceding Tahoe glaciation that peaked about 160,000 years ago. Both of these glacial intervals involved multiple advances and retreats that reflect the climatic oscillations of the Pleistocene. There is evidence of earlier glaciations, but the precise chronology and extent for them has yet to be determined with certainty.

- Valley glaciers in the mountainous regions carved U-shaped valleys and sculpted arêtes, horns, and cirques in bedrock exposures. Moraines and glacial erratics can also be identified in most mountains above 2,000 meters in elevation.

- In the lower elevations of the Basin and Range and Great Valley, pluvial lakes expanded and declined in rhythm with the Pleistocene climatic cycles. The largest of these Ice Age lakes submerged the floor of the entire Great Valley between 600,000 and 800,000 years ago.

- Sea level rose and fell multiple times during the Pleistocene Ice Ages. During times of glacial advances, sea level fell as much as 120 meters. Under such conditions, San Francisco Bay was a river valley through which rivers carried water to the Pacific coast located about 30 kilometers west of the modern shore. When the California glaciers melted during warm interglacial intervals, sea level rose. The modern estuary in San Francisco Bay formed about 8,000 years ago as glaciers in the interior mountains were melting under the influence of a warming climate.

Northern California.9
Northern California Earthquakes

Earthquakes in California are an inevitable consequence of the interactions between tectonic plates along the forward edge of a west-moving continent. The forces generated along the transform boundary between the Pacific and North American plates, and at the convergent boundary between the Juan de Fuca and North American plates, result in thousands of earthquakes every year. The ground literally shakes continuously in California, but most of these tremors go unfelt by people and rarely cause damage. Nonetheless, the seismic history of northern California includes some devastating events that provide a chronic reminder of the natural hazards facing those who live near a plate boundary. The infamous 1906 San Francisco earthquake (estimated magnitude 7.8 on the Richter scale) occurred along a segment of the San Andreas fault more than 400 kilometers long, ignited fierce building fires, and claimed more than 2,000 lives. In 1989, the magnitude 6.9 Loma Prieta earthquake in the Santa Cruz Mountains left 62 people dead in the San Francisco Bay area and caused nearly $10 billion in damages. In the dynamic tectonic environment of northern California, where millions of residents live in close proximity to active faults, future seismic disasters are as certain as those of the past have been tragic.

Earthquakes in northern California are associated with three different plate tectonic settings. The San Andreas fault system, including scores of subsidiary faults, generates many tremors in the San Francisco Bay area and the northern Coast Ranges. These earthquakes release stress associated with the northward motion of the Pacific plate along the edge of the North American plate. The earthquakes generated along this plate boundary tend to have very shallow foci, less than 15 kilometers deep (Figure NC. 53), and can be very powerful. In the densely populated urbanized area around San Francisco Bay, many buildings have been constructed on soils or bedrock that are not very resistant to seismic vibrations. In both the 1906 San Francisco and the 1989 Loma Prieta earthquakes, intense ground shaking, liquefaction, and fire all contributed to scope of the disaster and were related to the poor response of weak soil or rock to earthquake vibrations.

The second area of notable seismic activity is associated with the Mendocino triple junction along the northern coast, where each year 80 to 100 earthquakes greater than magnitude 3 occur. Here, forces generated as the Juan de Fuca (Gorda) plate descends into the Cascadia subduction zone can produce powerful earthquakes in a zone that dips eastward beneath the continent. In this plate tectonic setting, earthquakes occur in both the oceanic plate and on land, including some that can be powerful. In July and August of 1991, four major earthquakes, ranging from magnitude 6.3 to 7.1, occurred near the Mendocino triple junction, and there is good geologic evidence for even larger prehistoric events, possibly as great as magnitude 9. In addition to the impacts of local communities, large coastal earthquakes in northern California have the potential to produce **tsunamis** that could affect distant regions bordering the Pacific Ocean.

Along the eastern side of the Sierra Nevada, active normal faulting associated with the continuing uplift of the mountain block produces a third zone of seismic activity from the Lake Tahoe area south to Owens Valley. The eastern Sierra earthquakes are mostly generated along normal faults, but several prominent strike-slip faults in the region indicate that both tension and shear stresses exist in this area. Although the majority of eastern Sierra earthquakes are small, several greater than magnitude 6 have occurred

▶ **Figure NC.53** Distribution of earthquake foci along the San Andreas Fault from San Francisco to Parkfield. Note that most earthquakes occur within 15 km of the surface. The seismic gap filled by the 1989 Loma Prieta earthquake is represented on the bottom panel.

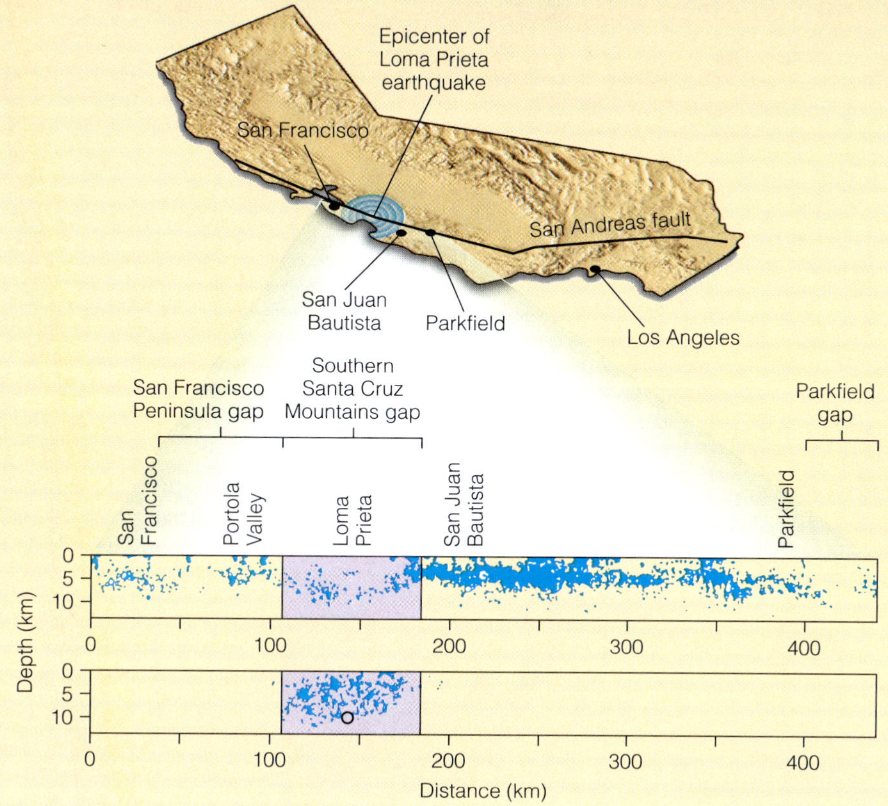

in the past 40 years. In Lone Pine, along the southern portion of the Sierra Nevada fault system, an 1872 earthquake destroyed all but seven houses and killed no fewer than 27 people. This earthquake occurred before seismometers were developed, but it may have been California's most powerful historic temblor. A recurrence of such an earthquake along the Sierra Nevada fault system would have major impacts on the resort communities in the Reno-Tahoe area, and regions farther south.

Earthquake Forecasts in Northern California: Even though scientists have been intensely studying earthquakes in California for more than a century, no reliable means to predict their occurrence with precision has been developed. Instead, geologists can forecast the probabilities of future earthquake occurrences based on measurements of ground motion and strain, historic patterns of seismic activity, and geologic evidence of fault activity. In recent years, the analysis of **seismic gaps**, seismically "quiet" segments along active faults, has helped to identify places most prone to future earthquakes. Because no two faults are exactly the same, the forecasts for major earthquake potential vary across the state, from very unlikely in some places to almost certain in others.

The time frame is a key consideration in making and using earthquakes forecasts. Over the long span of geologic time, the probability of a major earthquake in California is 100%, but the likelihood of such an event tomorrow is very low. Geologists generally try to forecast earthquakes over periods of 20 to 50 years to coincide with the time frames most widely used by planning commissions, disaster response agencies, and the insurance industry. In 1989, the U.S. Geological Survey developed the first widely accepted earthquake forecast for the San Andreas fault system. This forecast estimated the probabilities for major earthquakes between 1988 and 2018 within the entire fault zone. The probabilities varied from 10% to 90% at various locations along the fault system. Interestingly, the Loma Prieta earthquake occurred in an area where a seismic gap had been identified in this study (see Figure NC.53).

In 2007, an interdisciplinary team of scientists known as the Working Group on California Earthquake Probabilities developed a new updated earthquake forecast called the Uniform California Earthquake Rupture Forecast, Version 2 (UCERF). The new forecasts involve a broader array of data and techniques for estimating probabilities than did earlier assessments. Statewide, the UCREF estimates the probability of a magnitude 6.7 earthquake in the next 30 years at more than 99%. Californians also face a 46% chance of experiencing a magnitude 7.5 or larger earthquake in the same time frame.

▶ **Figure NC.54** Major faults of the San Andreas system in the San Francisco Bay Area. This map from the United States Geological Survey also identifies features resulting from recent offset along these faults by the circled numbers.

Map courtesy of the U.S.Geological Survey.

In northern California, the probabilities are lower than for the state as a whole. UCERF estimates that there is a 93% chance of a magnitude 6.7 or greater earthquake in the northern part of the state, with a 63% chance of the temblor occurring in the San Francisco area (Figure NC.54). The faults most likely to produce this large earthquake are the Hayward-Rogers Creek (32% probability), San Andreas (21% chance), and Calaveras faults (18% probability). The UCERF estimates only a 2% probability for an earthquake of magnitude 8 in northern California, illustrating that very large earthquakes are rare, even along the San Andreas fault system. This may seem comforting, but we should not forget how disastrous earthquakes in the magnitude 6 to 7 range have been in California. Earthquakes in this range of magnitude have resulted in more than $65 billion in property losses in California since 1971.

Living with Earthquakes in Northern California: Although human activities such as underground fluid injection can initiate earthquakes, such practices to reduce the earthquake threat in northern California would be extremely risky. There is no way to guarantee the size, location, or duration of such an induced earthquake, and even a moderate temblor could have disastrous consequences in densely populated areas. In the absence of any safe mechanism to control seismic events in California, careful planning and preparation for them makes good sense.

The California Geological Survey, the United States Geological Survey, and the California Office of Emergency Management have collaboratively developed forecast models and seismic hazard maps that are used to plan communities and make preparations for coordinated responses to earthquakes. In addition, in the most sensitive areas, stringent building codes have been developed that result in more earthquake-resistant structures. Local zoning ordinances in many areas limit residential and commercial development in locations where strong ground shaking or liquefaction are anticipated. Personal preparation is also important, and checklists are available at the aforementioned agencies that provide guidelines for preparing homes and apartments for a potential earthquake. The combination of continuing earthquake research, careful planning, stringent building codes, wise land use, and personal preparation is the best way to limit the property damage and loss of life in future northern California earthquakes.

Section Northern California.9 Summary

- Northern California earthquakes occur as a consequence of convergent plate interactions in the Cascadia subduction zone, in relation to the transform boundary between the Pacific and North American plates to the south, and along the normal fault zone at the base of the eastern escarpment of the Sierra Nevada. The San Andreas fault zone, part of the transform plate boundary, has historically produced the most large earthquakes, but destructive earthquakes could occur in the other plate tectonic settings in northern California as well.

- Earthquakes in populated regions of northern California have resulted in intense ground shaking leading to collapse of structures, liquefaction, landslides, and fires that have claimed many lives and caused billions of dollars of property damage. The 1906 San Francisco and the 1989 Loma Prieta earthquakes were the two most destructive earthquakes in northern California history.

- Geologists can formulate earthquake forecasts along active faults on the basis of ground motion and strain measurements, historic seismicity including "quiet" segments known as seismic gaps, and geological evidence of prehistoric fault activity. The most recent forecasts estimate the probability of a magnitude 6.7 or greater earthquake in northern California at 67%. Faults in the San Francisco Bay area within the San Andreas fault zone are most likely to produce this future earthquake.

- Because it is currently impossible for humans to prevent or weaken earthquakes, preparation for them is essential in reducing their impacts on people. In northern California, continued monitoring of active faults coupled with wise land use regulations, stringent building codes, and effective response plans will substantially reduce the impacts of future earthquakes.

Northern California.10 Living on the Edge: Coastal Hazards in Northern California

With 1,800 kilometers (1,120 miles) of ocean frontage, California has more coastline than all but two other states, Florida and Alaska. The northern portion of the California coast trends northwest, directly facing the enormous and tectonically active Pacific Ocean basin. Winter storms approach the exposed northern coast from the open ocean, bringing with them powerful wind-generated waves. In addition, normal wind swells, longshore currents, and occasional tsunamis also strike the north coast with full force. In contrast, the east-to-west alignment of much of the southern California coast partially protects the shoreline from the vigorous erosion experienced in the north. Consequently, the northern California coast is rugged and rocky rather than broad and sandy (Figure NC.55). Some locations along the north coast, such as Big Sur and Mendocino, are world famous for their spectacular coastal scenery.

The California coast also is unique in its recent emergence and in the variety of rock types exposed along the steep seashore. Recall from our earlier discussions that mountains of the northern Coast Ranges have been elevated over just the past several million years and, in fact, are still rising. The coast of northern California is therefore an **emergent** coast, where the land is being lifted by tectonic forces relative to sea level. As bedrock rises from the seafloor, it is exposed to the various agents of coastal erosion. In most places, the uplift exposes rocks of the heterogeneous Franciscan complex or the Salinian terrane. These rock assemblages include both easily eroded materials such as marine shale or sandstone and more durable igneous and metamorphic rocks such as granite, basalt, and greenstone. The erosion of such mixed

▶ **Figure NC.55** The steep and rugged shore in Point Reyes Nationa Seashore is typical of the scenic coast of northern California. Note the nearly flat marine terrace above the sea cliffs.

Photo courtesy of the U.S. Geological Survey.

▶ **Figure NC.56** A pocket beach developed between projecting headlands on the Monterey Peninsula.

Photo courtesy of the U.S. Geological Survey, Center for Coastal Geology.

▶ **Figure NC.58** Sea stacks along the northern Califonria coast at Shell Beach.

Sue Monroe

bedrock assemblages produces an irregular coastline, with coves and **pocket beaches** developing where soft rocks exist and **headlands** (Figure NC.56) projecting where harder rocks are exposed.

Coastal Landforms of Northern California: Because of the pervasive coastal emergence, the shoreline landforms in northern California are mostly of erosional origin. Sea cliffs and wave-cut platforms are actively forming in many locations, and elevated marine terraces are commonly observed above the coastal bluffs (Figure NC.57). Because the marine terraces originated as wave-cut platforms near sea level, geologists can estimate the rate of uplift along the coast by establishing the age of the terraces using fossils preserved on their surfaces. Radiometric ages for fossil material from the terraces suggest that the most prominent surfaces rose to their present elevation at rates varying from about 28 to 45 centimeters per thousand years in the Santa Cruz and Monterey Bay areas. Farther north, in the Mendocino area, the rate of emergence is even higher. This helps explain why the coast of California shows such strong emergent features, even though sea level has been rising for the past 20,000 years. The land simply is rising faster than the sea.

▶ **Figure NC.57** A marine terrace along the northern California coast with remnant sea stacks. Note the younger sea stacks forming along the modern coast.

James S. Monroe

Sea cliffs along the north coast are continuously undercut by waves generated during powerful winter storms. Because waves approaching a projecting headland are refracted around it, the wave-cut notches in some areas can develop into **sea caves** on opposite sides of the bedrock exposure. Continued wave erosion can deepen the caves until they connect, forming a **sea arch** (see Figure NC.26). **Sea stacks** (Figure NC.58), isolated exposures of bedrock, represent remnants of sea arches that have collapsed as a consequence of a continued wave erosion. The long-term effect of the formation of wave-cut platforms, sea caves, sea arches, and sea stacks along an emergent coast is the steady landward retreat of the cliffs. This has placed many coastal developments at risk along the northern California shoreline.

In addition to erosional features, depositional landforms have also developed in some areas along California's north coast. In places where streams discharge into bays or coves, **spits** (Figure NC.59) and **baymouth bars** (Figure NC.60) can develop from the accumulation of sediment transported by both rivers and longshore currents. **Tombolos**, sand spits that extend from the shore to a sea stack, can also be seen in several places along the northern California coast (Figure NC. 61). Tombolos are generally rare, but the abundance of sea stacks along the emergent northern California coast has created many places where the conditions exist under which tombolos form.

Coastal Hazards in Northern California: The majority of California's population lives within 100 kilometers (60 miles) of the coast. In northern California, the rugged beauty of the seashore has resulted in extensive residential and recreational development along the rapidly eroding emergent coast. Geologists have determined that coastal erosion is causing some sea cliffs in northern California to retreat at rates as high as 1.1 meters (3 feet) per year. In spite of recent regulatory action to restrict coastal development, there are

▶ **Figure NC.59** A spit at the mouth of the Russian River.

James S. Monroe

▶ **Figure NC.60** A baymouth bar in Marin County, California.

James S. Monroe

still hundreds of miles of coastline in northern California where people and property are threatened by retreating cliffs and unstable ground.

The principal hazard along the northern California coast is **mass wasting** triggered by wave undercutting, heavy rainfall, or earthquakes. Rock slides and slumps have repeatedly caused extensive damage to structures perched above sea cliffs (Figure NC.62). These problems are most severe where the bedrock is soft, sheared, or water saturated. Unfortunately these conditions are common along the tectonically active north coast, which is underlain by the highly deformed rocks such as those comprising the Franciscan complex. Mass wasting activity along the northern coast also extends offshore and down the continental slope. Turbidity currents initiated by earthquakes or storm events have scoured several impressive submarine. Monterey Canyon, the largest submarine canyon on the Pacific Coast, is deeper than the Grand Canyon and extends offshore for more than 400 kilometers (250 miles).

▶ **Figure NC.61** A small tombolo connecting the shore to a sea stack remnant in northern California

Sue Monroe

▶ **Figure NC.62** Mass wasting in Bodega Bay resulted from undercutting of the sea cliff by waves generated during winter storms of 1997–1998.

James S. Monroe

Because the northern California coast faces the Pacific Ocean basin where strong earthquakes occur frequently, tsunamis pose an additional hazard. Tsunamis can travel thousands of miles from their source, so even earthquakes in Japan or South America could affect people in northern California. In fact, the principal tsunami threat to northern California may be distant, not local, earthquakes. This is because the dominant strike-slip motion along the San Andreas fault system, much of which is on land, would not likely produce large vertical displacements on the seafloor. Subduction-related earthquakes generated in convergent plate boundaries more commonly result in the type of displacement that initiates tsunamis.

In April 1964, a 7-meter (23-foot) tsunami generated by the magnitude 9.2 Alaskan Earthquake struck Crescent City on the north coast, resulting in 11 deaths and an estimated $15 million in property damage. In the same town, a boat harbor suffered $10 million in damages from a tsunami that struck in November 2006. Other coastal towns in northern California to experience tsunami damages include Eureka, Half Moon Bay, and Santa Cruz. A major earthquake in the nearby Cascadia subduction zone is a primary concern in these coastal areas because it could potentially produce a much more disastrous tsunami than any experienced thus far. There is good geologic evidence of a major earthquake, perhaps as large as magnitude 9, in this area 300 years ago that may have caused a tsunami in Miho, Japan. Fortunately, most populated areas along the northern California coast have tsunami warning systems and evacuation plans in place.

Human Activities and Coastal Hazards: Although mass wasting and wave erosion are natural processes along California's north coast, they have been intensified and triggered in places by human activity. Whenever an ocean-facing slope or shore is modified for construction of roads or buildings, there is potential for otherwise stable slopes to fail. Highway 1, along the north coast, has experienced repeated closures due to mass wasting events. Property damage associated with mass wasting along this scenic highway has been significant in such coastal communities as Bolinas, Stinson Beach, and Pacifica (Figure NC.63). Farther south along the same highway at Half Moon Bay, a breakwater completed in 1961 to protect the harbor resulted in a fourfold increase in the rate of coastal erosion to the south. This was the natural response to the interruption of the southward flow of sand along the beach and the refocusing of wave energy produced by the breakwater. Many cities and planning commissions carefully review proposals for coastal development in an effort to reduce the negative impacts that result from altering natural slopes and shores. However, much of the development along the north coast was approved before the impacts of human activity were fully evaluated or understood. Coastal erosion will continue to be a concern for residents of California's scenic north coast.

▶ **Figure NC.63** Coastal erosion in northern California threatens thousands of homes such as this one near Bolinas.

Dick Hilton

Section Northern California.10 Summary

- The northern part of California's long coastline directly faces the Pacific Ocean and is vigorously eroded by wind waves, storm events, longshore currents, and occasional tsunamis. In addition, geologic forces are still lifting the Coast Ranges faster than sea level is rising so that new bedrock continuously appears along the emergent coast. In many coastal locations, the emerging bedrock is a mixed assemblage of hard and soft rock such as the Franciscan complex that does not respond uniformly to the agents of erosion. The combination of these factors produces an irregular and spectacularly rugged coast with many projecting headlands and small coves and pocket beaches.

- Erosional landforms such as wave-cut platforms, sea cliffs, sea caves, sea arches, sea stacks, and marine terraces dominate the scenery along the emergent coast of northern California. In places where rivers deposit sediments along the coast, or where longshore currents are active, sand can accumulate to form depositional features such as spits, baymouth bars, and tombolos.

- Many people live near the actively eroding coast in northern California. Coastal development is commonly at risk because of cliff retreat and mass wasting events that undermine support for structures. In addition, earthquakes in the Pacific Ocean basin can generate tsunamis that have affected several coastal communities. Human modifications, such as breakwaters, to protect portions of the coast can intensify erosion elsewhere and interrupt the transportation of sediment along the coast.

Review Workbook

ESSENTIAL QUESTIONS SUMMARY

Northern California.1 Introduction

■ *What are northern California's physiographic provinces?*
Physiographic provinces are regions of distinctive geology, landforms, climate, geomorphic trends, soils and vegetation, and drainage. In northern California, seven physiographic provinces are recognized: the Sierra Nevada, the Klamath Mountains, the Great Valley, the Coast Ranges, the Cascade Range and Modoc Plateau, and the Basin and Range. Collectively, these provinces endow northern California with extraordinary geologic diversity.

■ *What is the Farallon subduction zone?*
The Farallon subduction zone developed at the convergent plate boundary between the North American and Farallon plates about 200 million years ago. The entire subduction zone persisted until about 30 million years ago, when portions of it were disrupted by the collision between North America and the Farallon-Pacific spreading ridge. Many of the major geologic features of northern California resulted from the convergence of plates in the Farallon subduction zone.

■ *What two types of plate boundaries exist in northern California today?*
North of Cape Mendocino, the Juan de Fuca plate, a remnant of the Farallon plate, continues to be subducted under continental lithosphere to the east. South of Cape Mendocino, the transform boundary between the Pacific and North American plates shapes the geologic setting of northern California. The San Andreas fault system is directly related to the transform boundary, whereas northern California's volcanic centers are related to the subduction process.

■ *What are terranes, how do they originate, and why are they important in northern California?*
Terranes are large blocks of rock, typically bounded by faults, representing seamounts, oceanic rises, reefs, island arcs, or other oceanic features accreted to continents at convergent plate boundaries. Terranes are comprised of exotic rocks that become detached from a subducting plate and are sutured to the edge of the overriding continent. In northern California, many terranes accreted over the past 500 million years have been identified by geologists.

Northern California.2 The Sierra Nevada: California's Geologic Backbone

■ *What is the Sierra Nevada batholith?*
The Sierra Nevada batholith is a large body of plutonic rock that comprises the core of the Sierra Nevada, California's best known mountain system. It consists of more than 100 individual plutons emplaced mostly between 140 million and 80 million years ago.

■ *What kinds of rocks surround the Sierra Nevada batholith?*
Metamorphic rocks of Paleozoic and Mesozoic age surround the Sierra Nevada batholith. Most of these rocks are of oceanic origin and represent terranes accreted to North America before the intrusion of magma into the batholith during the Mesozoic Era.

■ *When and how was the modern Sierra Nevada uplifted?*
Though the Sierra Nevada region may have been elevated since the late Mesozoic Era, a pulse of rapid uplift 5 million to 10 million years ago lifted the mountain range to its present height. The recent ascent occurred via normal faulting and westward tilting along the eastern escarpment of the range.

■ *What types of gold deposits occur in the Sierra Nevada?*
Gold occurs in the Sierra Nevada regions as particles in modern river sediments (placer deposits), as flakes and nuggets in Eocene-age stream deposits known as "auriferous" gravels, and in association with quartz veins, or lodes, in Mesozoic-age basement rocks.

■ *What is the Mother Lode?*
The Mother Lode is a northwest-trending belt of gold mineralization in the Sierra Nevada foothills. Along this trend modern placer deposits, ancient gold-bearing river sediments, and lode deposits associated with quartz veins have all produced significant amounts of gold.

Northern California.3 The Klamath Mountains

■ *In what ways are the Klamath Mountains and the Sierra Nevada Similar?*
In both the Sierra Nevada and the Klamath Mountains bodies of Mesozoic-age granite have been emplaced in older metamorphic rocks representing numerous terranes accreted to North America. Gold mineralization is common to both areas as well, but slices of oceanic lithosphere (ophiolites) are more common in the Klamath Mountains than in the Sierra. In general, the Klamath Mountains represent a northwest continuation of the geological trends of the Sierra Nevada.

■ *What kinds of rocks comprise the ophiolites in the Klamath Mountains?*
Ophiolites represent disrupted fragments of oceanic lithosphere consisting of mafic and ultramafic igneous rocks associated with oceanic sediments. During emplacement of ophiolite sequences on land by plate convergence, the oceanic rocks generally became deformed and metamorphosed to greenstone, serpentinite, schists, and other altered materials.

Northern California.4 The Great Valley

■ *What factors have led to the formation of fertile soils in the Great Valley?*
The soils of the Great Valley have developed on alluvium derived from the weathering granitic and volcanic bedrock in the Sierra Nevada and adjacent ranges and deposited on a nearly flat surface by the Sacramento and San Joaquin River systems. These factors, along with the warm climate and abundant organic matter of the Central Valley, promote the formation of nutrient-rich, water-retaining soils that can support high agricultural productivity.

■ *What was the origin of the sedimentary rocks in the Great Valley Sequence?*
The Great Valley Sequence consists of sandstone, mudstone, and shale that accumulated in a forearc basin associated with the Mesozoic Farallon subduction zone. These rocks include sediments that were transported by turbidity currents to the deep seafloor where they built large submarine fans.

■ *What is the origin of the natural gas produced in the Great Valley?*
The natural gas resources of the northern Great Valley (Sacramento Valley) originate from organic matter trapped underground in the oceanic sediments of Mesozoic and Cenozoic age.

■ *Why is the Great Valley so prone to chronic flooding?*
The Great Valley is remarkably flat, surrounded by elevated terrain that receives abundant rain and snow, and has a climate that can produce both heavy winter rains and rapid spring melt of snow. The broad floodplain adjacent to the Sacramento and other rivers in the Great Valley has developed over thousands of years of recurring floods. The natural flooding behavior of rivers in the Great Valley is a serious threat to the populous cities, large farms, and industrial centers that have been established in the region.

Northern California.5 The Northern Coast Ranges

■ *What is the Franciscan complex and how did it form?*
The Franciscan complex is a complex assemblage of deformed rock consisting of oceanic sediments such as graywacke, shale, and chert mixed with metamorphosed igneous rocks. Such a complex rock mixture is known as a mélange, and represents an amalgamation of rock that formed in the Farallon subduction zone.

■ *How did the San Andreas fault system originate?*
The San Andreas fault system is the consequence of the development of a transform boundary between the North American and Pacific plates over the past 30 million years. The faults in the San Andreas system are mostly right-lateral strike-slip faults that accommodate the displacement between the Pacific plate (moving to the northwest) and the North American plate.

■ *What is the Salinian block?*
The Salinian block is a large terrane in the northern Coast Ranges that includes granitic and metamorphic basement rocks, similar to those of the Sierra Nevada, overlain by Cenozoic-age sediments of oceanic and terrestrial origin. From its original position near the south end of the Sierra Nevada, the Salinian block has been transported more than 300 kilometers northwest along the San Andreas fault system.

■ *How old are the northern Coast Ranges and what tectonic forces elevated them?*
The northern Coast Ranges are relatively young mountains, rising to their present elevations in the past 3 million to 4 million years. The forces that lifted these coastal mountains are probably related to compression between fault slices in the San Andreas system and the northward migration of the Mendocino triple junction.

Northern California.6 Volcanoes of the Cascade Range and Modoc Plateau

■ *What California volcanoes are part of the Cascade Range?*
Mount Shasta (elevation 4,319 meters/14.161 feet), Mount Lassen (elevation 3,188 meters/10,457 feet), and the Medicine Lake highland are the principal volcanic features of the Cascade Range in northern California.

■ *What is the Cascadia subduction zone?*
The Cascadia subduction zone generates the magma that sustains the volcanic activity in the Cascade Range. This subduction zone is created by the downward movement of the Juan de Fuca plate beneath the northwest edge of North America. The Cascadia subduction zone is only partly in northern California, extending northward from near Cape Mendocino to British Columbia.

■ *What kind of volcanic activity typifies the Cascade Range?*
Cascade volcanoes are mostly composite cone (or stratovolcanoes) that alternate between effusive eruptions of viscous andesitic lava and more violent explosive events that result in pyroclatic flows. The explosive phase of activity was most recently demonstrated by the 1980 eruption of Mount St. Helens in Washington, but the geologic record provides evidence of much more violent prehistoric activity in several Cascade Range locations.

■ *Why is Mount Lassen an especially interesting volcano in the Cascade Range?*
Mount Lassen is noteworthy among Cascade Range volcanoes because it is the most recently active volcano in California (it last erupted in 1914–1921) and is classified as a lava dome, not a composite volcano. Geothermal activity and ongoing seismic activity in the Mount Lassen region suggest that the underground magma system is still active.

Northern California.7 The Basin and Range of Northeast California

■ *What tectonic forces are responsible for the pattern of alternating mountains and valleys in the Basin and Range province?*
Tensional, or stretching, forces have broken the crust in the Basin and Range into hundreds of blocks bounded by normal faults. The blocks displaced downward (the hanging wall) along these faults are represented by the low valleys that separate the elevated ranges.

■ *What mountains and basins in northern California belong to the Basin and Range province?*
The Warner Range, more than 3,000 meters (9,870 feet) high, is the best example of a Basin and Range mountain system in northeast California. The adjacent Surprise Valley and the Lake Tahoe basin to the south both developed as a consequence of the downward displacement along normal faults.

Northern California.8 The Northern California Ice Ages

■ *During what time periods did northern California experience Ice Age conditions?*
Multiple cycles of climate change affected northern California during the Pleistocene Epoch, but the Tahoe and Tioga glaciations, about 160,000 and 20,000 years ago respectively, are the two best documented episodes.

■ *What landscape features resulted from the Pleistocene Epoch glaciations in northern California?*
Valley glaciers developed in the mountainous regions of northern California several times during the Pleistocene Ice Ages. Erosional features produced by these glaciers are widespread and include polished and striated bedrock surfaces, U-shaped glacial troughs, horns, arêtes, and cirques. In addition, sediments released from melting ice at the close of the glacial episodes left glacial erratics and moraine deposits in many locations.

■ *How did the Pleistocene landscape of northern California differ from the modern setting?*
During the periods of maximum glacial advance, the crest of the Sierra Nevada was buried under a thick ice cap. From this high mass of ice, valley glaciers extended down the major canyons, in some cases completely filling them with ice. Valley glaciers also existed in the Klamath Mountains and in the Cascade Range. In the lowland valleys, pluvial lakes expanded and withered repeatedly in rhythm with the Pleistocene climatic oscillations. The decrease in sea level that accompanied the glacial intervals exposed the floor of San Francisco Bay as a broad river valley several times. The Pacific shore was more than 30 kilometers west of its present location.

Northern California.9 Northern California Earthquakes

■ *What plate tectonic settings are associated with northern California earthquakes?*
Northern California earthquakes occur mostly along the San Andreas fault system (part of a transform plate boundary), in the Cascadia subduction zone (a convergent boundary), or along the eastern escarpment of the Sierra Nevada (a zone of developing plate divergence, or rifting).

■ *What is the likelihood of another major earthquake in northern California?*
Given the tectonic forces currently affecting the crust in northern California, another major earthquake is certain to occur eventually. Modern techniques for forecasting earthquake activity, including the analysis of seismic gaps, suggest that there is a 67% probability of an earthquake exceeding magnitude 6.7 in the next 30 years. The faults in the San Francisco Bay area, such as the Hayward-Rogers Creek, Calaveras, and San Andreas faults, are mostly likely to produce this earthquake. In this densely populated and intensively developed region, such an earthquake could have disastrous consequences.

■ *What might be the effects of a large northern California earthquake?*
Depending on the location of the epicenter and the magnitude, the ground shaking of a large earthquake in northern California could result in damaged or destroyed buildings, fires, collapsed bridges, disrupted communications and transportation networks, contaminated drinking water, liquefaction of soils, and tsunamis along the coast. Millions of people and billions of dollars in property are vulnerable to earthquake hazards in northern California.

■ *Given the severity of the potential hazards, how can the effects of earthquakes be minimized in northern California?*
Experience has shown that earthquakes hazards can be reduced through a combination of wise land use, continuous seismic monitoring, stringent building codes, and personal and public preparedness. Fortunately, most populated regions in northern California have plans in place for effective responses to a potential earthquake, and the regulated development of communities has reduced the number of people at risk.

Northern California. 10 Living on the Edge: Coastal Hazards in Northern California

■ *In comparison to the coast of southern California, why is the northern California shoreline so rugged and scenic?*
The northern California coast trends northwest, directly facing the waves, storms, and tsunamis generated in the Pacific Ocean basin. In addition, the emergent coast of northern California continuously exposes new bedrock to the vigorous erosion that results in such landforms as sea cliffs, sea caves, sea arches, and sea stacks. Finally, in most places, the bedrock of the northern California coast is a complex assemblage of hard and soft rock that erodes to an irregular shore with projecting headlands separated by recessed coves and pocket beaches.

■ *What coastal hazards exist in northern California?*
Coastal hazards in northern California are primarily related to erosion, mass wasting, or seismic activity. Erosion of sea cliffs by waves causes the cliffs to retreat at rates as high as a meter per year, potentially affecting structures overlooking the ocean. Mass wasting occurs readily along the steep slopes descending to the sea along the emergent coast, especially during the rainy season or where the rocks are weak and/or absorb water. Tsunamis generated by earthquakes in the Pacific basin have struck the California coast many times, some from very distant epicenters.

■ *How do human activities affect coastal hazards?*
Whenever humans modify the natural conditions along the coast, dynamic responses result that can sometimes intensify coastal hazards. For example, although breakwaters can protect one location from wave erosion, they also deflect the energy of approaching waves to other places where erosion increases. When coastal slopes are modified for the construction of roads or buildings, mass wasting can result if the balance between gravity and friction on a table stable slope is compromised. Today, in most coastal communities, awareness of these coastal hazards has led to restrictions on development, but there are still many places where people and property are at risk.

ESSENTIAL TERMS TO KNOW

accretionary terranes, terranes – a large block of rock with characteristics different from surrounding blocks. Terranes are typically bounded by large faults and are thought to represent seamounts, oceanic rises, reefs, island arcs, or other oceanic features accreted to continents at convergent plate boundaries.

alluvium – a general term for sediment transported and deposited by running water.

arête – a narrow jagged ridge separating two glacial valleys or cirques.

basalt – a dark-colored fine-grained igneous rock that forms from magma of mafic (45% to 52% silica) composition.

batholith – a large body of plutonic igneous rock with a surface exposure exceeding 100 square kilometers.

baymouth bar – a spit that extends across the mouth of a bay, closing it off from the open ocean.

blueschist – a foliated metamorphic rock that contains glaucophane, a bluish-colored mineral of the amphibole group.

breccia – a detrital sedimentary rock consisting of angular rock particles larger than 2 millimeters.

caldera – a large circular to oval depression that develops on the summit of a volcano in response to the partial evacuation of the underlying magma chamber.

chert – a fine-grained nonclastic sedimentary rock consisting of microcrystalline quartz, which may include the skeletons of silica-secreting microorganisms such as radiolarian.

cinder cone – a small steep-sided volcano consisting of a loose accumulation of volcanic cinder.

cirque – a generally circular depression that forms in the uppermost reaches of glacial troughs occupied by valley glaciers.

composite volcano – a volcano composed of lava flows, pyroclastic layers, and mudflow deposits; sometimes referred to as a stratovolcano.

dacite – an extrusive igneous rock intermediate in composition between rhyolite and andesite.

diorite – an intrusive igneous rock with nearly equal amounts of dark-colored and light-colored minerals. Diorite is intermediate in composition between granite (felsic) and gabbro (mafic).

emergent coast – a coast where land has risen with respect to sea level.

Farallon plate – an ancient lithospheric plate that separated the North America plate from the Pacific plate during the Mesozoic and Early Cenozoic Eras. Remnants of the Farallon plate include the modern Juan de Fuca, Gorda, Cocos, and Rivera plates in the eastern Pacific Ocean basin.

Farallon subduction zone – the east-dipping zone of subduction developed where the Farallon plate descended beneath the western edge of the North American plate during the Late Mesozoic and Early Cenozoic eras.

fault – a fracture in the Earth's crust that includes displaced rock masses.

felsic – a term describing the composition of magma or igneous rock that contains more than 65% silica and is rich in sodium, potassium, and aluminum.

forearc basin – a basin of sediment accumulation located between a volcanic arc mountain system and an offshore oceanic trench. Forearc basins are developed in association with subduction zones at convergent plate boundaries.

fumerole – a vent at the surface that emits volcanic gases.

gabbro – a dark-colored intrusive igneous rock of mafic composition.

glacial erratic – a rock carried by a glacier from its source to a surface of dissimilar bedrock.

glacial polish – a smooth and glossy bedrock surface produced by the movement of a glacier over it.

glacial striation – a linear scratch or shallow groove on a bedrock surface produced by the movement of a sediment-laden glacier over it.

graben – a term for the block of rock displaced downward as the hanging wall of a normal fault.

granite – a light-colored plutonic igneous rock of felsic composition.

greenstone – a metamorphic rock containing the greenish minerals epidote, chlorite, or amphibole that results from the metamorphism of mafic igneous materials.

headland – land that projects seaward from an irregular shore line.

horn – a steep and jagged peak having the form of a pyramid that developed between glacial cirques.

horst – a term describing the block of rock displaced upward as the footwall of a normal fault.

inverted topography – a term describing any landscape feature that originated at a low elevation, but is currently elevated above the surrounding terrain, or vice versa.

lahar – a volcanic mudflow composed of water, ash, and particles of volcanic rock.

limestone – a chemical sedimentary rock composed of calcite, $CaCO_3$.

lode – an ore deposit in which the valuable commodity is concentrated in a vein or pod within crystalline igneous or metamorphic rock.

mafic – a term describing the composition of magma or igneous rock that contains 45% to 52% silica and is rich in calcium, magnesium, and iron.

mass wasting – the downslope movement of material under the influence of gravity including rockslides, mudslides, rockfalls, and debris flows.

mélange – a deformed and sheared mass of metamorphic and sedimentary rocks that forms in seduction zones.

moraine – a ridge of mound of unsorted and nonstratified glacial sediment (till) deposited by melting glaciers.

mudstone – a fine-grained clastic sedimentary rock consisting of a mixture of sand, silt, and clay-size particles.

normal faults – a dip-slip fault in which the hanging wall has moved down relative to the footwall. Normal faults develop from tensional stress and are most common in areas of divergent plate boundaries

ophiolite – a sequence of mafic igneous rock representing oceanic lithosphere and upper mantle; ophiolites consist of pillow basalt, sheeted basalt dikes, layered and massive gabbro, and mantle periodite.

Pangaea – a supercontinent consisting of all the Earth's landmasses that existed at the end of the Paleozoic Era, about 250 million years ago.

peridotite –an ultramafic igneous rock thought to comprise much of the upper mantle.

physiographic province – a region of unique geology, landforms, drainage, soils, climate, and flora and fauna distinct from adjacent areas on the basis of these features.

pillow structures – rounded or bulbous structures that develop in lava erupted under water.

placer – a term applied to stream-transported sediment such as sand and gravel that contains significant quantities of a valuable mineral such as gold, silver, or platinum.

pluton – an intrusive igneous body that forms when magma cools and crystallizes below the surface and within the crust.

plutonic – a term describing the origin of an igneous rock that crystallized from magma within the crust.

pocket beach– a small beach along an irregular coast located in a recess between projecting headlands.

pyroclastic – a term that describes the fragmental texture or character of volcanic rock such as tuff and volcanic breccia produced during explosive eruptions.

roof pendant – a mass of metamorphic rock preserved above the top of an underlying pluton. Roof pendants represent remnants of the "country rock" into which subterranean magma intruded.

sea arch – an archlike exposure of coastal bedrock that results from wave erosion of sea caves along a projecting headland.

sea stack – a remnant of bedrock on a wave-cut marine terrace.

seismic gap – a portion of an active fault that has produced fewer and smaller earthquakes than other segments of the same fault.

Sierra Nevada batholith – a large composite pluton consisting of more than 100 individual masses of intrusive igneous rock making up the basement of the Sierra Nevada.

serpentinite – a greenish-black metamorphic rock that forms from mafic igneous rocks such as basalt, gabbro, and peridotite. Sepentinite occurs in the western metamorphic terranes of the Sierra Nevada, in the Klamath Mountains, and in the northern Coast Ranges. It is the official California state rock.

spit – a sandy projection of a beach into a body of water such as a bay or estuary.

superterrane – a large fragment of crustal rocks that is comprised of smaller terranes amalgamated during accretion to a continental margin.

syncline – a down-arched or concave-upward fold in a sequence of rock layers. (Note: the above definition is used to conform to that of Monroe, Wincander, and Hazlett)

tombolo – a spit that extends outward into the sea or a lake that connects an island or sea stack to land.

transform plate boundary – a boundary between plates that slide past one another and whose crust is neither produced or destroyed. The strike-slip faults of the San Andreas fault zone represent the transform boundary between the Pacific and North America plates.

triple junction – the point at which three lithosphere plates meet.

tsunami – a large sea wave that is produced when mass on the seafloor is suddenly displaced by earthquakes, volcanic eruptions, or submarine landslides.

tuff – a fine-grained pyroclastic igneous rock that consists of consolidated particles of volcanic ash.

turbidity currents – a dense mixture of sediment and water that flows downslope on the ocean floor.

ultramafic – a term describing the composition of magma or igneous rock that contains 40% or less silica and abundant magnesium and iron.

volcanic – a term describing the origin of igneous rocks from magma that cools rapidly after its eruption during a volcanic event.

volcanic arc – a chain of volcanoes that develops on the earth's surface where magma rises from an underlying subduction zone.

welded tuffs – a hard volcanic rock composed of tiny ash articles that are firmly welded together by the heat associated with an explosive volcanic eruption.

xenoliths – a type of inclusion in which a fragment of older rock is incorporated into younger igneous rock.

MORE ON NORTHERN CALIFORNIA GEOLOGY

California's stunning scenery, rich geologic heritage, and abundant natural resources have stimulated considerable interest from geologists for more than two centuries. Scientific studies in the state have resulted in a vast body of technical literature on California geology. Fortunately, there are many excellent summaries available that describe the geologic features of northern California in comprehensible terms and in the context of modern geologic concepts. Your college or local library may have copies of the following books that you may find valuable in learning more about the geology of northern California:

Assembling California, John McPhee, 2003, Farrar, Straus, and Giroux, New York, 304 pages.

Geology Of California, Deborah R. Harden, 2004, Pearson-Prentice Hall, Upper Saddle River, NJ, 552 pages.

Geology of Northern California, E.H. Bailey, editor, 1966, California Division of Mines and Geology (California Geological Survey) Bulletin 190, 508 pages.

Roadside Geology of Northern and Central Califorina, by David Alt and Donald Hyndman, 2003, Mountain Press Publishing, Missoula, MT, 384 pages.

The California Geological Survey is an excellent source of information on earthquakes, rocks and minerals, geologic hazards, and mapping programs in the Golden State. The Web site at http://www.conservation.ca.gov/cgs contains links to a vast array of information on the geology of California. In addition, the U.S. Geological Survey Web site (http://www.usgs.gov) offers a search feature that allows you to retrieve many USGS publications on specific aspects of California geology, water resources, and geologic hazards.

Finally, check with the geology department at your local college and university about other sources of information and additional opportunities to explore California geology. Most college geology programs offer field courses and introductory lecture courses that address the diverse geologic features of the state.

The *Geology of Northern California* provides an overview of the physiographic features and geological history of northern California's magnificent landscape. Written to accompanying any college-level earth science course, this module explores the rich geological heritage of northern California. The major geological features of the northern Sierra Nevada, the Klamath Mountains, the Cascade Range region, the northern Great Valley, the Basin and Range, and the northern Coast Ranges are explored with an emphasis on applying the fundamental concepts of modern geology to the interpretation of the scenic and varied terrain.

This module also surveys the affect of geological processes on humans in northern California. The origin of the rich mineral and fuel resources of the Golden State are explored in the context of hundreds of million of years of the tectonic evolution. In addition, the multiple hazards to people and property that result from California's on-going geological evolution are considered. We learn why future earthquakes, volcanic eruptions, slope failures, and floods *will* affect northern Californians, and how an understanding of these threats can help reduce their potential impacts.

Use this module to help make any introductory geology course more relevant to college students throughout northern California by providing local and regional examples of the consequences of dynamic geologic processes. Work with your Cengage Learning representative to learn how you can incorporate this or any of the modules from our Regional Geology Series into your course. Visit http://custom.cengage.com/regional_geology/ to learn more about other bonus content available and how to order it.

About the Author:

Frank DeCourten is Professor of Earth Science at Sierra College in Grass Valley, California, where he teaches courses in geology, geography, and environmental sciences. He has written four books and numerous research papers and popular articles on the geology of western North America.

Ambient Images/drr.net

CENGAGE Learning™

Visit Cengage Learning Custom Solutions online at www.cengagecustom.com

For your lifelong learning needs: www.academic.cengage.com

ISBN-13: 978-0-495-76382-6
ISBN-10: 0-495-76382-9

9 780495 763826

3

Earthquakes and Human Activities

It is useful to be assured that the heavings of the Earth are not the work of deities. These phenomena have a cause of their own.

Seneca, Roman statesman and philosopher (4 B.C.?–A.D. 65)

ASSOCIATED PRESS, AUGUST 25, 1999 "TURKEY'S PRIME MINISTER has promised stricter building rules to prevent the shoddy construction blamed for the thousands of deaths from last week's massive earthquake. Eight days after the powerful 7.4-magnitude temblor reduced much of western Turkey to rubble, an estimated 200,000 survivors remained camped out in parks and on vacant lots."

The cause of the earthquake was a sudden rupture along northern Turkey's Anatolia fault. The severe ground motion that was generated collapsed hundreds of buildings, many of which had been built below earthquake-resistant standards. Interestingly, the Anatolia fault and California's infamous San Andreas fault have some striking geographic similarities; see ◆Figure 1. Let's hope that is where the likeness ends.

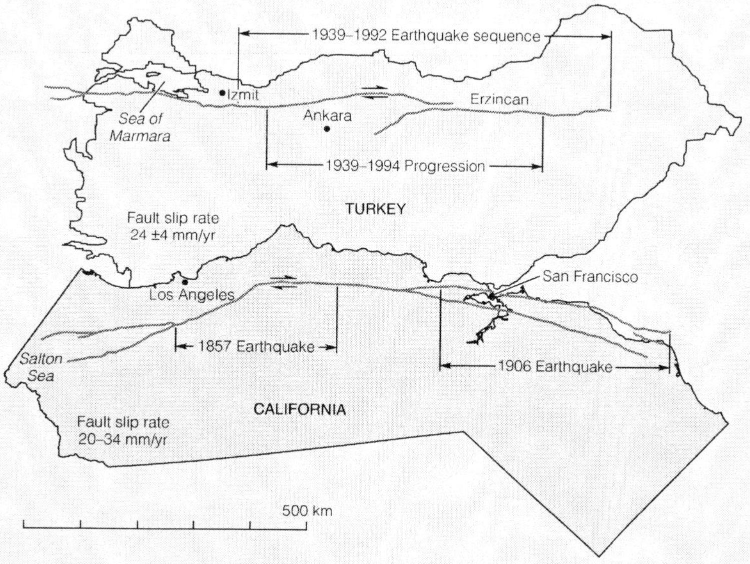

◆ FIGURE 1 The North Anatolian fault in Turkey and the San Andreas fault in California. The faults have the same sense of motion, are about the same age and length, and are capable of similar-magnitude earthquakes. The earthquake sequences shown on the North Anatolian fault represent events that occurred, one after the other from east to west, along the fault.

After experiencing an earthquake in Concepción, Chile, in **1835, Charles** Darwin noted that "A bad earthquake at once **destroys the old**est associations; the world, the very emblem **of all that is sol**id, had moved beneath our feet like crust over **a fluid." Darwin's** reflections are vivid, and many of us have felt the same way when experiencing strong earthquake motion. What Darwin did not know was the cause of earthquakes and the resulting ground motion.

The Nature of Earthquakes

Earthquakes are the result of abrupt movements on **faults**—fractures in Earth's lithosphere. The types of faults and the Earth forces that cause them are shown in ◈ Figure 4.1. The

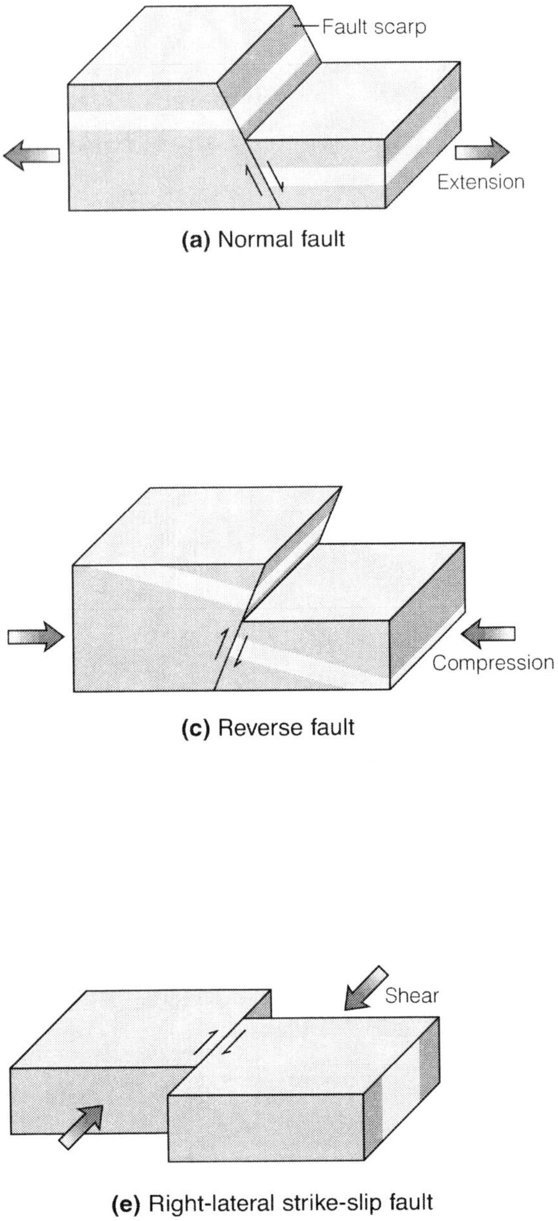

◈ FIGURE 4.1 (a) Normal fault geometry and (b) examples in a road cut on I-40 near Kingman, Arizona. The normal fault at the right is in relatively soft sedimentary rocks. Fault at the left is also a normal fault with a small displacement of white sandstone between it and a small fault that shows reverse displacement. (c) Reverse fault geometry and (d) low-angle reverse fault displacing sandstone strata at Wasatch Plateau near Salina, Utah. (e) Right-lateral strike-slip fault geometry and (f) a plowed field displaced by a strike-slip fault in the Imperial Valley, California, in 1979.

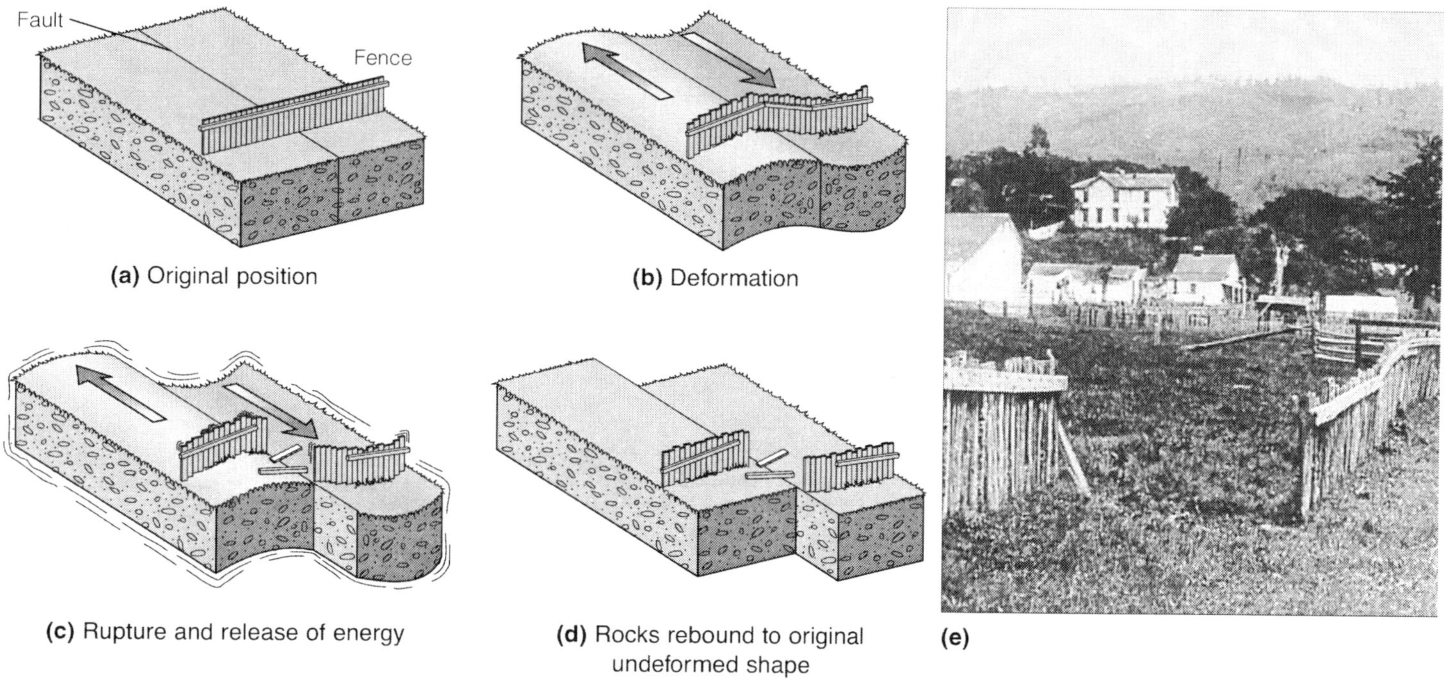

◈ FIGURE 4.2 (a–d) The cycle of elastic-strain buildup and release for a right-lateral strike-slip fault according to Reid's elastic rebound theory of earthquakes. At the instant of rupture, (c), energy is released in the form of earthquake waves that radiate out in all directions. (e) Right-lateral offset of a fence by 2.5 meters (8 ft) by displacement on the San Andreas fault in 1906; Marin County, California.

movements occur as Earth's crustal plates slip past, under, and away from one another. Because the **stress** (force per unit area) that produces **strain** (deformation) can be transmitted long distances in rocks, active faults do not necessarily occur exactly on a plate boundary, but they generally occur in the vicinity of one. The mechanism by which stressed rocks store up strain energy along a fault to produce an earthquake was explained by Harold F. Reid after the great San Francisco earthquake of 1906. Reid proposed a mechanism to explain the shaking, which resulted from movement on the San Andreas fault, known as the **elastic rebound theory** (◈Figure 4.2). According to this theory, when sufficient strain energy has accumulated in rocks, they may rupture rapidly—just as a rubber band breaks when it is stretched too far—and the stored strain energy is released as vibrations that radiate outward in all directions (◈Figure 4.3).

Most earthquakes are generated by movements on faults within the crust and upper mantle that do not produce ruptures at the ground surface. We can thus recognize a point within the Earth where the fault rupture starts, the **focus,** and the **epicenter,** the point on the earth surface directly above the focus (Figure 4.3). Elastic waves—vibrations—move out spherically in all directions from the focus and strike the surface. Damaging earthquake foci are generally within a few kilometers of Earth's surface. Deep-focus earthquakes, on the other hand, those whose foci are 300–700 kilometers (190–440 mi) below the surface, do little or no damage. Earthquake foci are not known below 700 kilometers. This indicates that the mantle at that depth behaves *plastically* due to high temperatures and confining pressures, deforming continuously as ductile substances do, rather than storing up strain energy.

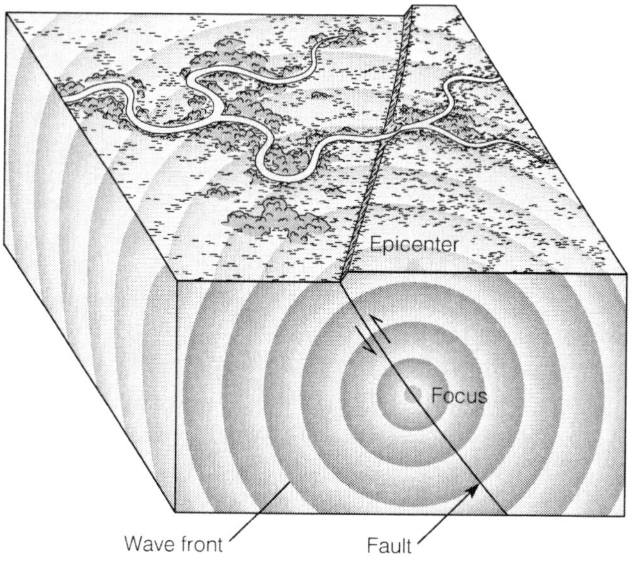

◈ ThomsonNOW™ ACTIVE FIGURE 4.3 The focus of an earthquake is the point in the earth where fault rupture begins, and the epicenter is directly above the focus at the ground surface. Seismic-wave energy moves out in all directions from the focus. From *Essentials of Geology,* 2nd ed., by R. Wicander and J. Monroe, p. 228, Brooks/Cole, 1999

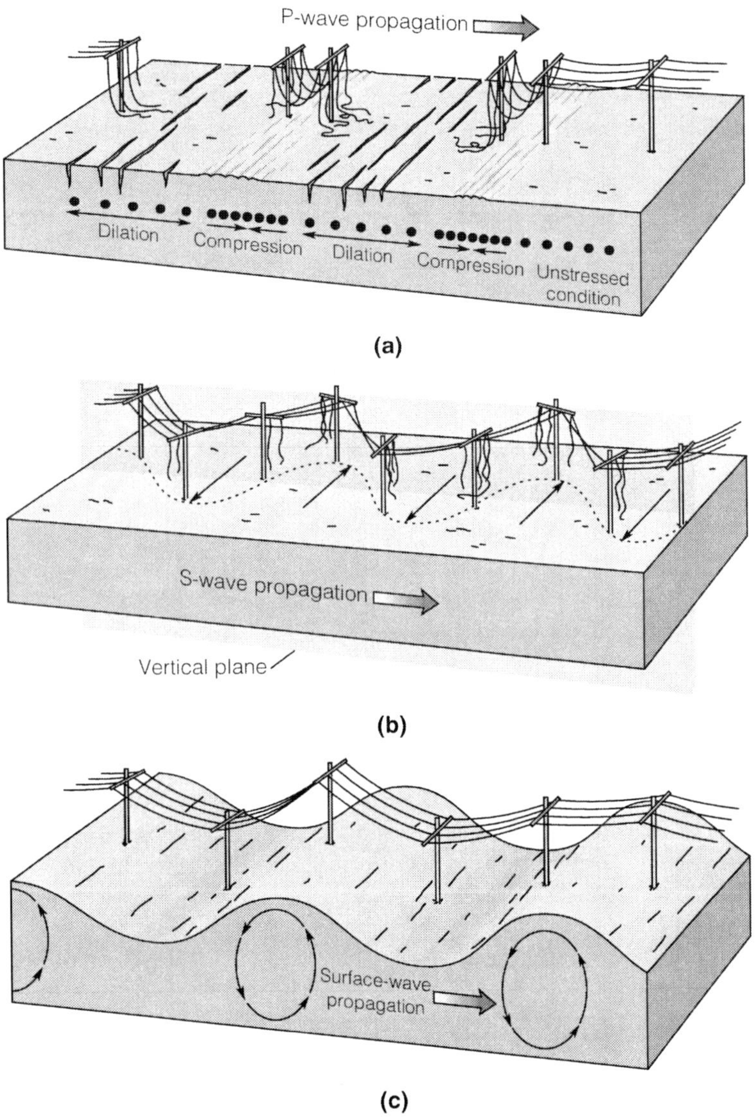

◈ ThomsonNOW™ ACTIVE FIGURE 4.4 Ground motion during passage of earthquake waves. (a) P-waves compress and expand the earth. (b) S-waves move in all directions perpendicular to the wave advance, but only the horizontal motion is shown in this diagram. (c) Surface waves create surface undulations that result from a combination of the retrograde elliptical motions of Rayleigh waves (shown) and Love waves, which move from side to side at right angles to the direction of wave propagation.

The vibration produced by an earthquake is complex, but it can be described as three distinctly different types of waves (◈Figure 4.4). Primary waves, **P-waves,** and secondary waves, **S-waves,** are generated at the focus and travel through the interior of Earth; thus, they are known as **body waves.** They are designated P- and S-waves because they are the first (*p*rimary) and second (*s*econdary) waves to arrive from distant earthquakes. As these body waves strike the planet's surface, they generate surface waves that are analogous to water ripples on a pond.

P-waves (◈Figure 4.4a) are **longitudinal waves;** the solids, liquids, and gases through which they travel are alternately compressed and expanded in the same direction the waves move. Their velocity depends upon the resistance to change in volume (in compressibility) and shape of the material through which they travel. P-waves travel about 300 meters (1,000 ft) per second in air, 300–1,000 meters (1,000–3,000 ft) per second in soil, and faster than 5 kilometers (3 mi) per second in solid rock at the surface. P-wave velocity increases with depth, because the materials composing the mantle and core become less compressible with depth. This increase causes the waves' travel paths to be bowed downward as they move through the Earth. P-waves speed through the Earth with a velocity of about 10 kilometers (6 mi) per second. At the ground surface they have very small amplitudes (motion) and cause little property damage. Although they are physically identical to sound waves, they vibrate at frequencies below what the human ear can detect. The earthquake noise that has been reported could be P-waves of a slightly higher frequency or some other, related vibrations whose frequencies are in the audible range.

◆ FIGURE 4.5 Seismometer (*left*) and seismogram (*right*) used to demonstrate method of detecting and recording earthquakes. The record on the seismogram is not an earthquake, but represents the seismometer's detection of vibrations from students' footsteps amplified about 2,000 times.

S-waves (◆Figure 4.4b) are **transverse (shear) waves;** they produce ground motion perpendicular to their direction of travel. This causes the rocks through which they travel to be twisted and sheared. These waves have the shape produced when one end of a garden hose or a loosely hanging rope is given a vigorous flip. They can travel only through material that resists shearing—that is, material that resists two forces acting in opposite directions in different planes. Thus, S-waves travel only through solids, because liquids and gases have no shear strength (try piling water in a mound). S-wave ground motion may be largely in the horizontal plane and can result in considerable property damage. S-wave velocity is approximately 5.2 kilometers (3.2 mi) per second from distant earthquakes; hence, they arrive after the P-waves.

Unexpended P- and S-wave energy bouncing off Earth's surface generates the complex surface waves (◆Figure 4.4c). These waves produce a rolling motion at the ground surface. In fact, slight motion sickness is a common response to surface waves of long duration. The waves are the result of a complex interaction of several wave types, the most important of which are *Love waves,* which exhibit horizontal motion normal to the direction of travel, and *Rayleigh waves,* which exhibit retrograde (opposite to the direction of travel) elliptical motion in a plane perpendicular to the ground surface. Surface waves have amplitudes that are greatest in near-surface unconsolidated layers and are the most destructive of the earthquake waves.

Locating the Epicenter

A **seismograph** is an instrument designed specifically to detect, measure, and record vibrations in the Earth's crust (◆Figure 4.5). It is relatively simple in concept (◆Figure 4.6), but some very sophisticated electronic systems have been developed. Seismic data are recorded onto a **seismogram** (◆Figure 4.7). Because seismographs are extremely sensitive to vibrations of any kind, they are installed in quiet areas such as abandoned oil and water wells, cemeteries, and parks.

When an earthquake occurs, the distance to its epicenter can be approximated by computing the difference in P- and S-wave arrival times at various seismograph stations.

◆ FIGURE 4.6 The first earthquake detector, invented about A.D. 130 by Chinese scholar Chang Heng. Balls held in the dragons' mouths were aligned with a pendulum inside the vase. When an earthquake occurred the pendulum swung and pushed balls into the mouths of the frogs aligned with the pendulum's swing. If the frogs facing north and south contained balls, for example, Chang Heng could say that the earthquake epicenter was north or south of the instrument.

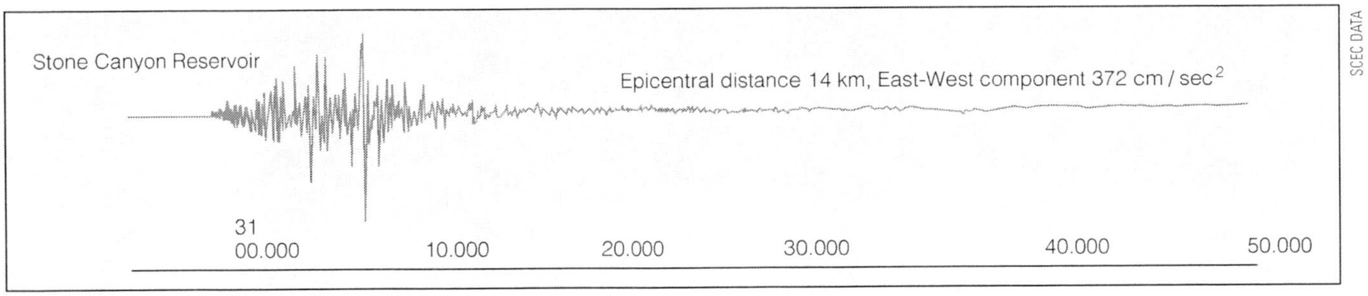

◈ FIGURE 4.7 Seismogram of the main shock of the Northridge earthquake, January 17, 1994. The time in minutes and seconds after 4:00 A.M. appears at the bottom. Distance from the epicenter and direction and ground acceleration at the recording site are also indicated. (Acceleration is explained in the next section.)

Although the method actually used by seismologists today is more accurate and determines the depth and location of the quake's focus and its epicenter, what is described here serves to show how seismic-wave arrival times can be used to determine the distance to an epicenter. Because it is known that the two kinds of waves are generated simultaneously at the earthquake focus and that P-waves travel faster than S-waves, it is possible to calculate where the waves started.

Imagine that two trains leave a station at the same time, one traveling at 60 kilometers per hour and the other at 30 km/h, and that the second train passes your house one hour after the first train goes by. If you know their speeds, you can readily calculate your distance from the train station of origin as 60 kilometers. The relationship between distance to an epicenter and arrival times of P- and S-waves is illustrated in ◈Figure 4.8. The epicenter will lie somewhere on a circle whose center is the seismograph and whose radius is the distance from the seismograph to the epicenter. The problem is then to determine where on the circumference of the circle the epicenter is. To learn this, more data are needed: specifically, the distances to the epicenter from two other seismograph stations. Then the intersection of the circles drawn around each of the three stations—a method of map location called *triangulation*—specifies the epicenter of the earthquake (◈Figure 4.9).

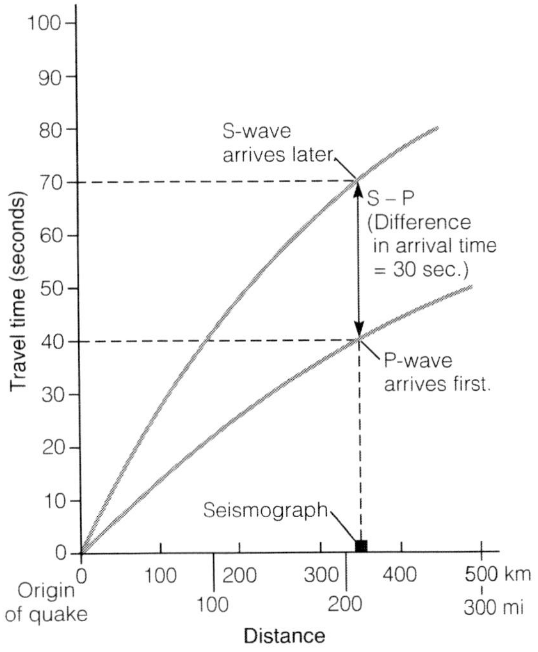

◈ FIGURE 4.8 Generalized graph of distance versus travel time for P- and S-waves. Note that the P-wave has traveled farther from the origin of the earthquake than the S-wave at any given elapsed time. The curvature of both wave paths is due to increases in velocity with depth (distance from epicenter).

CONSIDER THIS

You are shopping for a house in earthquake country. How could an isoseismal map of an earthquake whose epicenter was near a house you think you would like to buy be useful? What patterns would you look for when examining the isoseismal map?

Earthquake Measurement

Intensity Scales

The reactions of people (geologists included) to an earthquake typically range from mild curiosity to outright panic. However, a sampling of the reactions of people who have been subjected to an earthquake can be put to good use. Numerical values can be assigned to the individuals' perceptions of earthquake shaking and local damage, which can then be contoured upon a map. One **intensity scale** developed for measuring these perceptions is the 1931 **modified Mercalli scale *(MM)*.** The scale's values range from *MM* = I (denoting not felt at all) to *MM* = XII (denoting widespread destruction), and they are keyed to specific U.S. architectural and building specifications. (See Appendix 4.) People's perceptions and responses are compiled from returned questionnaires, and then lines of earthquake intensity, called **isoseismals,** are plotted on maps. Isoseismals enclose areas of equal earthquake damage and can indicate

◆ ThomsonNOW™ ACTIVE FIGURE 4.9 An earthquake epicenter can be closely approximated by triangulation from three seismic stations.

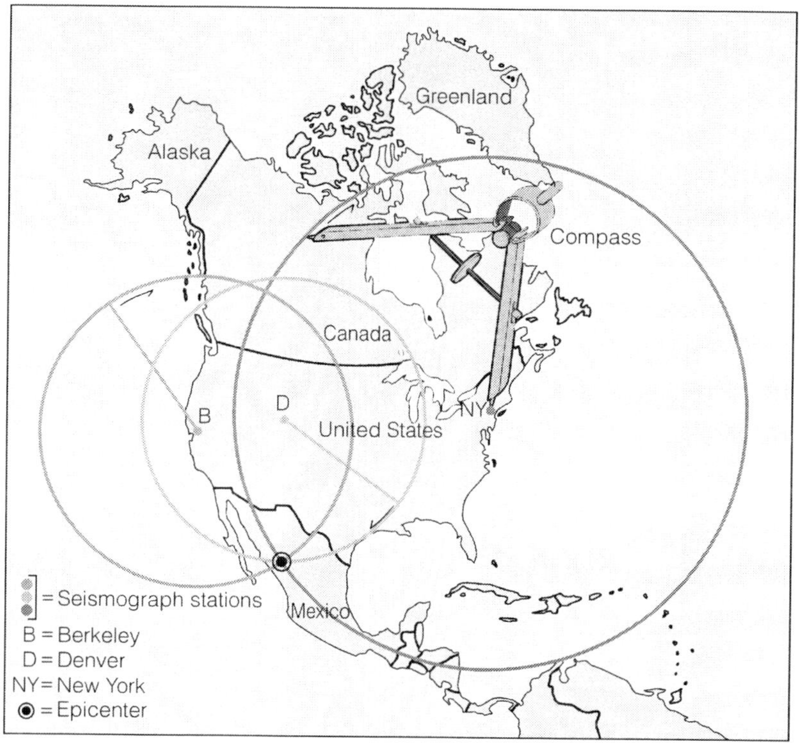

areas of weak rock or soil as well as areas of substandard building construction (◆Figure 4.10a). Such maps have proved useful to planners and building officials in revising building codes and safe construction standards.

Recently a new method of plotting shaking intensities has evolved that is faster and that compares favorably with the questionnaire method. It is the Community Internet Intensity Maps (CIIM), an example of which is shown in Figure 4.10b. The CIIM takes advantage of the Internet, and the time to generate intensity maps drops from months to minutes. The responses, which can take place within 3 minutes of the event, are summarized by computer, and an intensity number is assigned to each ZIP code area. This method is particularly useful in areas with sparse seismograph coverage. Although perhaps not as colorful as the map generated by the traditional *MM* intensities obtained by using postal questionnaires, the CIIM values agree well and have actually proved more reliable in areas of low shaking. However, neither method allows comparison of earthquakes with widely spaced epicenters because of local differences in construction practices and local geology.

Richter Magnitude Scale—The Best-Known Scale

The best-known measure of earthquake strength is the **Richter magnitude scale,** which was introduced in 1935 by Charles Richter and Beno Gutenberg at the California Institute of Technology. It is a scale of the energy released by an earthquake and thus, in contrast to the intensity scale, may be used to compare earthquakes in widely separated geographic areas. The Richter value is calculated by measuring the maximum amplitude of the ground motion as shown on the seismogram using a specified seismic wave, usually the surface wave. Next, the seismologist "corrects" the measured amplitude (in microns) to what a "standard" seismograph would record at the station. After an additional correction for distance from the epicenter, the Richter **magnitude** is the common logarithm of that ground motion in microns. For example, a magnitude-4 earthquake is specified as having a corrected ground motion of 10,000 microns ($\log_{10}$ of $10{,}000 = 4$) and thus can be compared to any other earthquake for which the same corrections have been made. It should be noted that because the scale is logarithmic, each whole number represents a ground shaking (at the seismograph site) 10 times greater than the next-lower number. Thus a magnitude-7 produces 10 times greater shaking than a magnitude-6, 100 times that of a magnitude-5, and 1,000 times that of a magnitude-4. Total energy released, on the other hand, varies logarithmically as some exponent of 30. Compared to the energy released by a magnitude-5 earthquake, a $M = 6$ releases 30 times (30^1) more energy; a $M = 7$ releases 900 times (30^2) more; and a $M = 8$ releases 27,000 (30^3) times more energy (◆Figure 4.11a).

The Richter scale is open-ended; that is, theoretically it has no upper limit. However, rocks in nature do have a limited ability to store strain energy without rupturing, and no earthquake has been observed with a Richter magnitude greater than 8.9—yet.

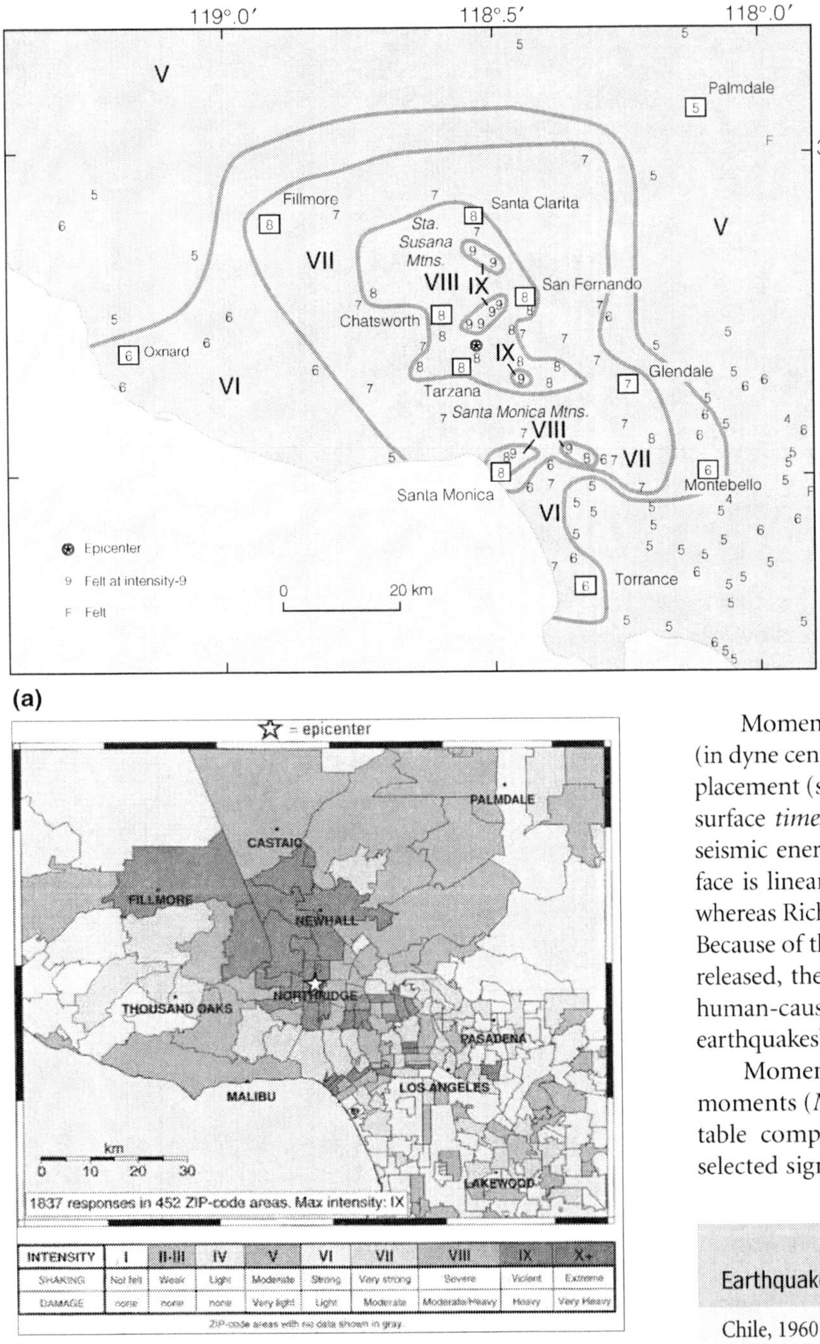

(a)

(b)

◆ FIGURE 4.10 (a) Isoseismal map of the near-field (close to the epicenter) modified Mercalli scale intensities of the 1994 Northridge, California, earthquake. Intensity-IX values are assigned where there are spectacular partial collapses of modern buildings, destroyed wood-frame buildings, and collapses of elevated freeways (see Appendix 4). (b) Community Internet Intensity Map (CIIM) for the same earthquake. The main difference in appearance is due to the fact that the intensities were not contoured but assigned to ZIP code areas of the respondents. (a) SCEC data (b) USGS

Moment Magnitude—The Most Widely Used Scale

Seismologists have abandoned Richter magnitudes in favor of **moment magnitudes** (**M_w** or **M**) for describing earthquakes. The reason is that Richter magnitudes do not accurately portray the energy released by large earthquakes on faults with great rupture lengths. The seismic waves used to determine the Richter magnitude come from only a small part of the fault rupture and, hence, cannot provide an accurate measure of the total seismic energy released by a very large event.

Moment magnitude is derived from seismic moment, M_0 (in dyne centimeters), which is proportional to the average displacement (slip) on the fault *times* the rupture area on the fault surface *times* the rigidity of the faulted rock. The amount of seismic energy (in ergs) released from the ruptured fault surface is linearly related to seismic moment by a simple factor, whereas Richter magnitude is logarithmically related to energy. Because of the linear relationship of seismic moment to energy released, the equivalent energy released by other natural and human-caused phenomena can be conveniently compared to earthquakes' moment magnitudes on a graph (◆Figure 4.11b).

Moment magnitudes (M_w) are derived from seismic moments (M_0) by the formula: $M_w = (\frac{2}{3} \log M_0 - 10.7)$. This table compares the two most commonly used scales for selected significant earthquakes:

Earthquake	Richter Magnitude	Moment Magnitude
Chile, 1960	8.3	9.5
Alaska, 1964	8.4	9.2
New Madrid, 1812	8.7 (est.)	8.1 (est.)
Mexico City, 1985	8.1	8.1
San Francisco, 1906	8.3 (est.)	7.7
Loma Prieta, 1989	7.1	7.0
San Fernando, 1971	6.4	6.7
Northridge, 1994	6.4	6.7
Kobe, Japan, 1995	7.2 JMA*	6.9

*The Richter scale is not used in Japan. The official earthquake scale there is the scale developed by the Japanese Meteorological Agency (JMA).

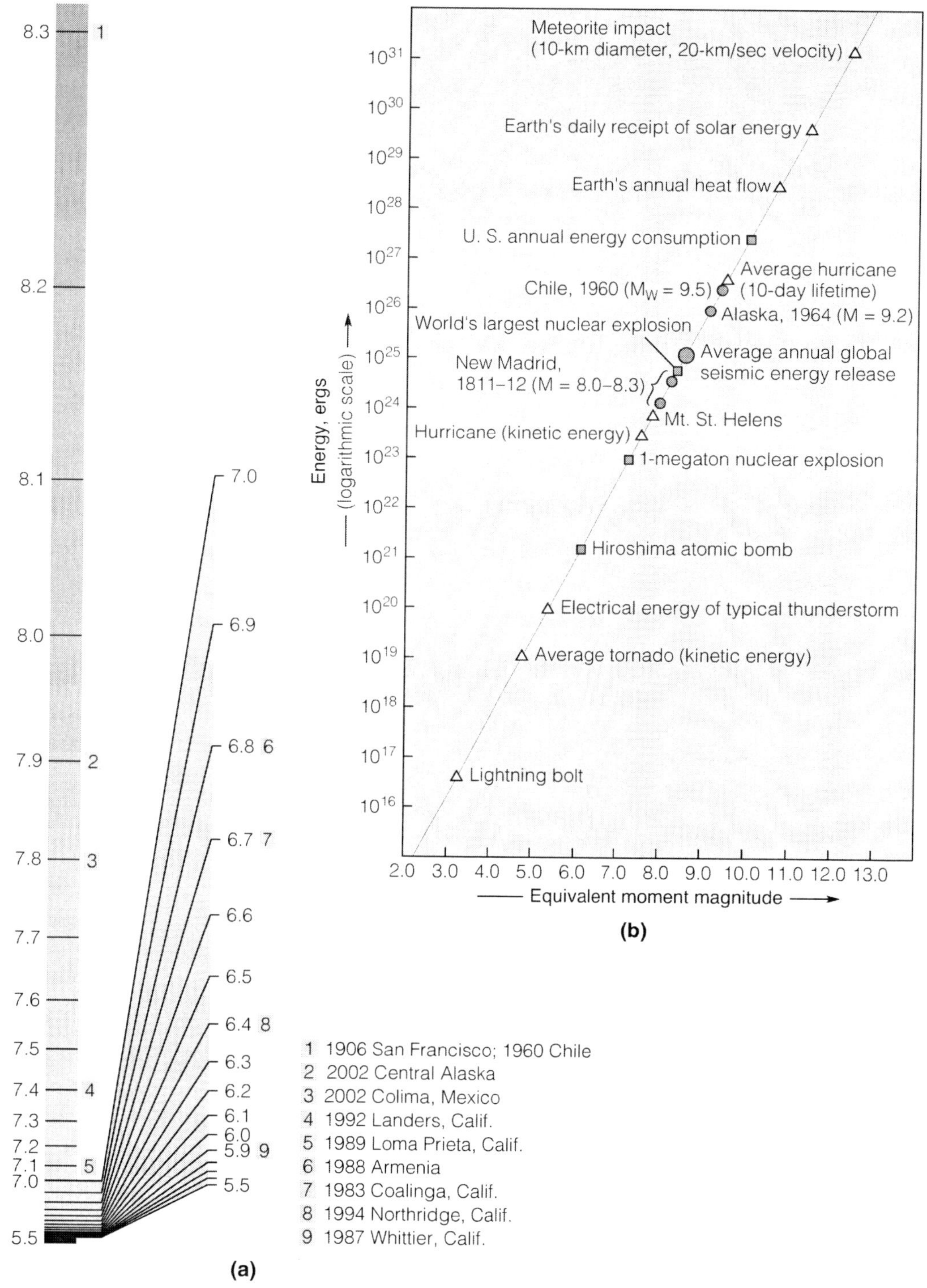

◆ FIGURE 4.11 Earthquake magnitude and energy. (a) Richter magnitudes of nine selected earthquakes, 1906–1994. (b) Equivalent moment magnitudes of energetic human-caused events (*squares*), natural events (*triangles*), and large earthquakes (*circles*). The erg is a unit of energy, or work, in the metric (cgs) system. To lift a one-pound weight one foot requires 1.4 x 10^7 ergs.

Confusion arises because some of the earthquake magnitudes are reported as Richter magnitude (M_L, teleseismic body wave magnitude (m_b), duration magnitude (m_d), surface wave magnitude (M_s), or moment magnitude (M_w or M). This is simplified somewhat because most newspapers and the Internet report earthquakes as either Richter or moment magnitude.

At the close of the twentieth century the Chilean earthquake of 1960 ($M_w = 9.5$) still held the record for the greatest seismic moment and energy release ever measured; that is, it had the longest fault rupture and the greatest displacement. The Alaskan earthquake of 1964 had the highest Richter magnitude of the 1900s and theoretically more severe ground motion than the event in Chile. (See ■ Case Study 4.2 on page 103.)

CONSIDER THIS

"Then the Lord rained upon Sodom and Gomorrha brimstone and fire. . . . And he overthrew those cities, and all the plain about . . ." (Genesis 19: 24–25). Sodom and Gomorrha are believed to have been located near the modern industrial city of Sedom in Israel at the south end of the Dead Sea. Could some geological event have caused the destruction of these towns? If so, what events might have been possible?

ThomsonNOW™
Click **Geology Interactive** to work through an activity on Earthquakes and Space and Time through Earthquakes and Tsunamis.

Fault Creep, the "Nonearthquake"

Some faults move almost continuously, or in short spurts, and do not produce detectable earthquakes. This type of movement, called **fault creep,** is well known but poorly understood. The Hayward fault, which is part of the San Andreas fault system, is an example of a creeping fault. It is just east of the San Andreas fault and runs through the cities of Hayward and Berkeley, California. (See the map in Figure 4.33.) Creep along this fault causes displacements of millimeters per year, and is one way that this plate boundary accommodates motion between the Pacific and North American plates. An alternative "accommodation" of plate motion on the San Andreas system is the storage of the strain energy and periodic release as a large displacement on a fault that generates a damaging earthquake. The obvious benefit of a creeping fault does not come without a price, as the residents of Hollister, California, can testify. The "creeping" Calaveras fault runs through their town and, as in Hayward, it gradually displaces curbs, sidewalks, and even residences.

A good strategy in areas subject to fault creep is to map the surface trace of the fault so that it can be avoided in future construction. The University of California–Berkeley Memorial Stadium was built on the Hayward fault before the hazard of fault creep was recognized. The fault creep caused damage to the drainage system that requires periodic maintenance. It is ironic that creep damage occurs at an institution that installed the first seismometer in the United States.

Forensic Use of Seismic Records

Seismic records can be used, in conjunction with other methods, to distinguish between earthquakes and nuclear explosions. The basis for the seismic method is that nuclear explosions are more concentrated in space and time and therefore excite more short-period seismic waves. This method requires extensive analysis of seismograms and is one of the tools that will be used to measure compliance by signers of the Comprehensive Test Ban Treaty. The following are select anomalous seismic events from 1995 to 2003, some of which were believed to be nuclear explosions. Note that the August 12, 2000 event marks the tragic explosion and sinking of the Russian submarine *Kursk,* and the February 1, 2003 event marks the *Columbia* space shuttle disintegration. Although not on this list of special studies compiled by Lynn Sykes of Lamont–Doherty Earth Observatory of Columbia University, the collapse of the Twin Towers on September 11, 2001 was also recorded on seismographs at Lamont–Doherty.

Location	Date	Type of Event
Novaya Zemlya	January 13, 1996	earthquake
Kola Peninsula	September 19, 1996	chemical explosion
Germany	September 11, 1996	mine collapse
East Kazakhstan	August 03, 1997	chemical explosion
India	May 13, 1998	alleged nuclear explosions
Novaya Zemlya	September 23, 1999	alleged nuclear explosion
Kursk, Barents	August 12, 2000	chemical explosion
Novaya Zemlya	February 23, 2002	earthquake
U.S. space shuttle *Columbia*	February 01, 2003	explosion 100 seconds long

Seismic Design Considerations

Ground Shaking

Ground movements, particularly rapid horizontal displacements, are most damaging during an earthquake. Shear and surface waves are the culprits, and the potential for them must be evaluated when establishing design specifications. The design objective for earthquake-resistant buildings is relatively straightforward: structures should be designed to withstand the maximum potential horizontal ground acceleration expected in the particular region. Engineers call this acceleration *base shear,* and it is usually expressed as a percentage of the *acceleration of gravity* (g). On Earth, g is the acceleration of a falling object in a vacuum (9.8 m/sec^2, or 32 ft/sec^2). In your car it is equivalent to accelerating from a dead stop through 100 meters in 4.5 seconds. An acceleration of 1 g downward produces weightlessness, and a fraction of 1 g in the horizontal direction can cause buildings to separate from their foundations or to collapse completely. An analogy is to imagine rapidly pulling a carpet on which a person is standing; most assuredly the person will topple.

The effect of high horizontal acceleration on poorly constructed buildings is twofold. Flexible-frame structures may be deformed from cube-shaped to rhomb-shaped, or they may be knocked off their foundations (◆Figure 4.12). More rigid multistory buildings may suffer "story shift" if floors and

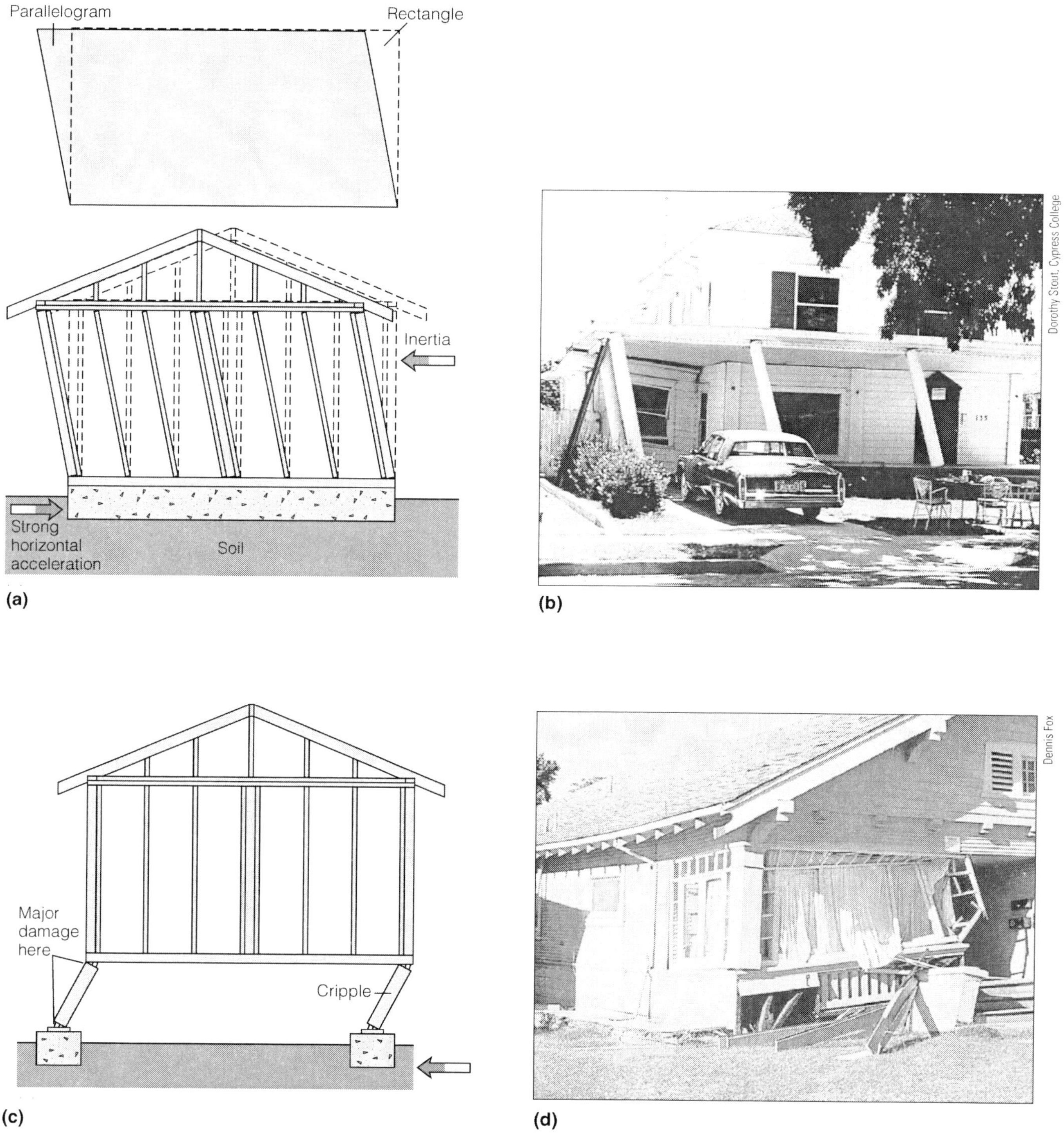

◆ FIGURE 4.12 (a) Strong horizontal motion may deform a house from a cube to a rhomboid or knock it from its foundation completely. (b) A Coalinga, California, frame house that was deformed by the magnitude-6.3 earthquake in 1983. (c) A cripple-wall consists of short vertical members that connect the floor of the house to the foundation. Cripple-walls are common in older construction. (d) A Watsonville, California, house that was knocked off its cripple-wall base during the Loma Prieta earthquake of 1989. Without exception, cripple-walls bent or folded over to the north relative to their foundations.

◆ FIGURE 4.13 Total vertical collapse as a result of "story shift"; Mexico City, 1985. Such structural failures are not survivable.

walls are not adequately tied together (◆Figure 4.13). The result is a shifting of floor levels and the collapse of one floor upon another like a stack of pancakes. Such structural failures are not survivable by inhabitants, and they clearly illustrate the adage that "earthquakes don't kill people; buildings do."

Damage due to shearing forces can be mitigated by bolting frame houses to their foundations and by *shear walls.* An example of a shear wall is plywood sheeting nailed in place over a wood frame, which makes the structure highly resistant to deformation. Wall framing, usually two-by-fours, should be nailed very securely to a wooden sill that is bolted to the foundation. Diagonal bracing and blocking also provide shear resistance (◆Figure 4.14). L-shaped structures may suffer damage where they join, as each wing of the structure vibrates independently. Such damage can be minimized by designing *seismic joints* between the building wings or between adjacent buildings of different heights. These joints are filled with a compressible substance that will accommodate movement between the structures.

Wave period is the time interval between arrivals of successive wave crests, or of equivalent points of waves, and it is expressed as T in seconds. It is an important consideration when assessing a structure's potential for seismic damage, because if a building's natural period of vibration is equal to that of seismic waves, a condition of resonance exists. **Resonance** occurs when a building sways in step with an oscillatory seismic wave. As a structure sways back and forth under resonant conditions, it gets a push in its direction of sway with the passage of each seismic wave. This causes the sway to increase, just as pushing a child's swing at the proper moments makes it go higher with each push. Resonance also may cause a wine glass to shatter when an operatic soprano sings just the right note or frequency.

Low-rise buildings have short natural wave periods (0.05–0.1 seconds), and high-rise buildings have long natural periods (1–2 seconds). Therefore, high-frequency (short-period) waves affect single-family dwellings and low-rise buildings, and long-period (low-frequency) waves affect tall structures. Close to an earthquake epicenter, high-frequency waves dominate, and thus more extensive home and low-rise damage can be expected. With distance from the epicenter, the short-period wave energy is absorbed or dissipated, resulting in the domination of longer-period waves.

CONSIDER THIS

Over time, tectonic stresses strain the rocks along a fault and increase the chance of sudden rupture and a potentially damaging earthquake. For this reason, it is generally believed that a region that has had many small earthquakes (magnitude <4.0) has a low probability for a large earthquake. Question: For a magnitude-7.0 earthquake with a 50-year recurrence interval, how many magnitude-4.0 earthquakes would be required to "balance the energy books"? How many small earthquakes a day is this? Is this belief realistic in a seismically active region based upon your calculations?

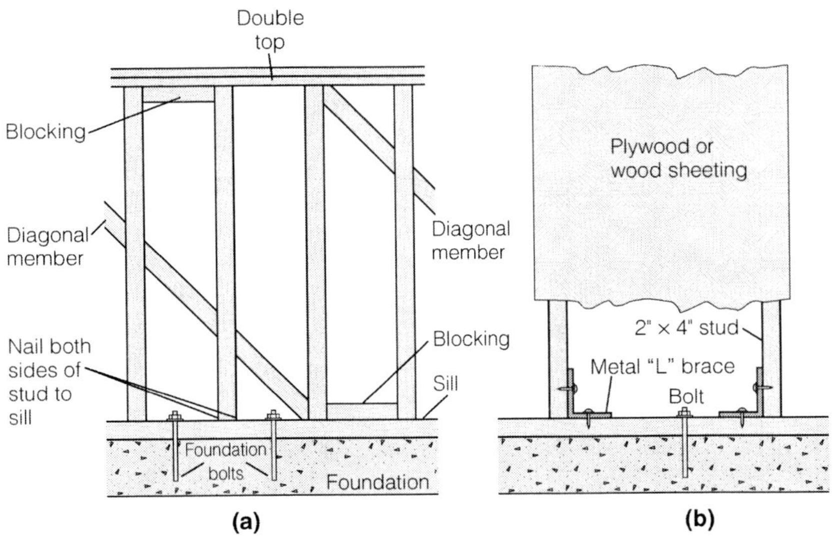

◆ FIGURE 4.14 Methods of reinforcing structures against base shear. (a) Diagonal cross-members and blocks resist horizontal earthquake motion (shear). (b) Plywood sheeting forms a competent shear wall, and metal "L" braces and bolts tie the structure to the foundation.

B. Pipkin

◆ FIGURE 4.15 Pictured is the site of the earthquake-induced rockslide at Red Mountain. The Madison River is just off the picture in the foreground. The rockslide was triggered by an earthquake and took place in schist with foliation planes inclined parallel to the mountain slope (see Chapter 7).

Landslides

Thousands of landslides are triggered by earthquakes in mountainous or hilly terrain; there were an estimated 17,000 during the Northridge earthquake alone (see ■ Case Study 4.1 on page 103). The 1989 Loma Prieta earthquake caused landsliding in the Santa Cruz Mountains and adjacent parts of the California Coast Ranges. Such areas are slide prone under the best of conditions, and even a small earthquake will trigger many slope failures. In the greater San Francisco Bay area there was an estimated $10 million damage to homes, utilities, and transportation systems because of landslides and surficial ground failures resulting from the Loma Prieta earthquake.

One of the worst earthquake-triggered tragedies in the United States occurred in August 1959, a short distance from Yellowstone Park. A landslide, in reality a massive rockslide (see Chapter 7), originated on Red Mountain above the Madison River in Montana. The rockslide was triggered by a moderate earthquake that caused the metamorphic rock (schist) composing the mountain to slide down its foliation planes, which were inclined parallel to the hill slopes (◆Figure 4.15). However, the rockslide occurred above a popular campground, and 26 people were killed by the falling rock. The slide generated a terrific shock wave of air that lifted cars, trees, and campers off the ground. In addition, the Madison River was dammed by the slide and a lake formed, later named "Quake Lake."

Ground or Foundation Failure

Liquefaction is the sudden loss of strength of water-saturated, sandy soils resulting from shaking during an earthquake. Sometimes called **spontaneous liquefaction,** it can cause large ground cracks to open, lending support to the ancient myth that the earth opens up to swallow people and animals during earthquakes. Shaking can cause saturated sands to consolidate and thus to occupy a smaller volume. If the water is slow in draining from the consolidated material, the overlying soil comes to be supported only by pore water, which has no resistance to shearing. This may cause buildings to settle, earth dams to fail, and sand below the surface to blow out through openings at the ground surface (◆Figure 4.16a). Liquefaction at shallow depth may result in extensive lateral movement or spreading of the ground, leaving great cracks and openings.

Ground areas most susceptible to liquefaction are those that are underlain at shallow depth—usually less than 30 feet—by layers of water-saturated fine sand. With subsurface geologic data obtained from water wells and foundation borings, liquefaction-susceptibility maps have been prepared for many seismically active areas in the United States.

Similar failures occur in certain clays that lose their strength when they are shaken or remolded. Such clays are called *quick clays* and are natural aggregations of fine-grained clays and water. They have the peculiar property of turning

NOAA

(a)

NOAA/NGDC 97

(b)

◆ FIGURE 4.16 (a) These apartment buildings tilted as a result of soil liquefaction in Niigata, Japan, in 1964. Many residents of the building in the center exited by walking down the side of the structure. (b) "House of cards" collapse of quick clay structure in Turnagain Heights, Anchorage, Alaska, in 1964. Total destruction occurred within the slide area, which is now called Earthquake Park.

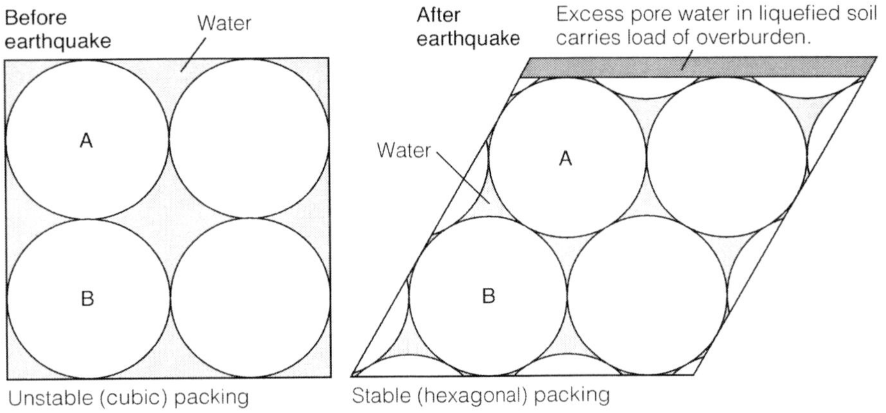

◆ FIGURE 4.17 Liquefaction (lateral spreading) due to repacking of spheres (idealized grains of sand) during an earthquake. The earthquake's shaking causes the solids to become packed more efficiently and thus to occupy less volume. A part of the overburden load is supported by water, which has no resistance to lateral motion.

from a solid (actually a gel-like state) to a liquid when they are agitated by an earthquake, an explosion, or even vibrations from pile driving. They occur in deposits of glacial-marine or glacial-lake origin and are therefore found mostly in northern latitudes, particularly in Scandinavia, Canada, and the New England states. Failure of quick clays underlying Anchorage, Alaska, produced extensive lateral spreading throughout the city in the 1964 earthquake (◆Figure 4.16b).

The physics of failure in spontaneous liquefaction and in quick clays is similar. When the earth materials are water-saturated and the earth shakes, the loosely packed sand consolidates or the clay collapses like a house of cards. The pore-water pressure pushing the grains apart becomes greater than the grain-to-grain friction, and the material becomes "quick" or "liquifies" (◆Figure 4.17). The potential for such geologic conditions is not easily recognized. In many cases it can be determined only by information gained from bore holes drilled deeper than 30 meters (100 ft). Because of this expense, many site investigations do not include deep drilling, and the condition can be unsuspected until an earthquake occurs.

Ground Rupture and Changes in Ground Level

Structures that straddle an active fault may be destroyed by actual ground shifting and the formation of a fault scarp (◆Figure 4.18). By excavating trenches across fault zones, geologists can usually locate past rupture surfaces that may reactivate. In 1970 the California legislature enacted the Alquist–Priolo Special Study Zones Act, which was renamed the *Earthquake Fault Zones Act* in 1995. It mandates that all known active faults in the state be accurately mapped and zoned for seismic safety. The act provides funds for state and private geologists to locate the youngest fault ruptures within a zone and requires city and county governments to limit land use adjacent to identified faults within their jurisdictions. Ironically, the faults responsible for the 1992 Landers earthquake (M_w = 7.3) were designated as Earthquake Fault Zones just prior to the June 28 event.

Changes in ground level as a result of faulting may have an impact, particularly in coastal areas that are uplifted or downdropped. For instance, during the 1964 Alaskan event, parts of the Gulf of Alaska thrust upward 11 meters (36 ft), exposing vast tracts of former tidelands on island and mainland coasts.

Dave Johnson

◆ FIGURE 4.18 The 1992 Landers earthquake (M_w = 7.3) in the Colorado Desert of southern California was felt from Phoenix, Arizona, to Reno, Nevada. Right-lateral offset of 4.27 meters (14 ft) on the Johnson Valley fault created a 2-meter (6.5-ft) vertical scarp due to lateral offset of a ridge. The geologist is 185 cm tall.

Fires

Fires caused by ruptured gas mains or fallen electric power lines can add considerably to the damage caused by an earthquake. In fact, most damage attributed to the San Francisco earthquake of 1906 and much of that in Kobe, Japan, in 1995 was due to the uncontrolled fires that followed the earthquakes. The Kobe quake hit at breakfast time. In neighborhoods crowded with wooden structures, fires erupted when natural-gas lines broke and falling debris tipped over kerosene stoves. Broken water mains made fire-fighting efforts futile. One of the principal "do's" for citizens immediately after a quake is to shut off the gas supply to homes and other buildings in order to prevent gas leaks into the structure. This in itself will save many lives (see Figure 4.41).

Tsunamis

The most myth-ridden hazard associated with earthquakes (and submarine volcanic eruptions and landslides) is tsunami (pronounced "soo-nah-mee"), or seismic sea waves. **Tsunami** is a Japanese word meaning "great wave in harbor" and it is appropriate because these waves are impulsively generated and most commonly wreak death and destruction inside bays and harbors. They have nothing to do with the tides, even though the term "tidal wave" is commonly used in the English-speaking world. The Japanese written record of tsunamis goes back 200 years and their power is dramatically displayed in the well-known print by Hokusai (◆Figure 4.19). Tsunamis, their causes, and effects are treated in detail in Chapter 10.

ThomsonNOW™
Click **Geology Interactive** to work through an activity on Tsunamis.

Five Earthquakes That Make a Point

Each year in the world, on average, there are at least two earthquakes of magnitude 8.0 or greater and 20 earthquakes in the magnitude 7.0 to 7.9 range. Release of seismic energy in the form of earthquakes has occurred throughout geologic time, and recorded history contains many references to strong earthquakes. More than 3,000 years of seismicity are documented in China, and Strabo's *Geography* mentions an earthquake in 373 B.C. in Greece. Thus, humans have probably always been subject to earthquakes (Table 4.1).

This section describes five earthquakes that occurred between 1994 and 2005. All five had very strong magnitudes but had quite different results because of differences in local foundation conditions, building standards, population density, and societal factors. The following is a synopsis of these events.

- A magnitude-7.7 earthquake struck Gujarat, India at 8:46 A.M. on January 26, 2001. It was 60 times more powerful than the Northridge, California, earthquake of 1994 but the death toll was 400 times greater.
- A magnitude-7.9 earthquake struck central Alaska on November 3, 2002. It was on the Denali fault and was one of the largest strike-slip ruptures in Alaska of the past two centuries. One minor injury was reported and the dollar loss was minimal.
- A magnitude-7.8 earthquake struck the state of Colima, Mexico, on January 22, 2003. The destruction was widespread, but the death toll was only 29. The victims were mostly in dwellings that were poorly built and collapsed. Modern high-rise buildings faired well.

◆ FIGURE 4.19 "The Breaking Wave off Kanagawa," wood-block color print by Katsushika Hokusai (1760–1849) from the series "Thirty-six Views of Mount Fuji," 1826–33.
Metropolitan Museum of Art, H. O. Havemeyer Collection (JP 1847)

TABLE 4.1 Selected Significant* World Earthquakes, in Chronological Order

Location	Year	Richter Magnitude**	Impact
San Francisco	1906	8.3	700 killed, $7 million damage, fire
Messina (Sicily)	1908	7.5	160,000 killed
Tokyo, Japan	1923	8.3	140,000+ killed, fire
Assam, India	1950	8.4	30,000 killed
Chile	1960	M_w 9.5	5,700 killed, 58,000 homes destroyed, tsunami
Alaska	1964	M_w 9.2	131 killed, tsunami
T'ang-shan, China	1976	7.9	330,000 killed
Mexico City	1985	8.1	10,000+ reported killed
Armenia	1988	6.8	55,000 killed
Loma Prieta (California)	1989	M_w 6.9	67 killed, $8 billion damage
Northridge, California	1994	M_w 6.7	60 killed, $25 billion damage
Kobe, Japan	1995	M_w 6.9	5,378 killed, $100 billion damage
Izmit (Kocaeli), Turkey	1999	M_w 7.4	15,370 killed, $10–$20 billion damage
Chi-Chi, Taiwan	1999	M_w 7.6	2,300+ killed
Gujarat, India	2001	M_w 7.7	Reported 20,000+
Alaska	2002	M_w 7.9	Little impact, no casualties
Colima, Mexico	2003	M_w 7.8	Widespread destruction, 27 fatalities
Bam, Iran	2003	M_w 6.5	30,000 believed killed, 30,00 injured
Simeulue, Sumatra	2004	M_w 9.3	230,000 killed, tsunami
Kashmir, Pakistan	2005	M_w 7.6	73,000 killed

*A "significant" earthquake is defined as one that registers a moment magnitude (M_w) of at least 6.5 or a lesser one that causes considerable damage or loss of life. The world averages 60 significant earthquakes per year.
**Estimated prior to 1935. M_w = Moment magnitude.
Sources: U.S. Geological Survey: *Earthquakes and Volcanoes;* California Division of Mines and Geology; *Academic American Encyclopedia* (1990, Grolier Electronic Publishing Company).

- A magnitude-9.3 earthquake struck the Andaman Islands and Sumatra on 26 December 2004. Despite the great shaking loss of life was due almost entirely to a tsunami.
- A magnitude-6.7 struck the San Fernando Valley at the community of Northridge on January 17, 1994. It is recognized as the second most costly natural disaster in U.S. history, second only to Hurricane Andrew (see Chapter 10).

Gujarat, India, 2001

A headline in a well-known American newspaper stated "Bad quake, worse buildings." This just about sums up the impact of this strong event in the state of Gujarat (◆Figure 4.20). It was a reverse-fault earthquake and the closest plate boundary lies many hundreds of kilometers away. It was felt 2000 km away and over an area 16 times that of the M_w = 7.8 San Francisco earthquake of 1906. Field investigation in Gujarat revealed no ground rupture, which is unusual for an earthquake of this magnitude. Liquefaction was widespread because of the high water table and thickness of unconsolidated sediment in the region. The quake occurred in an old rift system, which is a well-known seismic zone that is very similar to parts of the central United States. One doesn't have to be a seismologist to know that an earthquake of this magnitude has real potential to cause devastation. Here, the lack of strict building codes governing construction methods, some of which date back to the British Colonial period, exacerbated the destruction that led to the loss of more than 20,000 lives. Here also, one can become an architect or building contractor without any special education or license. The damage covered an area over 500 km wide from the large city of Ahmedabad in the east to the Arabian Sea in the west. Severe damage was reported within an area 50 km by 70 km in which most high-rises (anything more than 3 stories) and low-rise cobble-stone structures collapsed.

What is surprising to geologists is that this earthquake occurred in an area of relative flat topography without the usual tectonic landforms suggestive of faulting and seismic activity. Strong earthquakes in India usually occur as the Indian plate moves northward against the Himalayas. However, a similar event occurred in 1819 in the Rann of Katchchh, also in the state of Gujarat, just to the southeast of the 2002 earthquake epicenter. This suggests a greater rate of tectonic (earthquake) activity than is indicated by the land forms there.

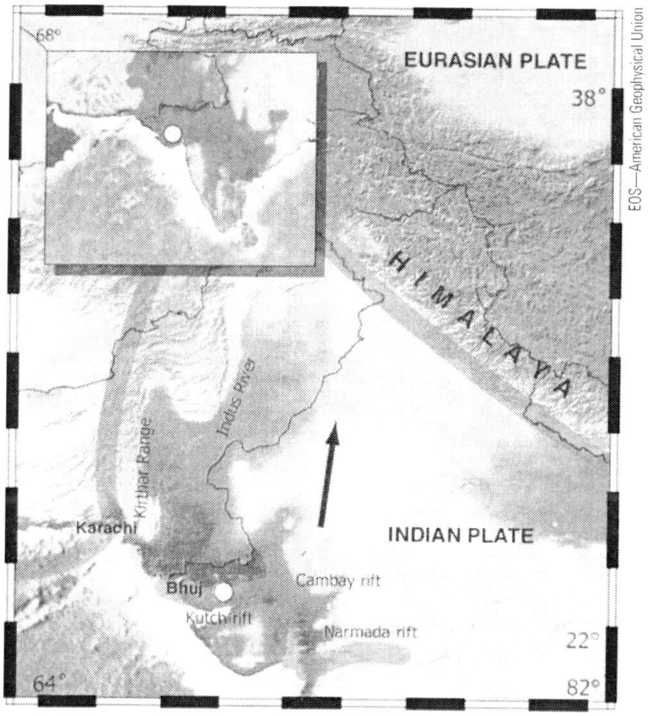

(a)

(b)

◆ FIGURE 4.20 (a) The regional geography and plate tectonic setting of the Gujarat (Bhuj) earthquake. The epicenter is shown by the yellow dot. Bhuj lies 400–500 km from a plate boundary and a greater distance from the Himalayan range. Also shown are buried rift basins that are seismically active and very similar to those of the central United States (see page 94 on New Madrid). (b) Collapsed houses in the town of Ratnal, in the epicentral region of the Bhuj earthquake.

Geologists from the United States are particularly interested in this event because it may serve as an analog for midcontinent earthquakes and the New Madrid, Missouri, seismic zone (Figure 4.30). Major earthquakes are rare in intraplate regions like New Madrid and Gujarat, far from plate edges and the usual seismicity associated with the world's tectonic-plate boundaries. As pointed out, there is little evidence at the ground surface left by these earthquakes. Thus, each intraplate earthquake is an opportunity to study and better understand the hazards posed by these events.

Alaska, 2002

On November 3 a magnitude-7.9 earthquake occurred some 90 miles south of Fairbanks along the Denali fault (◆Figure 4.21). Denali is the native Tanana word for "high one," and the Denali fault is the best-known and studied active fault in the state. This earthquake was among the strongest ever recorded in the United States. It shut down the Alaska pipeline, and caused lakes to ripple in Iowa and water to slosh out of swimming pools as far south as Louisiana. The earthquake was shallow and its energy went directly to the surface to produce the distant effects. One resident of Porcupine Creek is quoted as saying "A charging brown bear I can handle. This (the earthquake) scared the heck out of me." Alaska has a history of strong earthquakes, demonstrated in 1964 when the Denali fault ruptured and left 159 dead and many communities in ruins, mostly due to tsunami. This most recent event is a reminder that the fault has a great potential for damage.

Colima, Mexico, 2003

On January 22 a magnitude-7.8 earthquake struck near the village of Tecoman in the state of Colima (◆Figure 4.22). Twenty-seven people lost their lives in Colima and two more in the adjoining state of Jalisco. In this shallow-focus quake, the epicenter occurred near the junction of three tectonic plates: North American plate to the northeast, the Cocos plate to the south, and the Rivera plate to the northwest. Both the smaller plates are being subducted beneath the North American plate, producing significant seismic activity in the region. Twenty of the fatalities occurred in the collapse of adobe-brick houses. Over 150 houses were reported destroyed in the city of 200,000 that dates back to sixteenth-century Spanish colonial times. Mexican structural engineers are very aware of the seismic hazards that exist in this part of the country and are quite good at designing buildings to be earthquake-resistant. As a result, no damage occurred to modern high-rise buildings from this powerful earthquake. However, it was felt strongly and caused some to panic in Mexico City, where an earthquake in 1985 with its epicenter in the same region took over 10,000 lives. Two 20-story buildings in the capital swayed so much that they actually collided, but did little damage.

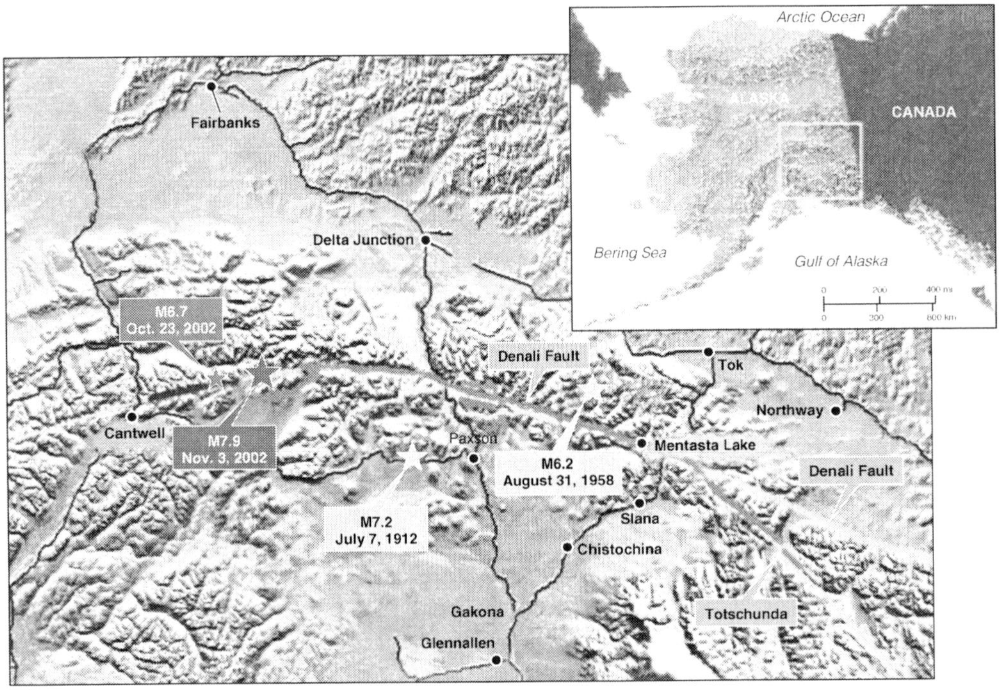

(a)

(b)

◆ FIGURE 4.21 (a) The magnitude-7.9 Denali fault earthquake of November 3, 2002, resulted from predominantly right-lateral offset along portions of the Denali and Totschunda fault systems in Alaska. Total length of the surface rupture was about 320 kilometers (200 mi). The western 49 km of the rupture shows mainly low-angle thrust offset as much as 1.5 m, with the northwest side up. Shown here are the epicenters of the November 3 event (red) and the magnitude-6.7 foreshock of October 23, 2002 (blue), as well as two previously recorded large shallow earthquakes (yellow) in the vicinity of the fault. (b) Aerial view of the Trans-Alaska Pipeline System (TAPS) line near the Denali fault, looking west. This is where the pipeline is supported by rails on which it can move freely in the event of fault offset. Here the line has slid toward the west end of the rails. Alyeska Pipeline Service Company reported no breaks to the line and therefore no loss of oil. Out of view to the left (south) is the 2.5 m (ft) right-lateral offset of the highway where it crosses the fault. Photo courtesy of Rod Combellick, Division of Geological and Geophysical Surveys, State of Alaska

Sumatra, 2004

The third deadliest and second most powerful earthquake in recorded history struck the northwest coast of Sumatra around breakfast time on December 26, 2004. In a mere six to eight minutes, the edge of the Indo-Pacific plate boundary flexed and lifted the coral-rimmed coasts of offshore islands closest to the 600-km-long fault rupture as much as 90 centimeters (35 in), while the shoreline of the main island of Sumatra sank below sea level, inundating important farmlands, roads, and built-up areas (◆Figure 4.23). An enormous tsunami radiated outward from the focal area and battered coasts as far away as eastern Africa, Sri Lanka, Myanmar, and Thailand. Because there was no early warning system in place to alert people of the approaching waves, nearly a quarter of a million people drowned. In one instance the sea waves derailed and washed away a crowded passenger train that happened to be traveling in the wrong place at the wrong time.

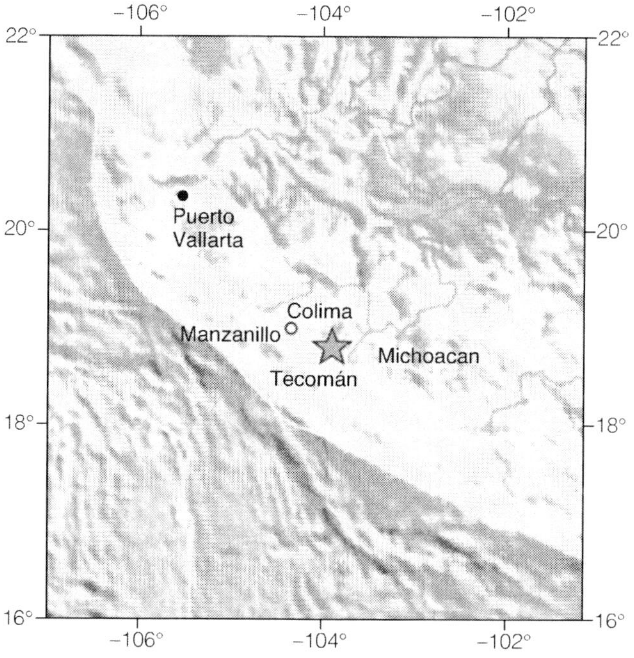

◆ FIGURE 4.22 Magnitude-7.8 earthquake near Tecomán, Colima, Mexico, on January 22, 2003. USGS

For geologists, the Great Sumatran earthquake was geophysically interesting because it set the whole Earth ringing like a giant bell struck with a hammer. Every location on the planet—including the remote poles and the flat plains of Kansas, rose and fell at least a centimeter in slow, gentle oscillations with periods ranging from every 20.5 to 54 minutes, too slow for people to notice without instrumental assistance. This wiggling continued with diminishing strength over a period of months. The quake was also strong enough to trigger aftershocks as far away as Alaska. Only a few months later, in March 2005, another great earthquake with a magnitude of 8.5–8.7—the biggest aftershock of all—struck the same area of Sumatra. Although it did not generate a large tsunami, 1300 people lost their lives, mostly on a single island.

Northridge, California, 1994

The largest earthquake in Los Angeles's short history occurred at 4:30 A.M. Monday, January 17, 1994, on a hidden fault below the San Fernando Valley. The $M_w = 6.7$ earthquake started at a depth of 18 kilometers (11 mi) and propagated upward in a matter of seconds to a depth of 5–8 kilometers (3–5 mi, ◆Figure 4.24). There were thousands of aftershocks, and their clustered pattern indicated the causative fault to be a reverse fault with a shallow dip of 35° to the south (Case Study 4.2 and Appendix 3). Such a low-angle reverse fault is called a *thrust fault,* and, because the thrust did not rupture the ground surface, it is described as

Kerry Sieh, Cal Tech

◆ FIGURE 4.23 Porites coral heads on the west coast of Simeulue that were uplifted about 90 centimeters (35 in) by the December 26, 2004, earthquake.

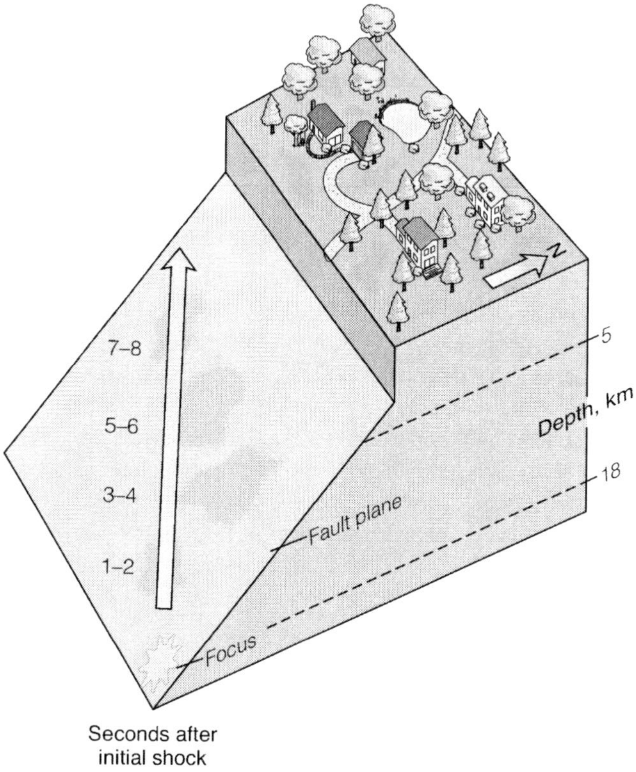

◆ FIGURE 4.24 In the 1994 Northridge earthquake, the fault rupture progressed up the fault plane from the focus at the lower right of the figure to the upper left in 8 seconds. Rather than rupturing smoothly, like a zipper opening, it moved in jerks along the fault plane, as shown by the pink patches. Total displacement was about 4 meters (12 ft).

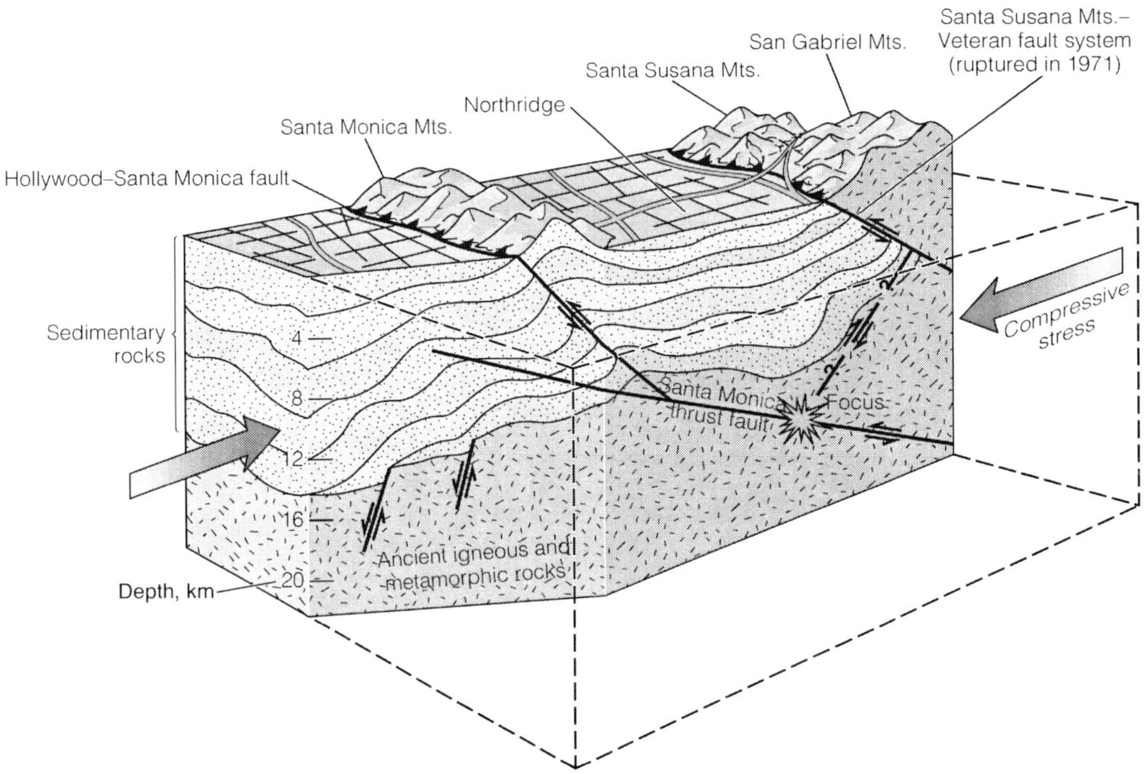

◈ FIGURE 4.25 Interpretation of the fault system in the San Fernando Valley and surrounding area of Southern California. Compressive stresses built up over a long period, causing the subsurface blind thrust fault to rupture. The ensuing Northridge earthquake impacted the entire valley, and extended from the Santa Susana Mountains on the north to the city of Santa Monica on the south. USGS DATA

"blind" (◈Figure 4.25). Blind thrusts were first recognized after the 1987 Whittier earthquake 40 kilometers (25 mi) southwest of Northridge, but they were not fully appreciated until after the Northridge event.

Blind thrust faults were a newly identified seismic hazard, and there is evidence that a belt of them underlies the northern Los Angeles basin. They are especially dangerous because they cannot be detected by traditional techniques such as trenching and field mapping (thus they do not fall within the Earthquake Fault Zones Act as it is written) and yet, as we know now, they can generate significant earthquakes. Just as the residents of earthquake country were beginning to feel confident that geologists knew the location and behavior of most active faults, blind thrusts made their presence known.

Field inspection of the epicentral region was a depressing experience. Thirteen thousand buildings were found to be severely damaged; 21,000 dwelling units had to be ordered evacuated; 240 mobile homes had been destroyed by fire; 11 major freeway overpasses were damaged at 8 locations (■ Case Study 4.3 on page 104). The reason for the extensive damage was the high horizontal and vertical accelerations generated by the earthquake. Accelerations of more than 0.30 *g* are considered dangerous, and the vertical acceleration of 1.8 *g* measured by an instrument bolted to bedrock in nearby Tarzana probably set a world record. The high ground acceleration explains why many people were literally thrown out of their beds and objects as heavy as television sets were projected several meters from their stands. Although the damage was not as visually spectacular as that at Mexico City in 1985, it was heartbreaking to many residents—brick chimneys, both reinforced with steel bars and unreinforced, came down; houses moved off their foundations; concrete-block walls crumbled; gaps broke open in plaster walls; the number of shattered storefront windows was beyond counting (◈Figure 4.26). The horizontal ground motion was directional; that is, it was strongest in the north–south direction. With few exceptions, block walls oriented east–west tipped over or fell apart, whereas those oriented north–south remained standing. Many two- and three-story apartment buildings built over open first-floor garages collapsed onto the residents' cars. The lack of shear resistance of these open parking areas led to failure of the vertical supports, and everything above came down.

California State University at Northridge sustained almost "textbook" earthquake damage. In its library is "Leviathan II," recognized as one of the most advanced automated book-withdrawal systems in the country. On this day it recorded a record withdrawal of about 500,000 books—all of them onto the floor of the library. Bottled chemicals fell off their storage shelves (a major concern at any university) and caused a large fire in the Chemistry

Greg Davis, USC

(a) (b) (c)

◆ FIGURE 4.26 (a) Two wings of an apartment building that collapsed toward one another. (b) A staircase that leads to nowhere. In reality, it was the second-floor breezeway that collapsed onto the ground floor. (c) Part of the collapsed parking structure of Fashion Center, Northridge. Such collapses instigated new design criteria for open structures. Note the outside staircase that separated from the main structure.

Building. In addition, one wall of a large open parking structure collapsed inward to produce the stunning art deco architecture seen in ◆Figure 4.26c.

What We Can Learn

These five earthquakes reinforce the adage that "earthquakes don't kill people, falling buildings and tsunamis kill people." The magnitudes were in the range where destruction is certain if construction methods are not adequate. In Gujarat high-rise structures and stone dwellings alike collapsed due to inadequate design leading to a high death toll. In Alaska, where the population density is about one person per square mile, the strong temblor took no lives and really did minimal damage. On the other hand, the example in Mexico illustrates that even when high population areas are struck by a powerful earthquake, recognition of the importance of seismic design criteria can reduce the loss of life as can an internationally organized tsunami early-warning alert network. The Northridge earthquake was an unusual case. There the combination of a blind thrust fault and very high ground accelerations, both vertical and horizontal, combined to cause the extensive damage in Southern California.

It should be noted that other factors beside construction, population density, and kind of fault movement can cause damage. Site effects are extremely important. During the 1989 Loma Prieta (San Francisco) earthquake the ground shaking around the bay area varied considerably. There was liquefaction and widespread destruction in the Marina District, where the two-level Interstate 880 was built on bay muds. Near Oakland it shook violently and failed (◆Figure 4.27) causing fatalities, whereas structures in San Francisco built on bedrock suffered much less or no damage. The seismograms shown in ◆Figure 4.28 illustrate this point (also see Case Study 4.3).

◆ FIGURE 4.27 The I-880 freeway failed because it was built on bay muds that reacted to seismic waves like a bowl of gelatin. Pictured is the Cypress Street viaduct with the upper deck collapsed upon the lower one. There was loss of life here.

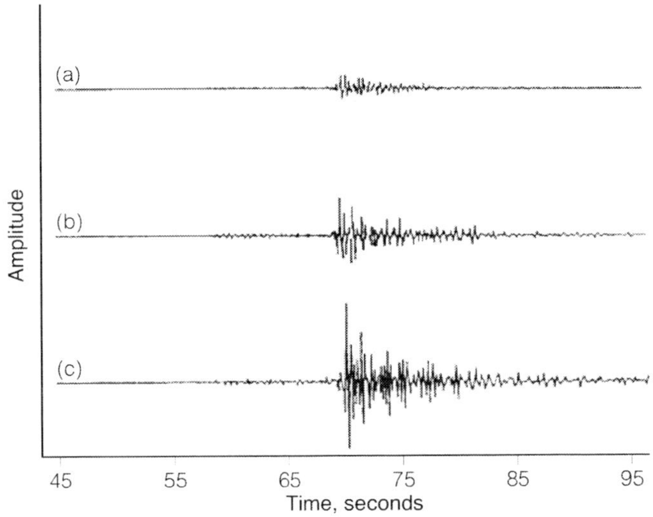

◆ FIGURE 4.28 Seismograms for a magnitude-4.1 aftershock as recorded on (a) firm bedrock, (b) alluvium (stream-deposited sediment), and (c) areas of fill and bay muds. This is where the most damage occurred in the San Francisco earthquakes of 1906 and 1989. Data from EOS—American Geophysical Union

Site effects were noted almost 200 years ago when an observer of the New Madrid sequence (next topic) noted "The convulsion was greater along the Mississippi, as well as along the Ohio, than in the uplands. The strata in both valleys are loose. The more tenacious layers of clay and loam spread over the adjoining hills . . . suffered but little derangement" (see Drake, 1815, in For Further Information).

Does Earthquake Country Include Idaho, Missouri, and New York?

Even though the vast majority of the world's earthquakes occur at plate boundaries, areas hundreds and even thousands of kilometers away are not free of seismic activity. In the United States the five most seismically active states between 1980 and 1991 were

	Earthquakes 1980–1991	
State	Number Recorded	Largest, M_w
Alaska	10,253	9.2
California	6,732	7.2
Washington	615	5.5
Idaho	536	7.3
Nevada	398	5.6

Although Alaska and California continue to lead the United States in the number of shakers, the earthquakes that have been felt over the largest area occurred in Missouri in 1811 and 1812, and significant seismic hazards are recognized in 39 states. No state is earthquake proof, as the seismic-risk map of the United States shows (◆Figure 4.29). It appears that U.S. residents who want to be seismically safe should move to Texas, Florida, or Alabama.

Why do such strong earthquakes occur *intraplate*—that is, far from a plate boundary (◆Table 4.2)? Intraplate earthquakes have several characteristics in common (see also the Gujarat earthquake, this chapter):

1. The faults causing them are deeply buried and have not broken the ground surface.
2. Because the rocks of the continental interior are stronger than those at plate boundaries, which are laced with faults, they transmit seismic waves better, causing ground motion over a huge area. In the United States this is usually many states (◆see Figure 4.29).
3. They do not appear to be random events.

The Charleston, South Carolina, earthquake of 1886 was larger than the 1989 San Francisco shaker. It killed scores of people, ruined the city, and slowed the South's recovery from the Civil War. The New Madrid (pronounced "mad′rid") earthquakes of 1811 and 1812 caused extensive topographic changes, locally reversed the course of the Mississippi River, and may have been felt over a larger area than any other earthquake in recorded history. Ground motion was felt as far away as Washington, D.C., where it caused church bells to ring and scaffoldings on the Capitol to collapse.

Recent studies suggest that intraplate earthquakes are concentrated in areas where normally stable crust has been stretched and faulted. Such zones form where continents have been split apart *(rifted)*, forming two continents, as occurs at divergent plate boundaries (see Chapter 3). When North America was separated from Africa at the Mid-Atlantic Ridge

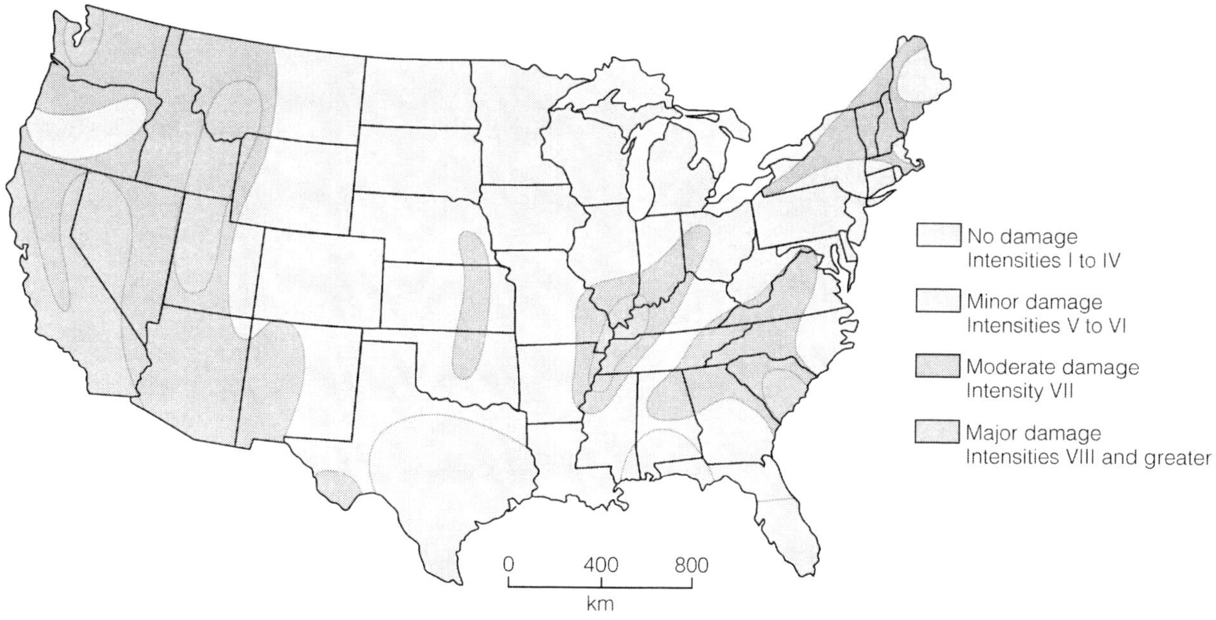

◆ FIGURE 4.29 Seismic-risk map of the 48 contiguous states based upon historical records and intensities collected by the U.S. Coast and Geodetic Survey. The Coast and Geodetic Survey gathers intensity data from questionnaires after earthquakes. From M. L. Blair and W. W. Spangle, USGS prof. paper 941-B, 1979

about 180–200 million years ago, the continental crust being rifted at the ridge was thinned, stretched, and faulted. The North American continent was transported westward with the plate, and the weakened part of the crust along its eastern edge was hidden by a thick cover of younger sedimentary rocks. Buried rifted crust under the Eastern seaboard states forms the intraplate earthquake zone from the Carolinas to Canada.

The Midwest earthquake zone through Arkansas, Missouri, and Illinois is believed to be where an ancient (pre-Pangaea) divergent boundary started to form but "failed" for some reason, leaving behind a significant buried fault zone. Known as the *Reelfoot Rift*, the buried block of crust is down-dropped between faults (the hachured lines on the map of ◆Figure 4.31). The rift is 60 kilometers wide and 300 kilometers long (roughly 40 mi × 190 mi) and formed at least 500 million years ago. The linear trend of earthquakes from Marked Tree, Arkansas, northeastward to Caruthersville, Missouri, reveals upwarped sedimentary rocks along the rift axis. Detailed analysis of the geology across the Reelfoot fault scarp by trenching revealed evidence of three large earthquakes within the past 2,000 years, which yields a recurrence-interval estimate of 600–900 years for the fault, but the recurring earthquakes are not necessarily of the same magnitude as the 1811–1812 events. In fact, scientists estimate that a magnitude-6–7 earthquake will occur in the New Madrid seismic zone between 2000 and 2050 with a 90 percent probability.

***TABLE 4.2 Selected North American Intraplate Earthquakes**

Location	Year	Moment Magnitude*	Impact
New Madrid, Missouri	1811	8.2	Reelfoot Lake formed in NW Tennessee.
New Madrid	1812	8.3	Elevation changes caused Mississippi River to reverse its course locally.
New Madrid	1812	8.1	
Charleston, South Carolina	1886	7.6	Felt from New York to Chicago; 60 killed.
Charleston, Missouri	1895	6.8	Damage in 6 states (see Figure 4.28).
Grand Banks, Newfoundland	1929	7.4	Submarine landslides broke trans-Atlantic cable, disrupting communications.

*Moment magnitude is used in reconstructing the strength of "preinstrument" earthquakes, because we know something of fault length, rupture length, and area felt.

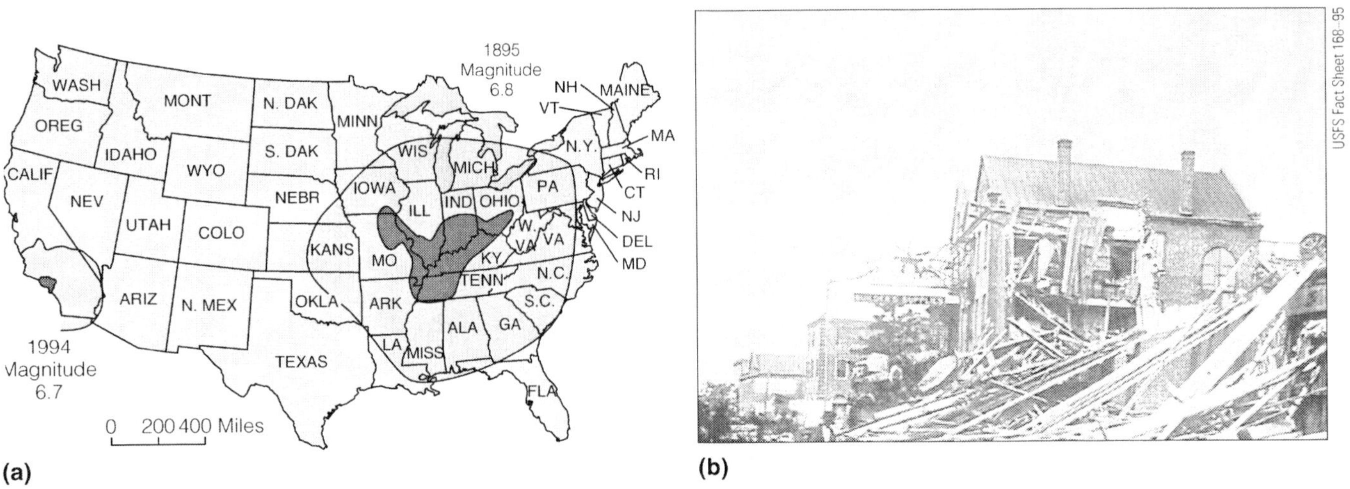

◆ FIGURE 4.30 Mid-continent earthquakes. (a) Comparison of the effects of two earthquakes of similar magnitude. The 1895 Charleston, Missouri, earthquake impacted a relatively huge area in the central United States, whereas the 1994 Northridge earthquake's effects were limited to Southern California. Red indicates areas of minor-to-major damage to buildings and their contents. Yellow indicates areas where shaking was felt but there was little or no damage to objects, such as dishes. Data from USGS. (b) Damage from the 1886 earthquake along East Bay Street in Charleston, South Carolina.

Nobody knows how many people died in the 1811–1812 earthquakes. An amateur observer in Louisville, Kentucky, recorded almost 2,000 shocks with a homemade pendulum. Liquefaction was widespread along the Mississippi River, and the town of New Madrid, Missouri, was completely destroyed when it settled from 8 meters (25 ft) above sea level to only 4 meters (12 ft). Subsidence caused some swamps to drain and others to become lakes; an example is Reelfoot Lake in northwest Tennessee, which today is more than 50 feet deep. The New Madrid seismic zone is a major geologic hazard in the United States, and efforts are being made to reduce the impact of a future large earthquake there.

Iben Browning, a scientist with a Ph.D. in physiology but who is best known for his work on climate, predicted there would be a repeat of the 1811 New Madrid earthquake on December 3, 1990. His prediction was based upon alignment of the planets and consequent gravitational pull, which he believed would be sufficient to trigger an earthquake on that date. Although the prediction was discounted by scientists, it created considerable anxiety in Arkansas, Tennessee, Alabama, and Missouri. Months before "the day," earthquake insurance sales boomed, moving companies were booked up, and bottled-water sales rose dramatically—as did sales of quake-related souvenirs. One church sold "Eternity Preparedness Kits," and "Survival Revivals" were held. On the day of the scientifically discredited prediction nothing earthshaking occurred.

Many earthquakes with epicenters in weakened intraplate continental crust have shaken parts of the Midwest, eastern Canada, New England, New York, and the East Coast of the United States. New York does not have a major fault, but it has many faults. In the 1980s an earthquake in Westchester County ($M = 4.0$) toppled chimneys and caused enough panic for the state to implement new seismic codes that added 2–5 percent to building costs. In April 2002 the state experienced a magnitude-5.1 shaker that centered 15 miles southwest of Plattsburgh near the Canadian border. It rattled dwellings from Maine to Maryland and left cracks in foundations and chimneys. There were no injuries. The point here is that a relatively small intraplate earthquake can be felt over a huge area. By the mid 1990s New York and Massachusetts were the only Northeastern states that had earthquake building codes.

The Pacific Northwest

In years past, earthquake hazards were considered minor in Oregon and Washington. During the 1980s, however, research changed the perception, revealing geologic evidence that "major" (≥7.0 but <8.0) or "great" (≥8.0 but <9.0) subduction-zone earthquakes have occurred in the past and that they can occur in the future. The Cascadia subduction zone (◆Figure 4.32), extending 1,200 kilometers (740 mi) from northern California to Vancouver Island in Canada, has destructive earthquake potential. Geological evidence, consisting of carbon-dated tsunami deposits and drowned red-cedar forests, suggests that great earthquakes strike the Pacific Northwest roughly every 500 years, the last one in A.D. 1700, approximately 300 years ago. Minor earthquakes occur daily, however, and the cities of Portland, Seattle, and Vancouver, British Columbia, are in earthquake country (◆Figure 4.33). The states now are working to minimize damage should the "big one" occur on the subduction zone or on the Seattle fault, which runs through downtown Seattle.

The year 2000 marked the tricentennial of the last great (equal to or greater than $M = 8.0$) earthquake generated along the Cascadia Subduction Zone. To commemorate this, almost 100 earth scientists and public officials gathered at

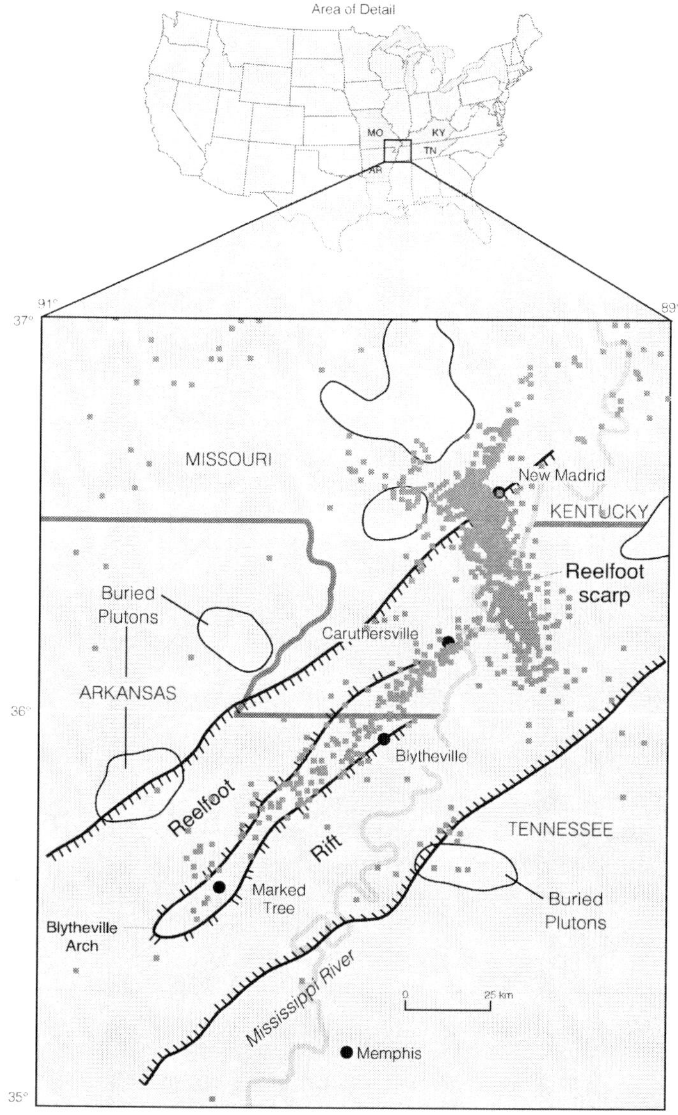

◆ FIGURE 4.31 The New Madrid seismic zone covers parts of four states and consists of an ancient, fault-bounded depression (rift) buried below 1.5 kilometers (almost a mile) of sedimentary rocks. An area of major seismic hazard, it is studied intensively. USGS DATA

Seaside, Oregon, to assess the seismic hazard posed by the Cascadia Subduction Zone. Besides ways to mitigate great loss of life, selected major points of agreement were:

- Most of the 1100 km of the subduction zone ruptured on January 26, 1700. A correlative tsunami struck Japan and the characteristics of the wave suggest a M_w of 9.0.
- Cascadia earthquakes generate tsunami, the most recent of which was about 10 meters high based on studies of the inundation zone.
- Strong ground shaking from a M_w −9.0 earthquake will last three minutes or more and will be damaging as far inland as Vancouver, Portland, and Seattle.

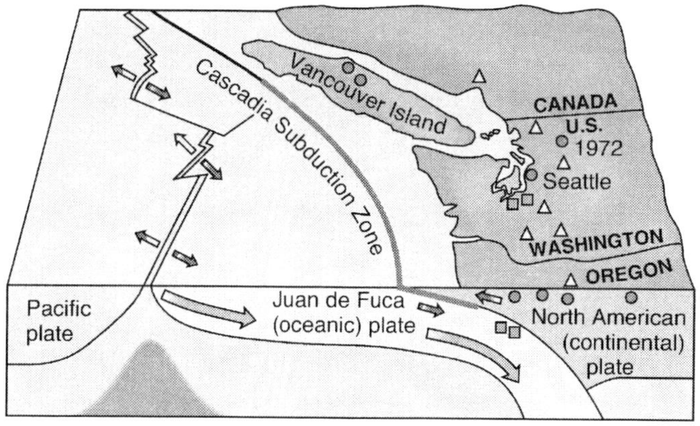

□ Deep earthquakes (>65 km or 40 mi deep)* occur when the oceanic plate descends beneath the continental plate. The largest deep earthquakes in recent times were in 1949 ($M = 7.1$) and in 1965 ($M = 6.5$).

● Shallow earthquakes (<17 km or 10 mi deep)* are caused by faults in the North American continent. Magnitude 7+ earthquakes have occurred along the Seattle fault in 1872, 1918, and 1946.

— Subduction earthquakes are huge quakes that occur when the subduction boundary ruptures. The most recent Cascadia Subduction Zone earthquake was 1700 and it sent a tsunami as far as Japan.

*Shallow and deep do not refer to focus of quake; rather, depth is subjective and is related to damage caused by faults in the Seattle area.

◆ FIGURE 4.32 Faults and epicenters in the Seattle, Washington and Vancouver, British Columbia area. USGS

- The recurrence interval for an event equivalent to the 1700 earthquake is 500–600 years.

Twenty years ago scientists were debating whether great earthquakes occurred at subduction zones. Now there are few doubters, and the 2000 meeting spent a great deal of time and effort in discussing methods of mitigating loss of life and damage. It is ironic that the conference hotel lies within the inundation zone of the 1700 tsunami.

Prediction

A guy ought to be careful about making predictions. Particularly about the future.
Yogi Berra, baseball player

Earthquake prediction has the great potential for saving lives and reducing property damage. A good prediction gives the location, time, and magnitude of a future earthquake with acceptable accuracy. Prediction was the hottest area of geophysical and geological research from the 1970s to the early 1990s, based upon the belief that measurable phenomena occurring before large and small earthquakes called **precursors** could be identified. Researchers studied earthquakes and seismograms where presumed precurors were seen. Such things as changes in the ratio of P-wave and S-wave velocities,

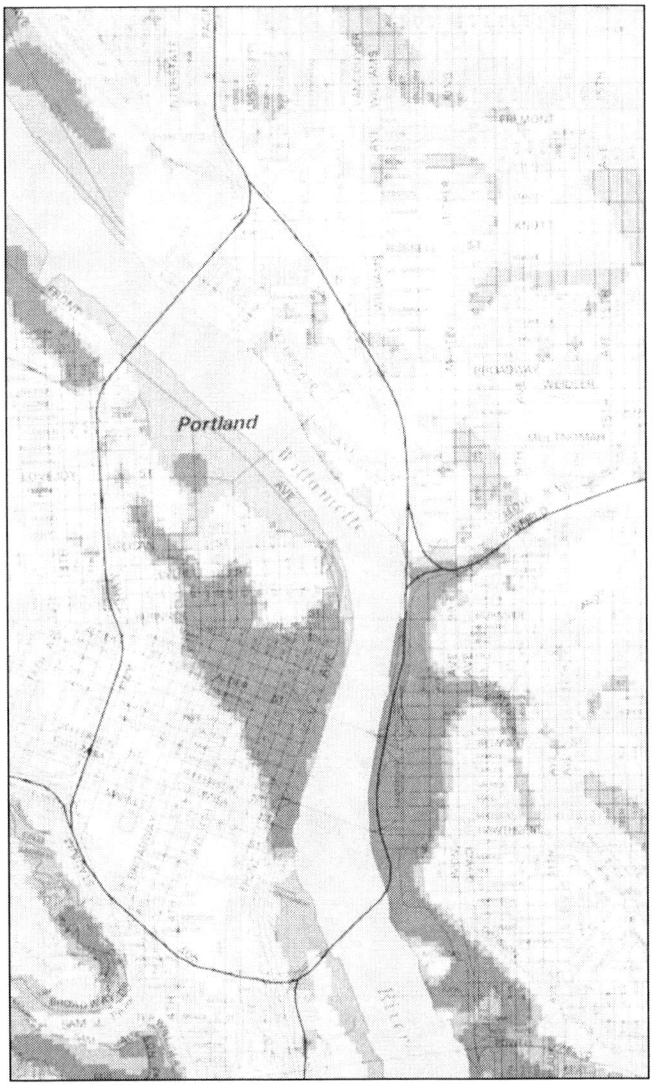

◆ FIGURE 4.33 Seismic-hazard map of downtown Portland, Oregon. The hazard zones are based upon liquefaction potential, landslide potential, and the probable degree of ground shaking. Red denotes greatest hazard; pale yellow, the lowest hazard. Oregon Dept. of Geology and Mineral Industries

ground tilt, water-well levels, and emission of noble gases in ground water were measured and piles of data were accumulated. Unfortunately, the hope of finding precursors that would lead to reliable predictions seems to have evaporated. In fact, the U.S. Geological Survey is on record as saying that the prospect of earthquake prediction is very dismal. In short, earthquakes cannot be predicted because the mechanics of earthquake generation are, with the present state of our knowledge, too complicated to predict.

Forecasts

We are familiar with weather forecasts and are aware that accuracy declines as the time span of the prediction becomes longer. We can usually rely on one-week forecasts, but a year ahead would be asking too much. The same can be said of earthquakes, only in reverse. There is an old saying in geology that "the longer it has been since the last earthquake, the sooner we can expect the next one."

A new approach is to evaluate the probability of a large earthquake occurring on an active fault during a given time period. This falls under the heading of long-term forecasting. An example of this is the U.S. Geological Survey and other scientists' conclusion that there is a 70 percent probability of at least one magnitude-6.7 or greater earthquake striking the San Francisco Bay region between 2000 and 2030. Such an event would be capable of causing widespread damage (◆Figure 4.34). Such forecasts are made on the basis of measured plate tectonic motion and slip on faults. The inexorable movement of the Pacific plate past the North American plate loads strain on the San Andreas network. Periodically this strain is released on one of the faults in the system and an earthquake occurs.

Statistical Approach

By compiling statistical evidence pertaining to past earthquakes in a region, we acquire basic data for calculating the statistical probability for future events of given magnitudes. These calculations may be done on a worldwide scale or on a local scale, such as the example in ◆Figure 4.35. Analysis of the graph indicates that for the particular area in Southern California, the statistical **recurrence interval**—that is, the length of time that can be expected between events of a given magnitude—is 1,000 years for a magnitude-8 earthquake (0.1/100 years), about 100 years for a magnitude-7 earthquake (1/100 years), and about 10 years for a magnitude-6 earthquake (almost 10/100 years). The probability of a magnitude-7 occurring in any one year is thus 1 percent, and of a magnitude-6, 10 percent. On an annual basis worldwide, we can expect at least two magnitude-8 earthquakes, 20 magnitude-7 earthquakes, and no less than 100 earthquakes of magnitude 6. Thus for seismically active regions, historical seismicity data can be used to calculate the *probability* of damaging earthquakes. Although these numbers are not really of predictive value as we defined it, they can be used by planners for making zoning recommendations, by architects and engineers for designing earthquake-resistant structures, and by others for formulating other life- and property-saving measures.

CONSIDER THIS

Humans have caused earthquakes by injecting fluids, such as oil-field waters and chemical warfare wastes, into deep wells. How does this open the possibility of controlling earthquake activity, and what might be a major deterrent to the effort?

Geological Methods

Research suggests that active faults and segments of long, active faults tend to have recurring earthquakes of characteristic magnitude, rupture length, and displacement. For

◆ FIGURE 4.34 Probabilities that one or more magnitude-6.7 or greater earthquakes will strike on specific faults in the San Francisco Bay region between 2000 and 2030. The San Andreas, Rogers Creek, and Hayward faults have the highest probabilities. Total probability for the region is computed as 70% (± 10%). These probabilities were developed by scientists with the U.S. Geological Survey, part of the U.S. Department of the Interior, and thus constitute official long-term *forecasts*. The message is that all communities in the Bay region should keep preparing for earthquakes.

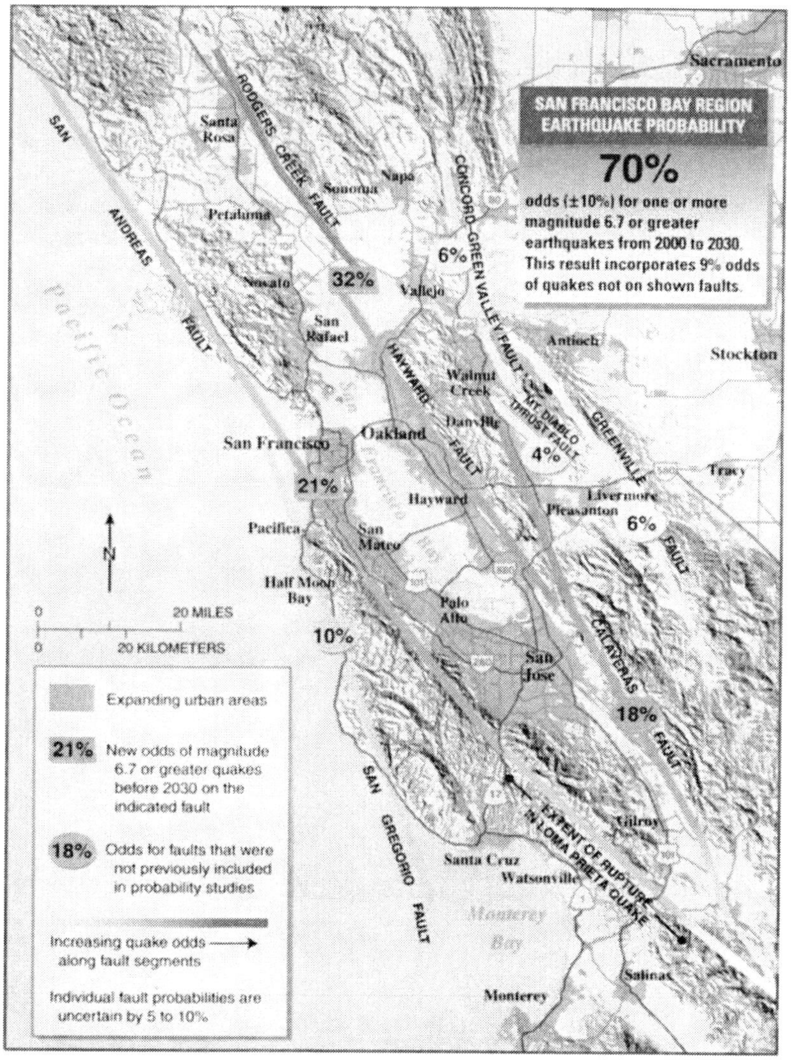

USGS, Modified and Updated

example, six earthquakes have occurred along the San Andreas fault at Parkfield since 1857 (◆Figure 4.36). They had Richter magnitudes of around 5.6, rupture lengths of 13–19 kilometers (8–12 mi), and displacements averaging 0.5 meter. These earthquakes have recurred about every 22 years since the earliest recorded earthquake. A 95 percent probability of an earthquake was predicted for this segment between 1988 and 1993, the first prediction in the United States to be endorsed by scientists and subsequently issued by the federal government. The prediction was a failure.

Geophysical and seismological precursors, as noted previously, have not proved to be reliable predictors of earthquakes. Seismic gaps, however, can indicate stretches of future fault activity. Seismic gaps are stretches along known active fault zones within which no significant earthquakes have been recorded. It is not always clear whether these fault sections are "locked" and thus building up strain energy, or if motion (creep) is taking place there that is relieving strain. Such gaps existed and were "filled" so to speak during the Mexico City (1985), Loma Prieta (1989; Figure 4.36), and Izmit (1999) earthquakes. Seismic gaps serve as alerts or warnings of possible future events and can be used as forecasting tools.

Since recorded history in North America is short, geologists need other means of collecting frequency data of large prehistoric earthquakes. One method is to dig trenches into marsh or river sediments that have been disrupted by faulting in an effort to decipher a region's **paleoseismicity,** its rock record of past earthquake events. Kerry Sieh of the California Institute of Technology has done this across the San Andreas in Southern California (◆Figure 4.37). Sieh found an intriguing history of seismicity recorded in disrupted marsh deposits at Pallett Creek and liquefaction effects extending from the seventh century to a great earthquake in 1857. Ten large events, extending from A.D. 650 to 1857, were dated using ^{14}C. The average recurrence interval for these ancient earthquakes is 132 years, but they are clustered in four groups. Within each cluster, the recurrence interval is less than 100 years, and the intervals between the clusters are two to three centuries in length.

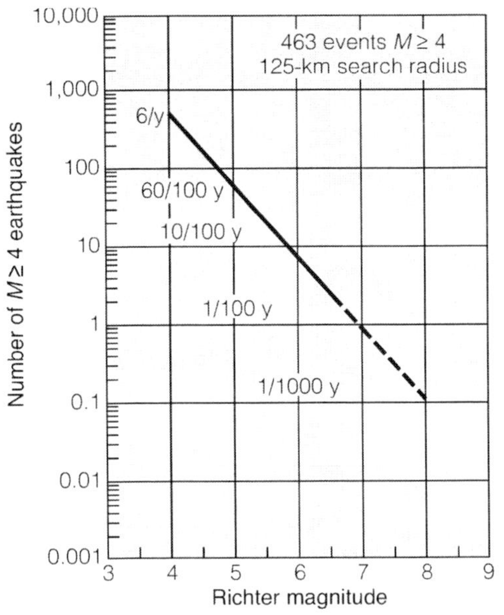

◆ FIGURE 4.35 A graphing of 463 earthquakes of magnitude 4 or greater in a small area near a nuclear reactor in Southern California over a period of 44 years. Statistically, the graph shows that 600 magnitude-4 earthquakes can be expected in 100 years, or 6 per year on average. The probability of a magnitude-8 earthquake in 100 years is 0.1, which would be 1 in 1,000 years. Such plots can be constructed for a region or for the world. The graphs are all similar in shape; only the numbers differ.

The last big one, also the last one of a cluster, was in 1857. Thus it appears that this section of the San Andreas may remain dormant until late in the twenty-first century or beyond.

Earlier in this chapter we quoted a geologist saying about earthquake prediction that "the longer it's been since the last earthquake, the sooner we expect the next one." Unfortunately, this is about the status of our present predictive ability. We remain uncertain that a quake will follow well-defined percursory phenomena, such as wave velocity changes or anomalous animal behavior (see ■ Case Study 4.4 on page 106). It seems that no one phenomenon is a predictor, and many changes will have to be monitored over a long period of time before our knowledge of fault behavior is refined enough for reliable prediction.

ThomsonNOW™

Click **Geology Interactive** to work through an activity on Seismic Risk USA through Earthquakes and Tsunamis.

Mitigation

Reducing earthquake risks is an admirable goal of scientists and lawmakers. Building codes provide the first line of defense against earthquake damage and help ensure the public safety. It has been shown that strict building codes in seismic regions, such as the Pacific Rim, reduce damage and loss of life. Laws passed after the 1933 (Long Beach) and 1971 (Sylmar) California earthquakes have proven the effectiveness of strict earthquake-resistant design. A good example is the magnitude-6.2 Morgan Hill (California) earthquake that

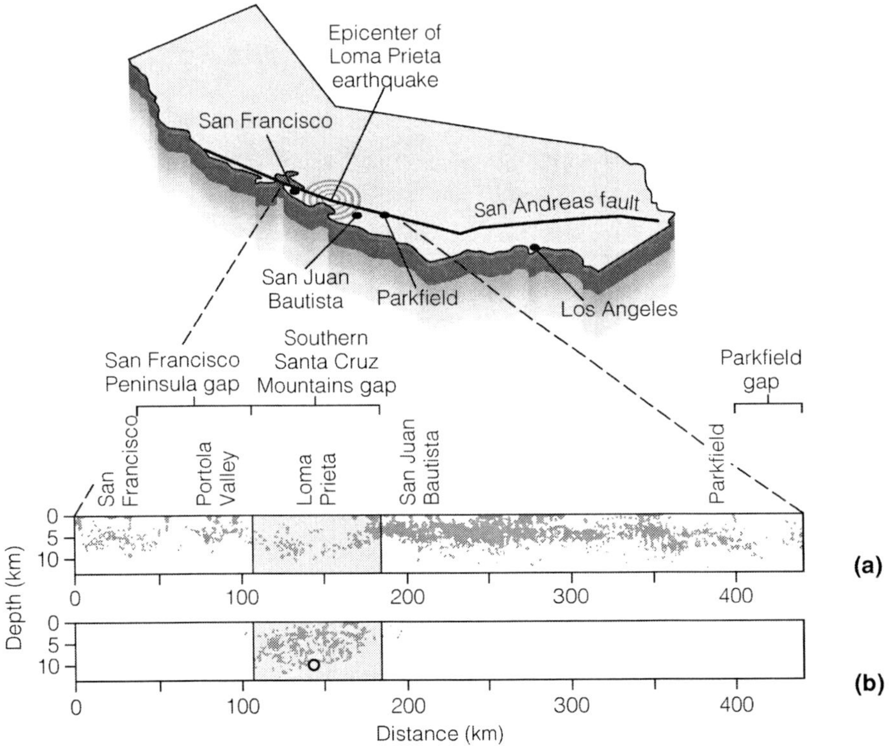

◆ FIGURE 4.36 (a) Historic seismic activity of about 450 kilometers (275 mi) of the San Andreas fault south of San Francisco before the 1989 Loma Prieta earthquake. The southern Santa Cruz Mountains seismic gap, where Loma Prieta is located, is highlighted. Other seismic gaps apparent in this historical record at the time were one between San Francisco and the Portola Valley and one southeast of Parkfield. (b) Seismicity record of the Southern Santa Cruz Mountains portion of the fault after the 1989 earthquake. The former seismic gap had been "filled" by the earthquake (*open circle*) and its aftershocks.

Kerry E. Sieh, Seismological Laboratory, California Institute of Technology

◆ FIGURE 4.37 Disrupted marsh and lake deposits along the San Andreas fault at Pallett Creek near Palmdale, California. Sediments range in age from about A.D. 200 at the lower left to A.D. 1910 at the ground surface. Several large earthquakes are represented here by broken layers and buried fault scarps.

shook West Valley College 20 miles from the epicenter. Seismic instruments on the gymnasium showed that the roof was so flexible that in a strong seismic event it could collapse (◆ Figure 4.38). Flexible roofs were permitted by the code at that time and many gyms and industrial buildings were built that way. As a result of the experience at West Valley College the Uniform Building Code was revised—this is the code used by hundreds if not thousands of municipalities across the country. The revision requires that the roofs be constructed to be less flexible and thus able to withstand nearby or strong distant earthquakes. Most large cities in strong earthquake zones have their own building codes, patterned after the Uniform Building Code, that require construction to modern seismic standards. For instance, ground response due to differing soil types is now more appreciated as contributing to quake damage. ◆ Figure 4.39 shows how the code has evolved from 1955 as we obtained more information. For example, soft clays react more violently with earthquake waves than do, say, granites, and this difference in response is now taken into consideration.

The primary consideration in earthquake design is to incorporate resistance to horizontal ground acceleration, or "base shear." Strong horizontal motion tends to topple poorly built structures and to deform more flexible ones. In California, high-rise structures are built to withstand about 40 percent of the acceleration of gravity in the horizontal direction (0.4 *g*), and single-family dwellings are built to withstand about 15 percent (0.15 *g*). *Base isolation* is now a popular design option. The structure, low- or high-rise, is placed upon Teflon plates, rubber blocks, seismic-energy dissipators that are similar to auto shock absorbers, or even springs, which allows the ground to move but minimizes building vibration and sway.

The ultimate earthquake resistance for large buildings is provided by "base isolation." The century-old Salt Lake City and County Building (◆Figure 4.40), scheduled at the time

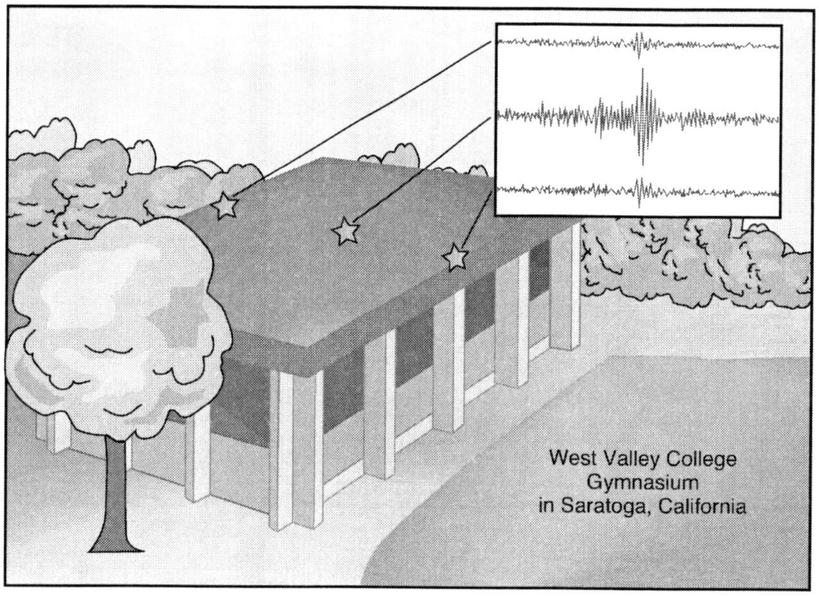

◆ FIGURE 4.38 Seismic records (upper right) obtained during the 1984 Morgan Hill, California, earthquake led to an improvement in the Uniform Building Code (a set of standards used in many states). The center of the gym roof shook sideways three to four times as much as the edges. The code has since been revised to reduce the flexibility of such large-span roof systems and thereby improve their seismic resistance. USGS

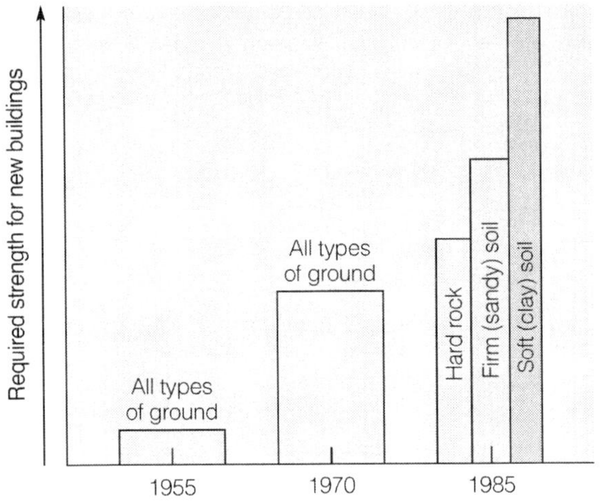

◆ FIGURE 4.39 Earthquake requirements in building codes have increased over time as scientists and engineers have obtained new information. Note that recent codes specify separate criteria for different ground types. USGS data

D. D. Trent

◆ FIGURE 4.40 Salt Lake City and County Building, Utah. Constructed in 1890, largely of brick, it is now seismically retrofitted to meet the modern building code.

for demolition, was cut from its foundation and retrofitted with 440 rubber base isolators. These permit the structure to move as a single unit during an earthquake while the ground shakes in all directions (◆Figure 4.41). About seventy-five percent of Utah's population live near the Wasatch Range and the fault responsible for the uplift. The retrofit of older public buildings is an example of the awareness of earthquake hazards by government officials and the public. In this case it saved an architectural treasure for future generations.

Survival Tips

Knowing what to do before, during, and after an earthquake is of utmost importance to you and your family.

Before an Earthquake:

◆Figure 4.42 provides the Federal Emergency Management Agency's (FEMA) suggestions for minimizing the possibility of damage, fire, and injuries in the home.

During an Earthquake:

- Remain calm and consider the consequences of your actions.
- If you are indoors, stay indoors and get under a desk, bed, or a strong doorway.

D. D. Trent

◆ FIGURE 4.41 Basement of the Salt Lake City and County Building. The building has been seismically retrofitted by cutting the massive structure from its foundation pillars (one shown on the far left, another in the center of the photograph), and lifting the building onto 440 large rubber isolators (one left of center, another at the extreme right of the photograph).

- If you are outside, stay away from buildings, walls, power poles, and other objects that could fall. If driving, stop your car in an open area.
- Do not use elevators, and if you are in a crowded area, do not rush for a door.

After an Earthquake:

- Turn off the gas at the meter.
- Use portable radios for information.
- Check water supplies, remembering that there is water in water heaters, melted ice, and toilet tanks. Do not drink waterbed or pool water.
- Check your home for damage.
- Do not drive.

(a)

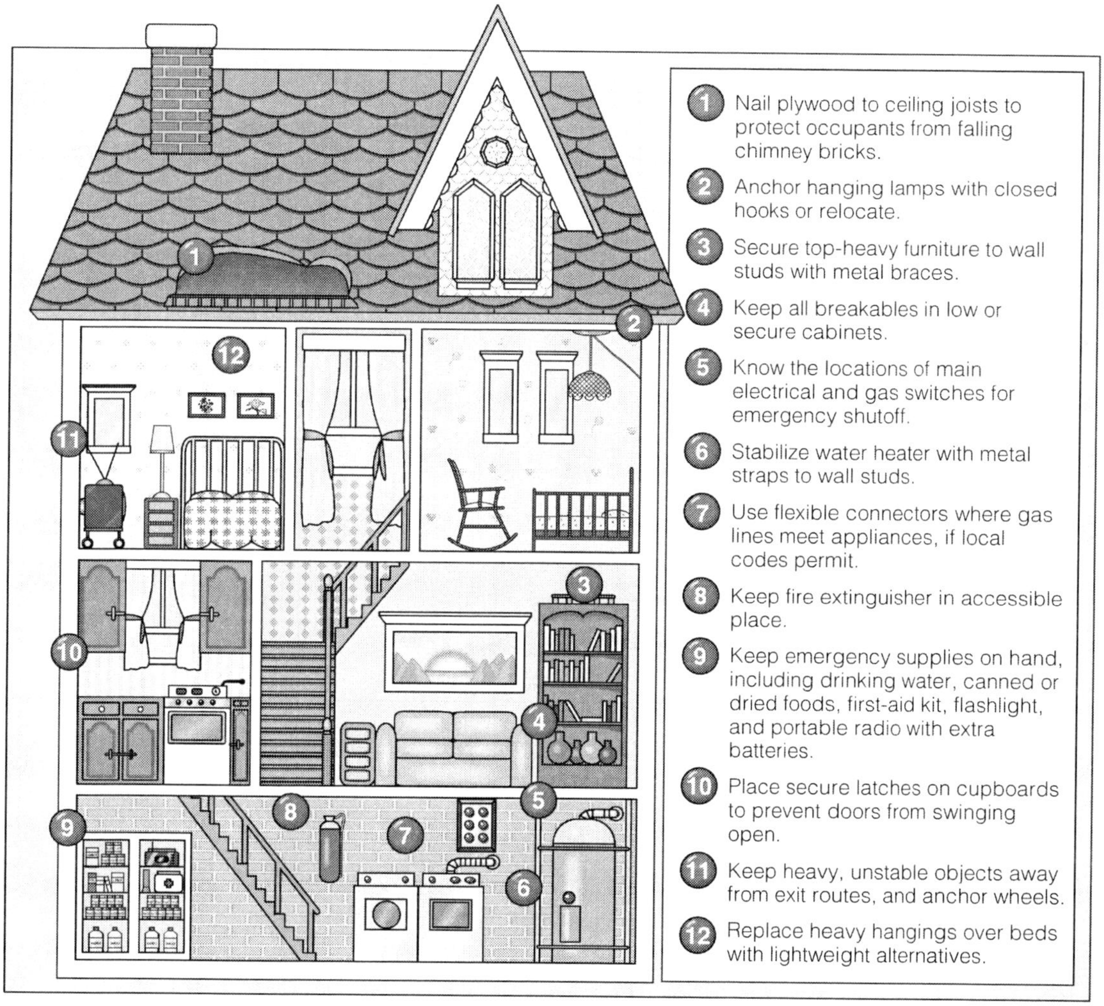

(b)

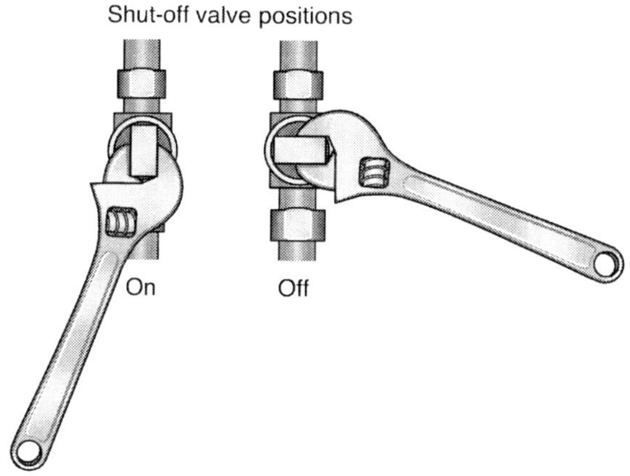

◆ FIGURE 4.42 (a) How to minimize earthquake damage in the home in advance. (b) Keep a small crescent wrench at the gas meter. Turn off the gas by turning the valve end 90°. FEMA

CASE STUDY 4.1

Earthquakes, Landslides, and Disease

Dynamically induced landslides are not particularly newsworthy, even though the 1994 Northridge earthquake caused 17,000 of them. However, the soil dislodged during sliding caused an outbreak of *coccidioidomycosis* (CM), commonly known as "valley fever." Endemic to the Southwest, the disease causes persistent flu-like symptoms and in extreme cases can be fatal. The victims breathe in airborne *Coccidioides immitis* spores that are released from the soil during sliding. From January 24 to March 15, 166 people were diagnosed with valley fever symptoms in Ventura County, up from only 53 cases in all of 1993. Most of the cases were reported from the Simi Valley, an area in which only 14 percent of the county's population resides (◆ Figure 1).

Large clouds of dust hung over the Santa Susana Mountains for several days after the earthquake, promoted by the lack of winter rains preceding the quake. During this period, pressure-gradient winds known locally as "Santa Ana" winds of 10–15 knots (11–17 mph) blew into the Simi Valley, carrying in spore-laden dust from the Santa Susana Mountains to the northeast. Researchers believe that the more "coherent" metamorphic rocks of the San Gabriel Mountains northeast of the San Fernando Valley probably account for the lack of cases reported in the epicentral region. This is the first report of an earthquake-associated outbreak of valley fever, even though many earthquakes have occurred in CM endemic areas.

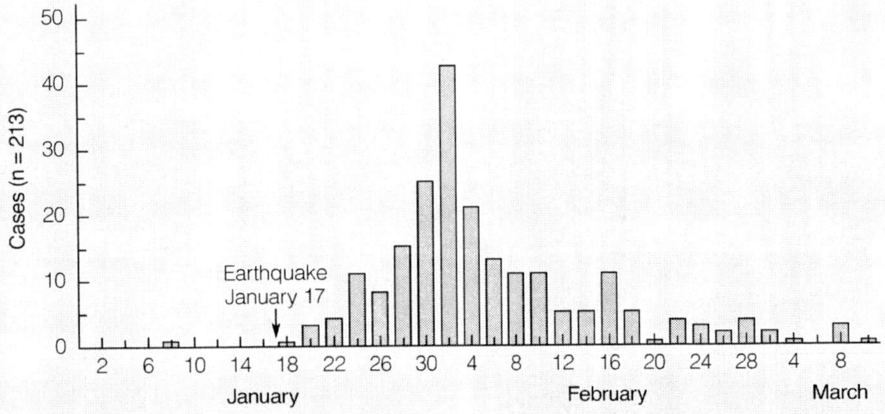

◆ FIGURE 1 Histogram of the almost-epidemic outbreak of coccidioidomycosis (valley fever) in Ventura County in early 1994 following the Northridge earthquake.

CASE STUDY 4.2

Predictable "Future Shocks"

A large earthquake is normally followed by thousands of smaller-magnitude earthquakes known as aftershocks (◆ Figure 1). If the main shock is small—say in the magnitude-4.0–5.0 range—aftershocks are small and nonintimidating. Following Northridge-size and bigger earthquakes, however, the strongest aftershocks ($M = 5.0$–6.0 at Northridge) can cause buildings damaged by the main earthquake to collapse and, even worse, can greatly increase anxiety in the already damaged psyche of the local citizenry. Aftershocks are caused by small adjustments (slips) on the causative fault or on other faults close to the causative one. For example, try this: push the eraser on the end of a lead pencil across a desktop. You'll find that it does not slide smoothly; it moves in jerky jumps and starts. This is called "stick–slip" and it is what happens along faults that are adjusting after a big earthquake.

Los Angeles experienced 2,500 aftershocks in the week following the Northridge earthquake. Three strong ones, magnitude

CONTINUED

5 or greater, occurred the first day. The largest ($M = 5.6$) occurred eleven hours after the main shock, causing concern among rescuers digging for victims beneath the rubble and the already traumatized citizens. Aftershocks follow statistically predictable patterns, as exemplified by the Northridge sequence. On the first day there were 188 aftershocks of magnitude 3 or greater, and on the second day only 56 were recorded (◆Figure 2). By fitting an equation to the distribution of aftershocks during the first few weeks, seismologists were able to estimate the number of shocks to be expected in the future. Statistically, there was a 25 percent chance of another magnitude-5 or greater aftershock occurring within the following year, but it did not happen.

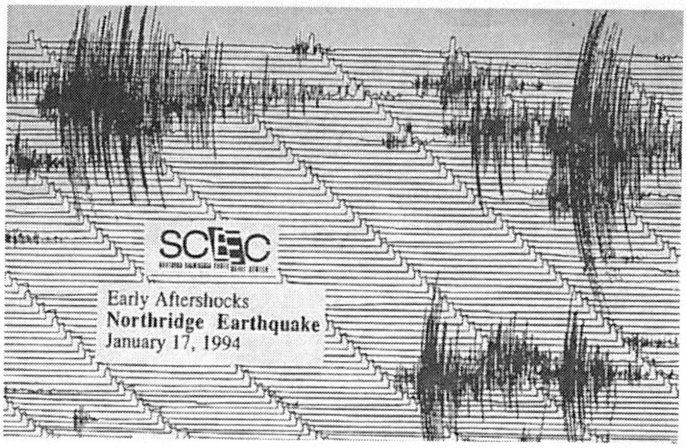

◆ FIGURE 1 Seismogram of early aftershocks following the Northridge earthquake, one of them a magnitude-5.6. Southern California Earthquake Center (SCEC)

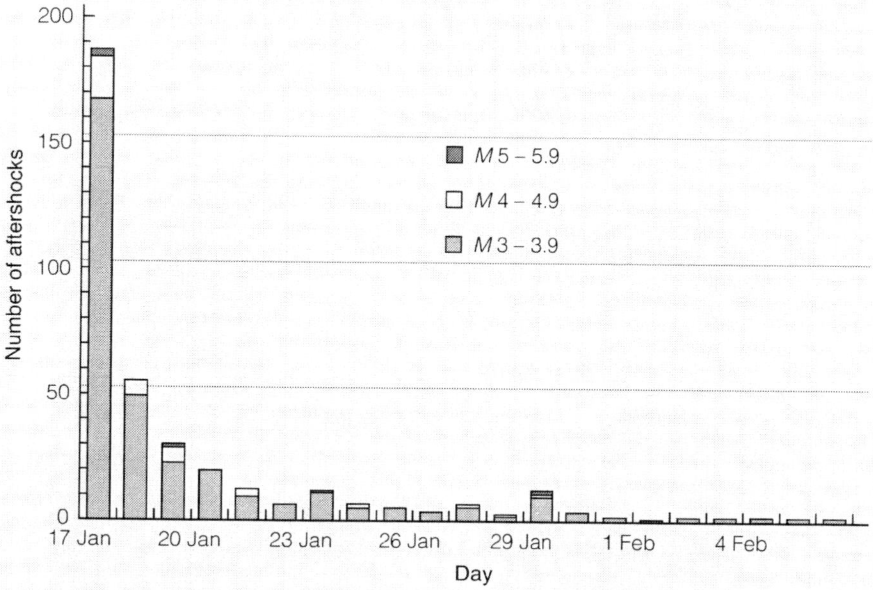

◆ FIGURE 2 Daily record of aftershocks of magnitude 3.0 to 5.9 during the three weeks following the main shock at Northridge. Note the sharp drop in aftershock frequency in the first four days. Redrawn from F. Harp and R. Jibson, USGS

CASE STUDY 4.3

Rx for Failed Freeways

Extensive damage to freeway bridges and overpasses typically accompanies earthquakes in large urban areas. Such damage has occurred in Alaska, California, and Japan. Overpass damage commonly is due to failure of the shorter columns, which lack the flexibility of longer ones. During an earthquake the tall columns supporting a bridge or overpass system bend and sway with the horizontal forces of the quake. Because the parts of the overpass system are tied together, the stresses are transferred through the structure to the short columns, which are designed to bend only a few centimeters (◆Figure 1). Failure causes the short columns to bulge just above ground level, which breaks and pops off the exterior concrete, exposing the warped, "birdcaged" steel in the interior. In addition,

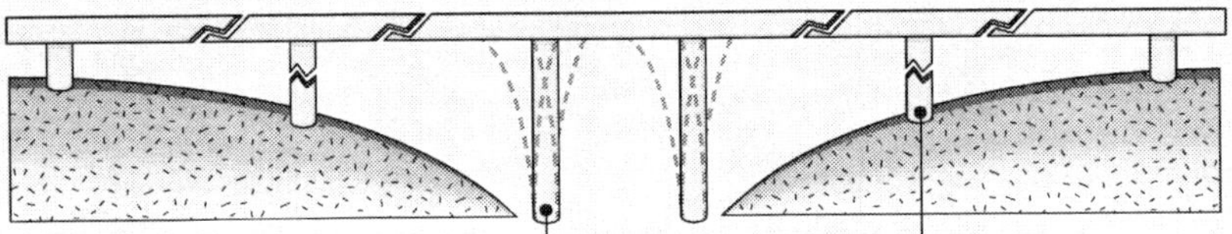

◆ FIGURE 1 The difference in flexibility of long and short columns results in failure of the short ones.

high vertical accelerations can cause some columns to actually punch holes through the platform deck. With excessive horizontal motion some deck spans may slip off their column caps at one end and fall to the ground like tipped dominoes.

The California Department of Transportation in 1971 decided to retrofit 122 overpasses to alleviate these problems. One aspect of the retrofitting was jacketing the short columns with steel or a composite substance and filling the space between the jacket and the original column with concrete (◆Figure 2a). This allows the columns to bend 12.5 centimeters (5 in), instead of 2.5 centimeters (1 in), without shattering. Another solution is to increase columns' horizontal strength by wrapping heavy steel rods around their vertical support bars, particularly on short columns. This allows the columns to bend but prevents "birdcaging" or permanent bending (◆Figure 2b). To prevent the decks from slipping off their supports and dropping to the ground, steel straps or cables are installed at the joints (◆Figure 2c). Of the 122 overpasses that CalTrans retrofitted, not one collapsed in the 1994 Northridge earthquake. Ten of the eleven that collapsed were slated for future retrofitting.

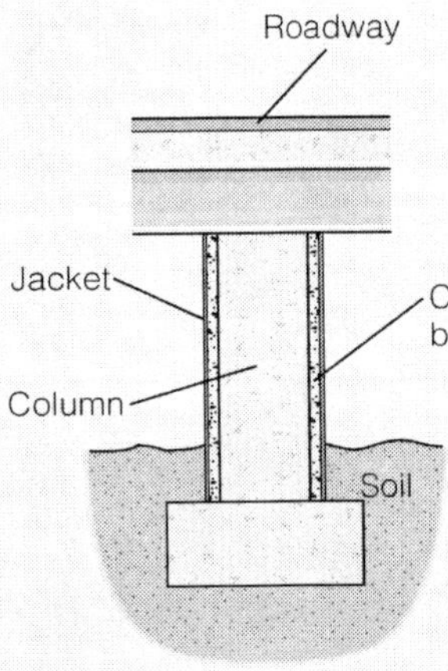

(a)

(b)

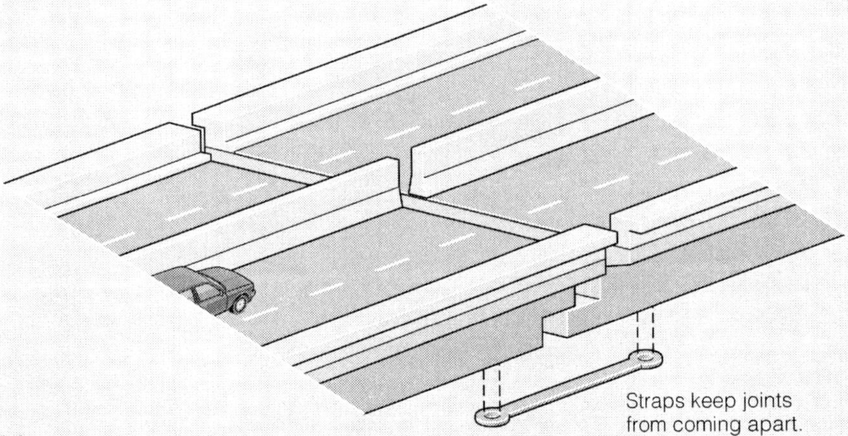

(c)

◆ FIGURE 2 Earthquake-resistant design for bridges and overpasses. (a) Short columns are retrofitted with steel or composite jackets that allow them to bend five times as much. (b) Old method of constructing highway support columns. The vertical steel supports have bent and "birdcaged." (c) Steel straps are used to hold deck sections together, preventing them from falling off their support columns. (a), (c) After *The Los Angeles Times*

CASE STUDY 4.4

Depressed Tigers, Restless Turtles, and Earthquakes!

Anomalous animal behavior preceding earthquakes is well documented. Domesticated animals such as barnyard fowl, horses, cats, and dogs have been known to behave so peculiarly before big events that they attracted the notice of people not knowledgeable in what is normal or abnormal animal behavior (◈Figure 1). Remember that although anomalous animal behavior precedes many earthquakes, it does not appear to precede every earthquake, and abnormal animal behavior is not always followed by an earthquake. Other natural phenomena, such as atmospheric disturbances, also can cause animals to behave strangely. Here are some reported incidents of unusual animal behavior noted before earthquakes:

- Tientsin Zoo, China, 1969: 2 hours before the magnitude-7.4 earthquake the tiger appeared depressed, pandas screamed, turtles were restless, and the yak would not eat.
- Haicheng, China, 1975: 1½ months before the magnitude-7.3 earthquake, snakes came out of hibernation; 1–2 days before, pigs would not eat and they climbed walls; 20 minutes before, turtles jumped out of the water and cried.
- Tokyo, Japan, 1855: 1 day before the magnitude-6.9 earthquake wild cats cried, and rats disappeared.
- Concepción, Chile, 1835: 1 hour and 40 minutes before the earthquake, flocks of sea birds flew inland, and dogs left the city.
- San Francisco, 1906: dogs barked all night before the magnitude-8.3 earthquake.
- Friuli, Italy, 1976: 2–3 hours before the magnitude-6.7 earthquake, cats left their houses and the village, mice and rats left their hiding places, and fowl refused to roost.

B. Pipkin

◈ FIGURE 1 Anomalous animal behavior. An earthquake may be in the offing if your dog dons a hard hat.

Earthquakes Are Hard on Cars

The sight of cars buried beneath a pile of bricks and rubble after an earthquake seems to be standard fare for the press. We captured a few of these scenes just by coming upon them and hope you feel, as we do, that there is a light side to most bad natural disasters. The other photos were just too unusual or amusing to pass up.

B. Pipkin

◆ FIGURE 1 This luxury car seems to be pursuing the garage in which it is normally kept. The house and garage slid down the hillside, but the car was parked with its rear wheels on the driveway, which was on stable ground. Nobody was hurt; Northridge, 1994

Frederick Larson, *SAN FRANCISCO CHRONICLE*

◆ FIGURE 2 A row of cars in need of body-and-fender work because their owners were asleep in the apartment building at the time of the earthquake; Northridge, 1994

James Williams

◆ FIGURE 3 Earthquakes have no respect for "No Parking" signs; Boys Market Sylmar, California, 1971

Yi-Ben Tsai, National Central Univ., Taiwan

◆ FIGURE 4 One hundred yard . . . dash? The September 21, 1999 earthquake (M_w = 7.8) in Taiwan was caused by movement on the Chelungpu fault. This running track at Kuang Fu High School was built across the fault that here had a vertical offset of 2.5 meters (8.1 ft). It is obvious that no 100-meter-dash records will be broken here.

Ted Reeves, Chaffey High School, Upland, Calif.

◆ FIGURE 5 The Hector Mine earthquake of October 16, 1999 was a real bomb! The magnitude-7.0 event occurred in California's remote Mojave Desert. The fault scarp, shown here, forms a gash across the desert for 40 kilometers (25 mi), but the earthquake did little damage. This area happens to be a U.S. Marine Corps bombing range, as you can see, which made geologists' field work interesting.

Summary

Earthquakes

Cause

Movements on fractures in the crust known as *faults* that result in three types of wave motion: P- and S-waves, which are generated at the focus of the earthquake and travel through the Earth, and surface waves.

Distribution

Most (but not all) large earthquakes occur near plate boundaries, such as the San Andreas fault, and represent the release of stored elastic strain energy as plates slip past, over, or under each other. Intraplate earthquakes can occur at locations far from plate boundaries where deep crust has been faulted, probably at "failed" continental margins.

Measurement Scales

Some of the scales used to measure earthquakes are the modified Mercalli intensity scale (based on damage), the Richter magnitude scale (based on energy released as measured by maximum wave amplitude on a seismograph), and the moment magnitude (based on the total seismic energy released as measured by the rigidity of the faulted rock, the area of rupture on the fault plane, and displacement).

Earthquake-Related Hazards and Mitigation

Ground Shaking

Damaging motion caused by shear and surface waves.

Ways to reduce effects—seismic zoning; building codes; construction techniques such as shear walls, seismic joints, and bolting frames to foundations.

Landslides

Hundreds of landslides may be triggered by an earthquake in a slide-prone area.

Ways to reduce effects—proper zoning in high-risk areas.

Ground Failure (Spontaneous Liquefaction)

Horizontal (lateral) movements caused by loss of strength of water-saturated sandy soils during shaking and by liquefaction of quick clays.

Ways to reduce effects—building codes that require deep-drilling to locate liquefiable soils or layers.

Ground Rupture/Changes in Ground Level

Fault rupture and uplift or subsidence of land as a result of fault displacement.

Ways to reduce effects—geologic mapping to locate fault zones, trenching across fault zones, implementation of effective seismic zonation like the Earthquake Fault Zones Act in California.

Fire

In some large earthquakes fire has been the biggest source of damage.

Ways to reduce effects—public education on what to do after a quake, such as shutting off gas and other utilities.

Tsunamis

Multidirectional sea waves generated by disruption of the underlying sea floor (see Chapter 10).

Earthquake Prediction

It is not feasible at this time to predict within useful limits the time, magnitude, and location of an earthquake. However, seismologists continue to search for clues to accurate prediction, which is what scientists do and what, we might add, science is.

Upside/downside—no proven large earthquakes predicted. Inaccuracies in short-term prediction have made the public wary.

Earthquake Forecasts

Based on plate movement and known fault movement there is a 70 percent chance there will be a magnitude-6.7 earthquake in the San Francisco Bay area between 2000 and 2030. This kind of forecast is the research thrust today and forecasts will vary for each seismic region.

Upside/downside—good for planning purposes but not for short-term warnings.

Statistical Methods

Tell us that worldwide there will be two $M = 8.0+$ and at least 20 $M = 7.0+$ earthquakes each year. Statistics can be applied for a smaller area in earthquake country.

Upside/downside—good for planning purposes but not for short-term warnings.

Geological Methods

Active faults are studied to determine the characteristic earthquake magnitudes and recurrence intervals of particular fault segments; sediments exposed in trenches may disclose historic large fault displacements (earthquakes) and, if they contain datable C-14 material, their recurrence intervals.

Upside/downside—useful for long-range forecasting along fault segments and for identifying seismic gaps; not useful for short-term warnings.

Key Terms

base isolation
body wave
elastic rebound theory
epicenter
fault
fault creep
focus (pl. *foci*)
forecasts
intensity scale
isoseismal
liquefaction
longitudinal wave

magnitude
modified Mercalli scale (*MM*)
moment magnitude (M_w or *M*)
paleoseismicity
precursor
P-wave
recurrence interval
resonance
Richter magnitude scale
seismogram
seismograph
spontaneous liquefaction
strain
stress
S-wave
transverse (shear) wave
tsunami (pl. *tsunamis*)
wave period (*T*)

Study Questions

1. What is "elastic rebound," and how does it relate to earthquake motion?
2. Distinguish among earthquake intensity, Richter magnitude, and moment magnitude. Which magnitude scale is most favored by seismologists today? Why?
3. Why should one be more concerned about the likelihood of an earthquake in Alaska than of one in Texas?
4. In light of plate tectonic theory, explain why devastating shallow-focus earthquakes occur in some areas and only moderate shallow-focus activity takes place in other areas.
5. What should people who live in earthquake country do before, during, and after an earthquake (the minimum)?
6. Describe the motion of the three types of earthquake waves discussed in the chapter and their effects on structures.
7. Why do wood-frame structures suffer less damage than unreinforced brick buildings in an earthquake?

For Further Information

Books and Periodicals

Atwater, Bryan, and others. 1999. *Surviving a tsunami: Lessons from Chile, Hawaii, and Japan.* U.S. Geological Survey circular 1187, 19 pp.

Bolt, Bruce. 1998. *Earthquakes, newly revised and expanded.* New York: W. H. Freeman.

Brocher, Thomas M., 2000, Urban seismic experiments investigate Seattle fault and basis, *EOS, Transactions of the American Geophysical Union,* vol. 81, no. 46, November 14.

Celebi, M., P. Spudich, Robert Page, and Peter Stauffer. 1995. *Saving lives through better design standards.* U.S. fact sheet 176–95.*

Drake, D. 1815. *Natural and statistical view, or picture of Cincinnati and the Miami County, illustrated by maps.* Cincinnati, Ohio: Looker and Wallace.

Ellis, Michael, Joan Gomberg, and E. Schweig. 2001. Indian earthquake may serve as an analog for New Madrid earthquakes, *EOS, Transactions of the American Geophysical Union,* vol. 82, no. 32, August 7.

Field, Edward H. 2000. Accounting for site effects in probabilistic seismic hazard analysis of Southern California. *Bulletin of the Seismological Society of America,* vol. 90, no. 6b, December.

Gomberg, Joan, and Eugene Schweig. 2002. *East meets Midwest: An earthquake in India helps hazard assessment in the central United States.* U.S. Geological Survey fact sheet 007-02.

González, Frank I. 1999. Tsunami. *Scientific American,* May: 56–65.

Gore, Rick. 1995. Living with California's faults. *National geographic,* April:2–34.

Hickman, Steve, and John Langbein. 2000. *The Parkfield experiment—Capturing what happens in an earthquake,* U.S. Geological Survey fact sheet 049-02.

Iacopi, Robert. 1971. *Earthquake country.* Menlo Park, Calif.: Lane Books.

Knopoff, L. 1996. Earthquake prediction: The scientific challenge. In *Earthquake prediction: Proceedings of the National Academy of Sciences,* pp. 3719–3720.

Kockelman, William J. 1984. *Reducing losses from earthquakes through personal preparedness.* U.S. Geological Survey open file report 84-765.

Koper, Keith D., and others. 2001. Forensic seismology and the sinking of the *Kursk, EOS,* Transactions of the American Geophysical Union, vol. 82, no. 4, January 23.

Kovachs, Robert. 1995. Earth's fury: An introduction to natural hazards and disasters. Englewood Cliffs, N.J.: Prentice-Hall.

Michael, Andrew J., and others. 1995. *Major quake likely to strike between 2000 and 2030.* U.S. Geological Survey fact sheet 151–99.

Mori, James J. 1994. Overview: The Northridge earthquake: Damage to an urban environment. *Earthquakes and volcanoes* 25, no. 1 (special issue).

Renwald, Marie, Tammy Baldwin, and Terry C. Wallace. 2003. Seismic analysis of the space shuttle *Columbia* disaster (abstract). Geodaze Geoscience Symposium, University of Arizona, Tucson, p. 75.

Richter, C. 1958. *Elementary seismology.* New York: W. H. Freeman.

Wuethrich, Bernice. 1995. Cascadia countdown. *Earth: The science of our planet,* October: 24–31.

Yeats, Robert S. 1998. *Living with earthquakes in the Pacific Northwest.* Corvallis, Oreg.: Oregon State University Press.

*Fact sheets are one- or two-page condensations of a geological problem and are free of charge. Some are found on the Web at http://quake.usgs.gov. Hard copies may be obtained from the U.S. Geological Survey, Mail Stop 977, 345 Middlefield Road, Menlo Park, CA 94025.

ThomsonNOW™

Assess your understanding of this chapter's topics with additional comprehensive interactivities at http://www.thomson.com/login, which also has up-to-date web links, additional readings, and exercises.

4

Energy and the Environment

> It is difficult for people living now, who have become accustomed to the steady exponential growth in the consumption of energy from the fossil fuels, to realize how transitory the fossil-fuel epoch will eventually prove to be when it is viewed over a longer span of human history.
>
> M. King Hubbert, 1971

Flaming Ice, Energy, and the Greenhouse!

SETTING ICE BALLS ON FIRE IS U.S. GEOLOGICAL SURVEY OCEANOGRAPHER Bill Dillon's idea of fun, and it never fails to amaze his visitors. It is not ordinary ice that flames. Dillon uses hydrated methane (CH_4) recovered from sea-floor sediments deep below the ocean surface. This flame has become a focus of the energy-hungry nations of the world (◆ Figure 1). What actually burns is methane gas released from a crystalline solid of methane molecules encased in a "cage" of ice (◆ Figure 2). Methane is the gas that constitutes the bulk of the "natural gas" that is piped to your home. Methane hydrates are so extensive in deep-sea sediments that they have the potential to supply a share

USGS

◆ FIGURE 1 A flame from ignited synthetic methane hydrate.

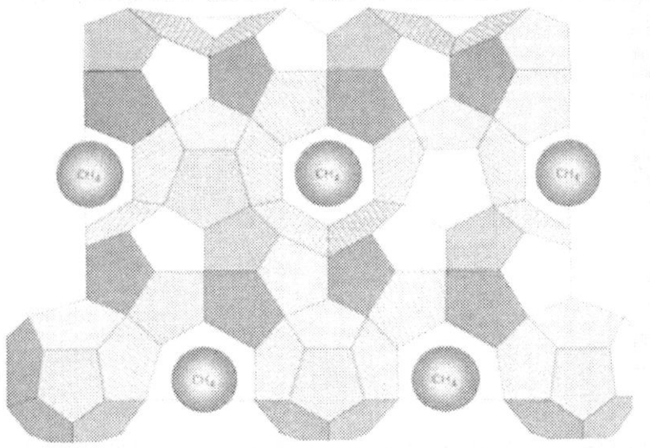

◆ FIGURE 2 Crystalline structure of methane hydrate. Methane gas (CH_4) generated by bacterial digestion of organic matter in sea-floor sediments is sometimes trapped in "cages" of frozen H_2O (the large polygons) to form methane hydrate. Peter McCabe and others, 1993, USGS circular 115, p. 46

of the world's future fuel needs, *when and if* practical mining techniques can be developed. India and Japan, at present energy poor, are most interested in developing methane hydrates as an energy source. Experts agree that marine gas hydrates harbor twice the carbon contained in all known natural gas, petroleum, and coal deposits on Earth—about 10,000 gigatons, or 10^{13} tons, of carbon. Extensive gas-hydrate deposits are currently known in hundreds of localities. Particularly interesting for potential development are those of offshore Oregon, Costa Rica, New Jersey, Japan, and Florida (◆Figure 3). Also intriguing are the significant deposits that have been encountered in oil drilling through Arctic permafrost areas in Canada, Alaska, and Siberia (see Chapter 6).

Gas hydrates in marine sediments have been known for several decades, but their full extent and potential were not realized until the 1990s (◆Figure 4). Conditions necessary for their formation are high water pressure (water depth >530 m in low latitudes and >250 m in high latitudes) and low ambient temperatures ($<7°$ C) at the water–sediment interface. These conditions are met on the sea floor's continental slope and rise. The exact origin of the gas is not known, but it is strongly suspected that bottom bacteria consume organic-rich detritus and generate methane as a waste product. Under particular conditions, the gas becomes trapped in ice in the sediment instead of being released to the overlying water column.

The bad news is that these hydrated molecular structures can release their caged methane into the atmosphere if there is a slight decrease in pressure (if sea level is lowered, for instance) or a slight warming of the surrounding seawater. Because methane is a powerful "greenhouse" gas linked to global warming (it is about 10 times as effective as carbon dioxide), melting of the hydrates and release of methane could have a drastic impact on Earth's surface temperature (see Chapter 11). One model suggests that as sea levels rose toward the end of the last ice age, the waters covered land that was underlain by permafrost containing the methane hydrates. As the warmer water temperature melted the permafrost ice, the caged greenhouse gas was released, increasing the global warming and the melting of glaciers.

Gas hydrates may also be a geologic hazard and may occupy much of the pore space in bottom sediments. When the hydrate-bearing sediment becomes deeply buried, the temperature may increase to the point at which gas hydrates are no longer stable. When a solid–gas mixture becomes a liquid–gas mixture, high pore pressure and unstable slopes result (see Chapter 7). This presents a danger to offshore drilling platforms, pipelines, communication cables, and other offshore structures. Current interest in methane hydrate is focused on it as an energy source and as a factor in global climate change.

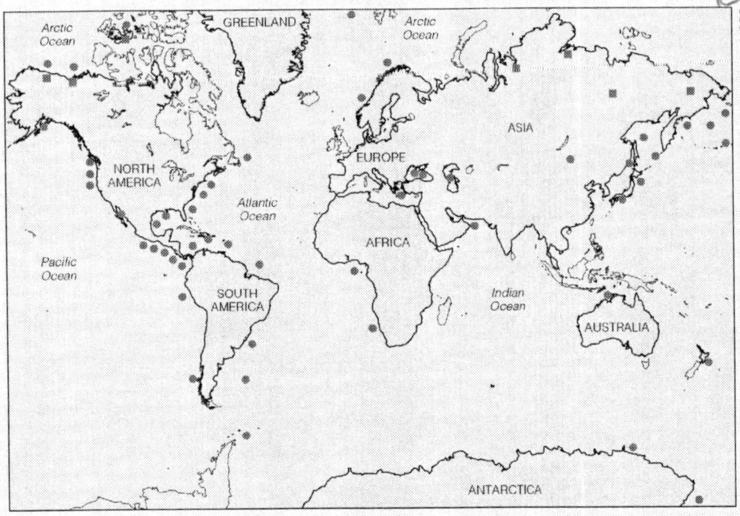

Keith A. Kenwolden, USGS

◆ FIGURE 3 Location of the world's known methane hydrate deposits, 1999. Dots are marine deposits; squares are permafrost deposits.

Dillon, USGS

◆ FIGURE 4 Methane hydrate nodule in marine sediment from the Blake Ridge off the Florida coast.

From the simplest algal scum to the most complex **ecosystem,** energy is essential to all life. Derived from the Greek **word** *energia,* meaning "in work," *energy* is defined as the **capacity** to do work. The units of energy are the same as **those** for work, and the energy of a system is diminished only by the amount of work that it does.

Prosperity and quality of life in an industrialized society such as ours depend in large part on the society's energy resources and its ability to use them productively. We may illustrate this in a semiquantitative fashion with the equation

$$L = \frac{R + E + I}{\text{population}}$$

where L represents quality of life, or "standard of living," R represents the raw materials that are consumed, E represents the energy that is consumed, and I represents an intangible we shall call *ingenuity.* As the equation expresses, when high levels of raw materials, energy, and ingenuity are shared by a small population, a high material quality of life results. If, on the other hand, a large population must share low levels of resources and energy, a low standard of living would be expected. Some highly ingenious societies with few natural resources and little energy can and do enjoy a high quality of life. Japan is a prime example. Some other countries that are self-sufficient in resources and energy, such as Argentina, are having difficult times. Thus we see that ingenuity, which is reflected in a country's political system, technologies, skills, and education, is heavily weighted in the equation, and can cancel out a lack of resources and a large population.

To a physicist there is no energy shortage, because she or he knows that energy is neither created nor destroyed; it is simply converted from one form to another, such as from nuclear energy to heat energy. Fuels of all kinds are warehouses of energy, which can be tapped by some means and applied in some way to do work. Coal and oil, for example, are fossil fuels that have been storing solar energy in the lithosphere for millions of years (◆ Figure 14.1).

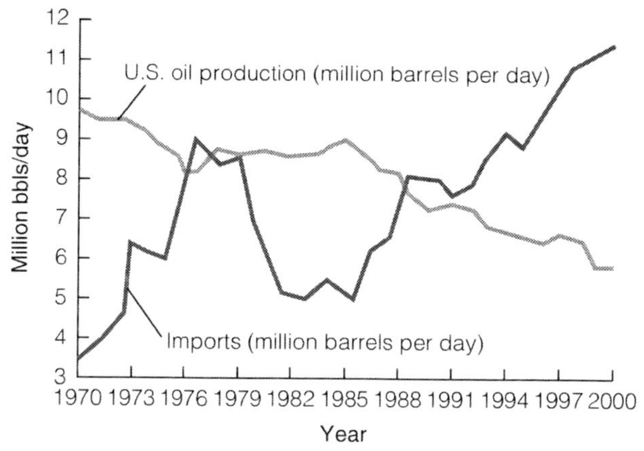

◆ FIGURE 14.1 Graph illustrating the relationship between oil production in the United States and the amount of oil imported by the United States. Adapted from Larry Fellows, Arizona Geological Survey

Some forms of energy are **renewable,** that is, they are replenished at a rate equal to or greater than the rate at which they are used. Examples include solar, water, wood, wind, ocean and lake thermal gradients, geothermal, and tidal energy. The energy in all of these resources except for geothermal and tidal (gravitational) energy was originally derived from the sun. Renewable resources are dependable only if they are consumed at a rate less than or equal to their rate of renewal. If they are overexploited, some period of time will be required to replenish them. Peat, a fuel used extensively for space heating and cooking in Ireland and Russia, is estimated to accumulate at a remarkable 3 metric tons per hectare per year (1.3 tons/acre/year). Nonetheless, the conversion from plant litter to peat may take a hundred years. Wood energy may renew in a matter of a few decades, and water and wind are renewed continuously.

Nonrenewable resources are not replenished as fast as they are utilized, and once consumed, they are gone forever. Crude oil, oil shales, tar sands, coal, and fissionable elements are nonrenewable energy resources. Their quantities are finite. Supplies of crude oil, for example, are within a few human generations of exhaustion. Oil underground was discovered almost fifty years before the first automobile was operational. Prior to then, gasoline was a minor by-product of refining oil for kerosene lamps. In 1885 Germans Gottlieb Daimler and Karl Benz independently developed gasoline engines, and in 1893 Massachusetts bicycle makers Charles and Frank Duryea built the first successful U.S. gasoline-powered automobile. By the turn of the century, production automobiles were hitting the roads, horses were being put out to pasture, and refineries were stepping up production to satisfy an increasing demand for gasoline. The transportation revolution to automobiles spawned the largest private enterprise on Earth: the exploration and production of petroleum. This has led to major geopolitical and economic problems, because the nations that consume the most do not produce in comparable amounts.

Petroleum

Although the carbon content of Earth's crust is less than 0.1 percent by weight, carbon is one of the most important elements to Earth's people. It is indispensable to life, and it is the principal source of energy and the principal raw material of many manufactured products. Crude oil, or petroleum (Latin *petra,* "rock," and *oleum,* "oil"), is composed of many **hydrocarbon compounds,** simple and complex combinations of hydrogen and carbon (◆ Table 14.1). Petroleum occurs beneath Earth's surface in liquid and gaseous forms and at the surface as oil seeps, tar sands, solid bitumen (gilsonite), and oil shales. In addition to its use as a fuel, hydrocarbon compounds derived from petroleum are used in producing paints, plastics, fertilizers, insecticides, soaps, synthetic fibers (nylon and acrylics, for example), and synthetic rubber. Carbon combines chemically with itself and with hydrogen in

***TABLE 14.1 Hydrocarbon Compounds Typically Found in Crude Oil**

Molecular Type: Name	Molecular Type: General Formula	Hydrocarbon Compound		Percentage of Weight in Medium-Grade Crude Oil
Paraffins	C_nH_{2n+2}	methane, CH_4	lighter	25
		ethane, C_2H_6	↑	
		propane, C_3H_8		
		butane, C_4H_{10}*		
		pentane, C_5H_{12}**		
Aromatics	C_nH_{2n-6}***	benzene, C_6H_6		17
Naphthenes	C_nH_{2n}	asphalt	↓	50
Asphaltenes	(solid hydrocarbons)	gilsonite	heavier	8
				100

* Butane gives gasoline quick-starting capability.
** Pentane gives smooth engine warm-up.
*** Aromatics improve mileage and "knock" resistance.

an infinite variety of bonding schemes; some two million hydrocarbon molecules have been identified to date. The manufacturing process of separating crude oil into its various components is known as *refining,* or *cracking.*

Origin and Accumulation of Hydrocarbon Deposits

Carbon and hydrogen did not combine directly to form petroleum; they were chemical components of living organisms before their transformation to complex hydrocarbons in crude oil. Porphyrin compounds found in petroleum are derived either from chlorophyll, the green coloring in plants, or from hemin, the red coloring matter in blood, and their presence is solid evidence for an organic origin for crude oil. The fact that large quantities of oil are not found in igneous or metamorphic rocks also rules out an inorganic source for oil.

Four conditions are necessary for the formation and accumulation of an exploitable petroleum deposit in nature:

- a source rock for oil,
- a reservoir rock in which it can be stored,
- a caprock to confine it, and
- a geologic structure or favorable strata to "trap" the oil.

Even where these geologic conditions are met, the human elements of exploration, location, and discovery still remain. Oil companies utilize the skills of geologists to interpret surface and subsurface geology, to locate the potential oil-bearing structure or stratum, and to specify the optimal location for a discovery well. Until a well is drilled, the geologists' interpretation remains in doubt, much like a medical doctor's diagnosis of an ailment subject to surgery; the diagnosis is tentative until the patient is opened up.

A **source rock** is any volume of rock that is capable of generating and expelling commercial quantities of oil or gas. Source rocks are sedimentary rocks, mostly shales or limestones, usually of marine (ocean) origin and sometimes of *lacustrine* (lake) origin. The biological productivity (biomass) of surface waters must have been high enough to generate a settling "rain" of dead organisms, and the bottom waters must have been low enough in oxygen to prevent the deposited organic matter from being oxidized or consumed by scavengers. Almost all source rocks are dark-colored, indicative of high organic-matter content, and some of them carry a fetid or rotten-egg odor.

Favorable marine environments are rich in microscopic single-celled plants known as *diatoms* (phytoplankton), which form the largest biomass in the sea. Where there are diatoms, we also find zooplankton—tiny protozoans and larvae of large animals. These together with diatoms provide the molecules that make up crude oil. As the rocks are buried, they are heated. The conversion from organic matter to petroleum takes place mostly between 50° C and 200° C. Thus a proper thermal history, ideally between 100° C and 120° C, is necessary to form liquid petroleum—too cool, and oil does not form; too hot, and the hydrocarbons "boil" away.

After petroleum has formed in a source bed, it is squeezed out and migrates through a simple or complex plumbing system into a **reservoir rock.** This migration is a critical element in the formation of an economically exploitable accumulation of oil. Reservoir rocks are porous and permeable (see Chapter 8). Commonly they are sandstones, porous limestones, or in some cases, fractured shales. Reservoir porosities range from 20 to 50 percent, meaning that for each cubic foot of reservoir rock we will find 1 to 4 gallons of oil. The unit of oil volume is the barrel (equal to 42 gallons), and a so-called giant field, such as the north slope of Alaska, will yield a billion barrels of oil in its lifetime. Using standard recovery techniques, however, as much as 40 to 80 percent of the oil may be left in pore spaces and as films on mineral grains.

An impermeable **caprock** prevents oil from seeping upward to form tar pits at the surface or dissipate into other rocks. Such seals are analogous to aquicludes in groundwater systems and are most commonly clay shales or limestones of low permeability. Our discussion of the fourth requirement, a suitable geologic structure for trapping the oil, requires a separate subsection.

Geologic Traps—Oil and Gas Stop Here

Structural Traps

An **anticline** is an ideal structure for trapping gas and oil. It is a convex-upward fold in stratified rock (◆ Figure 14.2a). Analogous to a teacup inverted in a pan of water that traps a layer of air inside it, an anticline holds a reservoir of gas and oil.

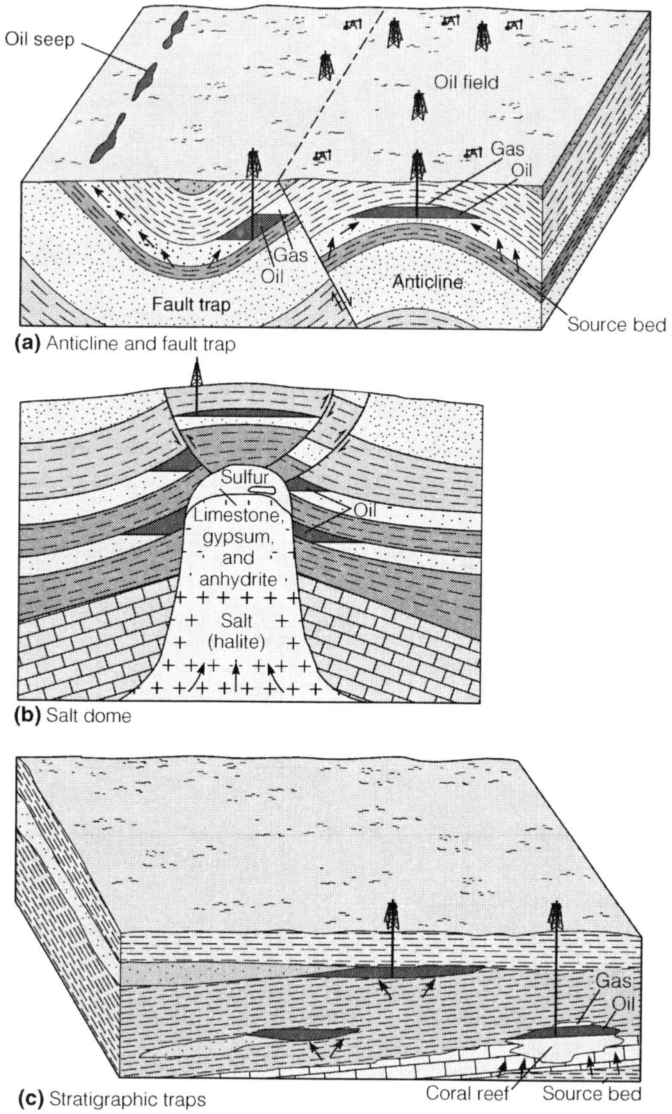

◆ FIGURE 14.2 (a and b) Common structural oil traps. (c) Stratigraphic oil traps.

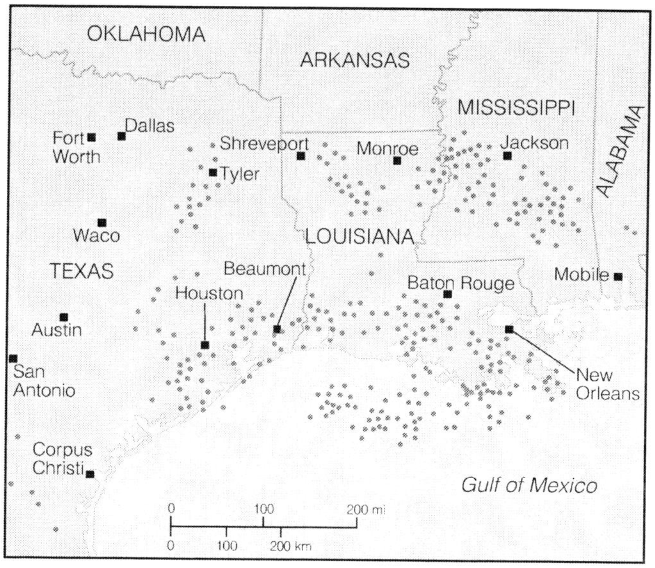

◆ FIGURE 14.3 Locations of salt domes of the U.S. Gulf Coast. More than 500 domes have been discovered on land and in the shallow parts of the Gulf of Mexico, and more are known to be in deep water offshore.

This occurs because crude oil floats on water and natural gas rises to the top of the reservoir. The anticlinal theory of oil accumulation was not developed until 1900, forty-one years after oil was discovered. Strata dip away from the central axis of an anticline at the ground surface, and most anticlines with surface geologic expression have been drilled. Today the search for oil is more difficult, because less obvious and geologically more complex traps need to be discovered. Many times faults form impermeable barriers to hydrocarbon migration, and oil becomes trapped against them (see Figure 14.2a). Faulted and folded stratified rocks in a single oil field may thus contain many isolated oil reservoirs.

Salt domes are of much interest to geologists (◆ Figure 14.2b). Not only do they create oil traps, but they are also valuable sources of salt and sulfur and may be potential underground storage sites for petroleum and hazardous waste. More than 500 salt domes have been located along the U.S. Gulf Coast, both offshore and on land (◆ Figure 14.3). They rise as flowing fingers of salt, literally puncturing their way through the overlying rocks and buckling the overlying shallow strata into a dome (◆ Figure 14.4). Some fingers rise as high as 13 kilometers (8 mi) above the "mother" salt bed,

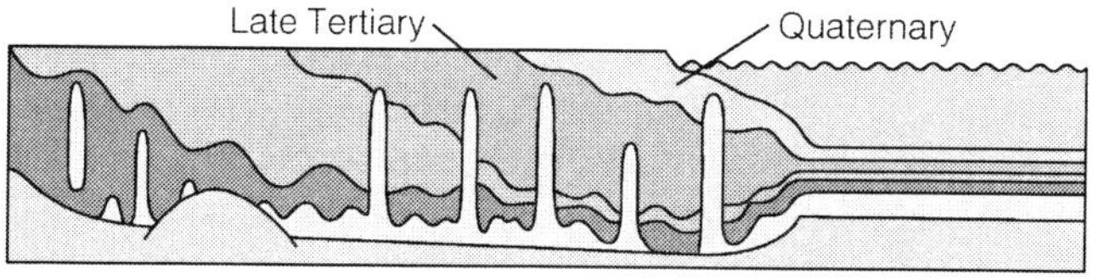

◆ FIGURE 14.4 Ancient salt deposits are buried deeply, and because salt is less dense than the overlying sediments, it rises buoyantly as pillars of salt, creating salt domes.

the Louann salt, and would be taller than Mount Everest if they were at the surface of Earth. Because the salt is considerably less dense than the overlying rock, buoyancy forces drive the salt upward—much as a blob of oil will rise through water. Brittle solids, such as salt or ice, will flow over long periods of time; many salt domes are still rising measurably. The "mother" salt was deposited in Jurassic time by evaporation of seawater when the embryonic Gulf Coast was connected to the open ocean by a shallow opening. The opening allowed seawater to enter the basin but did not allow the denser salt brines at the bottom to escape. Thus the brine became concentrated to the point of saturation, and the great thickness of salt that now underlies the entire Gulf Coast was deposited. Oil accumulates against the salt column and also in the dome overlying the salt. "Old Spindletop," a salt dome near Beaumont, Texas, that was discovered in 1901, is perhaps the most famous U.S. oil field. Soon after its discovery it was producing more oil than the rest of the world combined, and the price of oil plummeted to two cents a barrel.

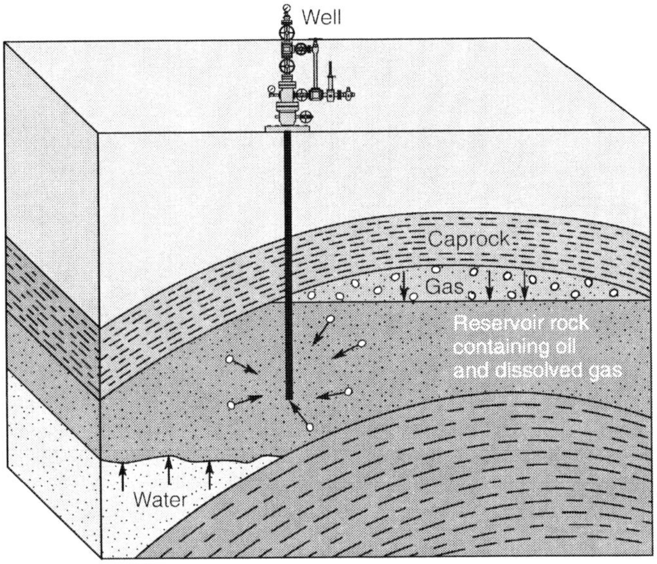

◆ FIGURE 14.5 Oil in an anticline is driven by gas pressure from above and by buoyant water pressure from below.

Stratigraphic Traps

Any change in sedimentary rock lithology (its physical character) that causes oil to accumulate is known as a **stratigraphic trap** (in contrast to anticlines and salt domes, which are structural traps). Thus, if a stratum changes laterally from a permeable sandstone to an impermeable shale or mudstone, oil may be trapped in the stratum (◆ Figure 14.2c).

Ancient coral reefs are ideal reservoirs, because they are porous and were biologically productive when they were living. Oil may be trapped in the porous, permeable debris on the flanks of the reef; production from such fields can be measured in thousands of barrels per day (bbl/d). Many oil fields of the Middle East are of this type, and their production potential is tremendous. A comparison of reef production with that of sandstone reservoirs such as those of California or Texas, which typically yield only a few hundred barrels per day, explains why the Middle East can control oil production and therefore price.

Oil Production

The first successful oil well was drilled in Titusville, Pennsylvania, in 1859 (■ Case Study 14.1 on page 438). In modern jargon, this well would be called a **wildcat well,** because it was the discovery well of a new field. There is 1 chance in 50 of a wildcat well being successful—less favorable odds than those of winning at roulette. The probability improves when we consider all wells drilled, including those in known oil fields. Between 1970 and 2003, 1 well in every 4 drilled was successful. Independent oil entrepreneurs take considerable risks; it is not a business for the faint of heart.

Most successful wells require pumping. If gas and water pressure are sufficient, however, oil may simply flow to the ground surface. High reservoir pressures develop from water pressing upward (buoyancy) beneath the oil and gas pressure pushing downward on the oil (◆ Figure 14.5). In some cases dissolved gases "drag" the oil along with them as they spew forth, as though from a bottle of champagne. If high pressures are not controlled, rocks, gas, oil, and even drill pipe may shoot into the air as a "gusher." High-pressure wells—749 in all—were set afire by Iraqi soldiers in Kuwait during Operation Desert Storm (◆ Figure 14.6). Fewer than a dozen wells were set ablaze in the 2003 battle to free Iraq. Professional oil-fire-fighting firms such as the well-known Red Adair Company and Boots and Coots were called in. With the help of Kuwaiti *roughnecks* (the nickname applied to drilling-rig workers), the fires were extinguished in nine months, three months earlier than the most optimistic estimate.

It is possible to drill a well so that the drill hole slants from the vertical to penetrate reservoir rocks far from the drilling site. This is desirable where the oil structure is offshore and must be "slant-drilled" from land or from a drilling platform. **Slant drilling** is also employed for tapping reservoirs beneath developed land, such as in Beverly Hills, where a large number of oil wells are slant-drilled under fashionable commercial areas. Modern methods use "smart" drill bits that are remotely controlled. Producers can thus withdraw oil from below a large area more economically and at the same time minimize the visual blight of drilling towers.

The ultimate drilling technology of the 1990s was horizontal drilling, a technique by which the drill bit and pipe follow gently inclined reservoir sands and limestones. By this means a single borehole can provide access to reservoirs beneath a much larger area than provided by vertical drilling or slant drilling. This method is now superseded by multilateral drilling, which enables operators to reach multiple oil-bearing reservoirs by means of lateral extensions from one

◆ FIGURE 14.6 The Kuwaiti night skies were illuminated by 749 blazing oil wells set afire by Iraqi troops during the 1991 Gulf War. The fires continued for nine months and burned off approximately 6 billion barrels of petroleum.

well (◆ Figure 14.7). It is worth noting that some offshore drilling platforms are in water more than a mile deep and have strings of drill pipe dangling below them 8.5 kilometers (5 mi), about the height of Mount Everest. There are plans to drill exploratory holes in water depths greater than 3,100 meters (10,000 ft). (See ■ Case Study 14.2).

Secondary Recovery

Secondary recovery methods extract oil that remains in the reservoir rock after normal withdrawal methods have ceased to be productive. As much as 75 percent of the total oil may remain. Secondary recovery methods can be grouped into three categories: thermal, chemical, and fluid-mixing (miscible) methods (◆ Figure 14.8). All of these methods require *injection* wells for injecting a fluid or gas and *extraction* wells for removing the remobilized oil.

Thermal methods include steam injection, which makes the adhering oil less viscous and thus more free to flow, and fire flooding, in which air is injected into the reservoir in order to set fire to the oil and thus produce gases and heat that will increase the flow of oil. Water injection is a chemical method that utilizes large-molecule compounds, which when added to water, thicken it and increase its ability to wash or sweep the adhering oil films and globules toward an extraction well. Fluids that mix with oil, that are **miscible,** are very effective in removing "stuck" oil from the reservoir. Miscible recovery methods use mixtures of water with propane or ethane extracted from natural gas or mixtures of CO_2 and water. Even after the use of these secondary recovery methods, as much as a fifth of the total oil may remain in the reservoir.

CONSIDER THIS

An acquaintance has offered you a share in a sure-shot wildcat oil-drilling venture in the Sierra Nevada. How should you respond to the offer?

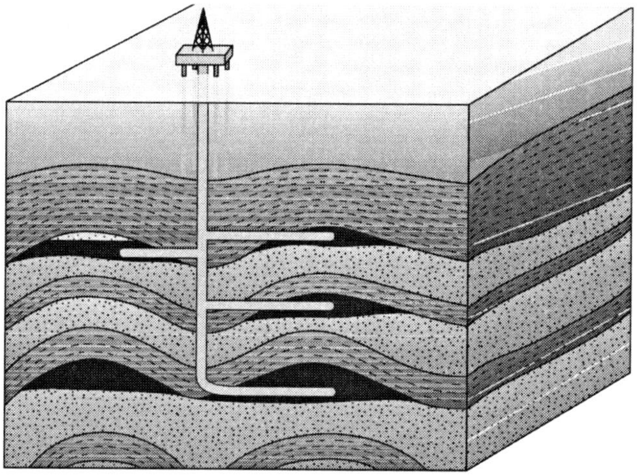

◆ FIGURE 14.7 Multilateral drilling from an offshore platform allows many oil-producing zones to be tapped from one platform. In this illustration, four zones are tapped by lateral horizontal pipes.

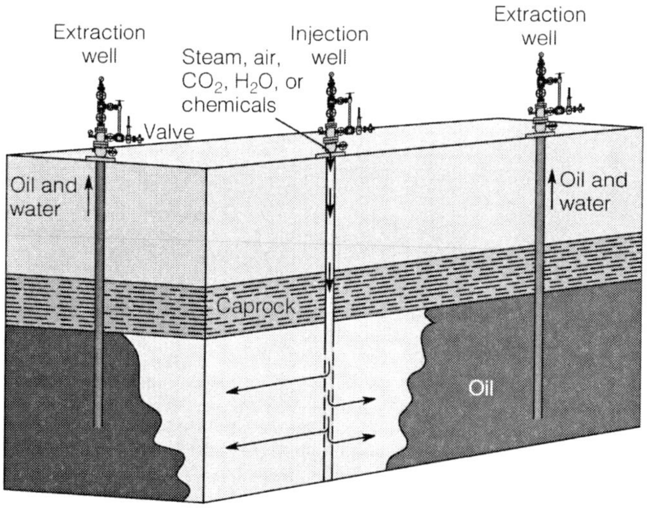

◆ FIGURE 14.8 Secondary recovery. Steam, air, carbon dioxide, or chemicals dissolved in water are injected into a sluggishly producing formation in order to stimulate the flow of oil to extraction wells.

*TABLE 14.2 Typical Composition of an A.P.I. Medium-Grade Crude Oil

Common Name	Number of Carbon Atoms*	Percentage of Crude-Oil Weight
Gasoline	5–10	27
Kerosene	11–13	13
Diesel fuel	14–18	12
Heavy gas oil	19–25	10
Lubricating oil	26–40	20
Heavy fractions (tars, asphalt)	>40	18
		100

* Volatility decreases as the number of carbon atoms increases, C_3 to C_{40}.

Quality and Price

The price of a barrel of crude (unrefined) oil varies with its grade—its quality—and with market demand at the time. Light oils bring higher prices, because they contain large proportions of paraffins and aromatic hydrocarbons, which are desirable for gasoline and bottled gases. Heavy crudes contain lower proportions of those components and greater proportions of the heavy, less valuable asphalts and tars (see Table 14.1). A scale of crude-oil quality has been established by the American Petroleum Institute based upon its weight. Light crude oil is very fluid and yields a high percentage of gasoline and diesel fuel. Heavy crude, on the other hand, is about the consistency of molasses. The percentages of fuels and lubricating oils yielded by a barrel of medium-weight crude oil are shown in Table 14.2.

Reserves and the Future for Oil

The time when we could count on cheap oil and even cheaper natural gas is clearly ending.

—Dave O'Reilly, Chairman, Chevron-Texaco, February, 2005

The United States has about 5 percent of the world's population, yet consumes 28 percent of the world's energy supplies. Daily, Americans use some 3,000 products, including such things as gasoline, lubricating oil, jet and other fuels, fertilizer, makeup, synthetic fabrics, and pharmaceuticals, which are derived from petroleum. Even with conservation efforts, our consumption of petroleum in 2005 was up 6 percent from 2000, averaging 21.11 million barrels a day out of some 80 million barrels a day used each year worldwide (Figures 14.1 and 14.10). (One barrel equals 42 gallons.) To put our oil consumption in a different context, of those 21.11 million barrels a day, about 12 million barrels are imported. Assuming a price of $50 per barrel, $600 million *each day* goes to other countries to pay for our thirst for oil. Thus, it's not surprising that the United States as a net importer must concern itself actively with global issues that affect the long-term oil-reserve outlook for energy supply and demand.

On a global scale, petroleum reserves in 2005 were estimated to be 1,200 billion barrels of oil (Figures 14.9 and 14.10; Table 14.3). (**Reserves** are defined as the amount of oil and gas that can be extracted at a profit. But to muddy the waters, there are several kinds of reserves: proved, probable, possible, developed, and undeveloped, and the use of these various terms is not universally consistent.) The 1,200 billion barrels may mean little to us until we realize that approximately two-thirds of the proved reserves are located in just six countries, and about 70 percent of reserves are in predominantly Islamic states (Figure 14.11; Table 14.3). Of these, Saudi Arabia has about one-third of the total, and Iran, Iraq, Kuwait, and the United Arab Emirates have about 100 billion barrels each of proved reserves. The world discovers less than one barrel of oil for every four consumed, and it's been running on deficit since 1981.

Let us look more closely at the significance of the reserve figures in Table 14.3. Most people overlook the margins of uncertainty concerning the actual size of proved reserves. For example, for apparent political reasons, some nations do not allow audits of their oil fields. This is especially true for nations that belonged to the former USSR and for the eleven

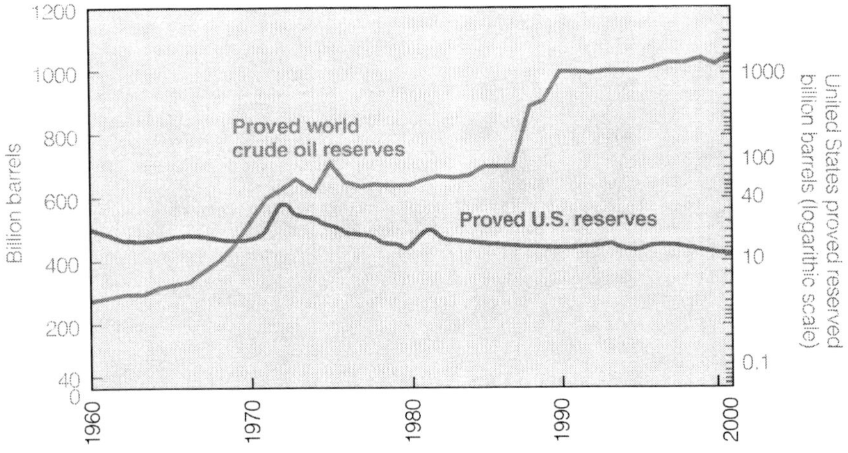

◆ FIGURE 14.9 World and U.S. proved petroleum reserves. Note the scale change for U.S. reserves and that proved world reserves increased from 1970 to 1990 and then leveled off from 1990 to 2000. In contrast, U. S. reserves have decreased overall since 1970.

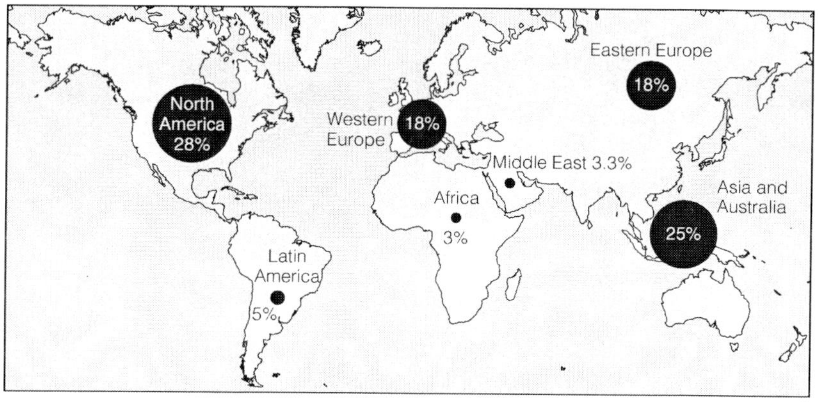

◆ FIGURE 14.10 World energy consumption.
Colorado Geological survey

★TABLE 14.3 Global proved reserves, 2002

Country	Proved Reserves (Billion Barrels)	Percentage of Total World
Saudi Arabia*♦	264	22.0
Iran*♦	137.5	11.5
Iraq*♦	115	9.6
United Arab Emirates*♦	97.8	8.1
Kuwait*♦	101.5	8.5
Venezuela♦	79.7	6.6
Russian Federation	74.4	6.2
Kazakhstan*	39.6	3.3
Libya*♦	39.1	3.3
Nigeria♦	35.9	3.0
United States	29.3	2.4
Canada	16.5	1.4
China	16.5	1.4
Qatar*♦	15.2	1.3
Mexico	13.7	4.1
Algeria*♦	12.2	1.0
Brazil	11.8	1.0
Norway	9.7	0.8
Angola	9.0	0.8
Oman*	5.6	0.5
India	5.9	0.5
Malaysia*	4.2	0.3
Indonesia*♦	4.3	0.4
United Kingdom	4.0	0.3
Yemen*	2.9	0.2
Rest of world	55.9	4.6
Total world	1,200.7	100.0
Total OPEC	838.9	70.0
Total for dominantly Islamic nations	577.5	48.0

* Dominantly Islamic nation
♦ OPEC nation
Source: British Petroleum Statistical Review of World Energy, 2006

members of OPEC. The quantity of oil exported from OPEC member countries is based on the portion of the country's remaining reserves, thus giving these countries an incentive to manipulate data. For example, in 1985 Kuwait increased the estimated reserves of their oil fields by 50 percent, and yet there were no new discoveries of consequence. This raised Kuwait's oil export quota. The other OPEC members quickly followed suit, and because reserves in these nations are state secrets, their official reserve reports are highly suspect. To quote Colin Campbell, an international petroleum-exploration geologist and former president of the Association for the Study of Peak Oil,

> It is noteworthy that Saudi Arabia reported 170 Gb [170,000 million barrels] in 1989, before the country had reason to inflate the number to 258 Gb in the following year to protect itself in the "OPEC quota wars." Engineers familiar with the [Saudi oil] fields think that a reasonable estimate of the oil-in-place is 600 Gb compared with the official estimate of 716 Gb. Applying a fairly generous recovery factor of 45 percent yields about 270 Gb recoverable, of which 104 Gb have been produced. It suggests that Saudi Arabia is reporting *original* and not *remaining* reserves [italics added for emphasis].

The reserve figure for Saudi Arabia shown on ★ Table 14.3 is the value used by British Petroleum in their 2006 annual summary of world reserves. All of the reserve estimates from the OPEC states are open to question.

The United States is the most energy-deficient of the largest oil-producing countries (Figure 14.11; ★ Tables 14.3 and 14.4). Overall U.S. production dropped 4.6 percent in 2005, the largest decline in 50 years. Production in 2005 from the nation's largest oil field at Prudhoe Bay dropped to about 75 percent of its peak in 1987. We imported approximately 60 percent of the 6.5 billion barrels of crude oil and natural gas condensates consumed in 2005. (Condensates are natural gas liquids that form when the gas depressurizes at Earth's surface.) At the current rate of production, the U.S. reserves are expected to last about 10 years. U.S. natural gas production is also declining at five percent per year despite frenetic drilling; and adding to the problem is that half of our nation's homes are heated with natural gas, and all electric power plants built after 1980 are designed to be fueled by gas.

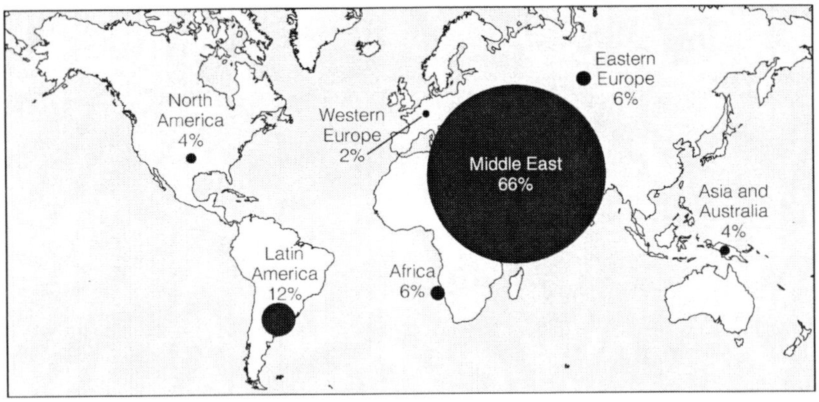

◆ FIGURE 14.11 World and U.S. proven petroleum reserves. Notice the scale change for U.S. reserves.
Colorado Geological survey

What about "all the oil" in the Arctic National Wildlife Refuge (ANWR)? In 2002 the U.S. Geological Survey estimated that there are between 5.7 and 16 billion barrels of probable and possible reserves oil in ANWR. The average of these estimates gives a median value of 10.4 barrels, a value that has been widely quoted in the media. The number, however, is a statistical fiction that represents the median value between a very large amount with only a five percent chance of being found and a more likely small amount with a 95 percent chance of being found. The very high (very low chance) estimate unduly pumps up the median, a figure that the uncritical public may not recognize. It should be noted that even if the refuge were opened tomorrow, it would be at least seven years before a single barrel reaches the lower 48 states. Even if 10.4 billion barrels exist, with our present thirst for oil amounting to 6 to 7 billion barrels a year, that amount of oil is equal to only about 1.5 year's supply. Some significant facts that should be kept in mind when evaluating annual per capita energy consumption are that each individual in the United States consumes

- 4 tons of oil
- 2.5 tons of coal
- 2+ tons of natural gas, and
- the individual's weight in oil—every 7 days.

Even though domestic and worldwide production rates will continue to decline, new reserves, such as the recent deep-water discoveries in the Gulf of Mexico (■ Case Study 14.2), may be discovered, and there will not be an abrupt end to the nation's production. Nevertheless, worldwide production of oil and gas liquids is on the declining limb of bell-shaped production curve (◆ Figure 14.12), and the gap between petroleum discovery and production continues to grow (◆ Figure 14.13). The future is not so bleak when we factor U.S. ingenuity into the energy equation. Some oil remains to be discovered using current exploration techniques, although a crude-oil price of at least $50/barrel may be required to make the expensive searches feasible. In addition, there are nonconventional sources of petroleum such as oil shale, tar sands, and coal from which liquid fuel can be derived; and coal is also an alternate source of fuel-grade natural gas. Complacency, however, is not warranted, because the availability of energy from these sources comes with a steep price tag for it takes considerable energy to process these alternate sources of oil.

TABLE 14.4 2002 U.S. Oil Production

State	Thousand Barrels/Day	Million Barrels/Yr	Percentage
Louisiana	1,513	552	25.7
Texas	1,400	511	23.8
Alaska	1,050	383	17.9
California	857	313	14.6
Oklahoma	193	70	3.3
New Mexico	176	64	3.0
Wyoming	167	61	2.8
North Dakota	90	33	1.5
Kansas	80	29	1.4
Other	355	130	6.0
Total	5,881	2,147	100.0
Offshore U.S. production			
La., Tex., Alaska, Calif.	4,820	1,759	82.0
Gulf of Mexico OCS	1,420	518	24.1

* Not including natural gas liquids
Source: Arizona Geological Survey, American Petroleum Institute (A.P.I.)

CONSIDER THIS

It has been forecast that the last drop of oil will not be pumped until some time beyond your life expectancy. In light of this, do you think we should go on consuming energy as usual, or should the world's peoples and their governments do something to delay the day when that last drop will be used?

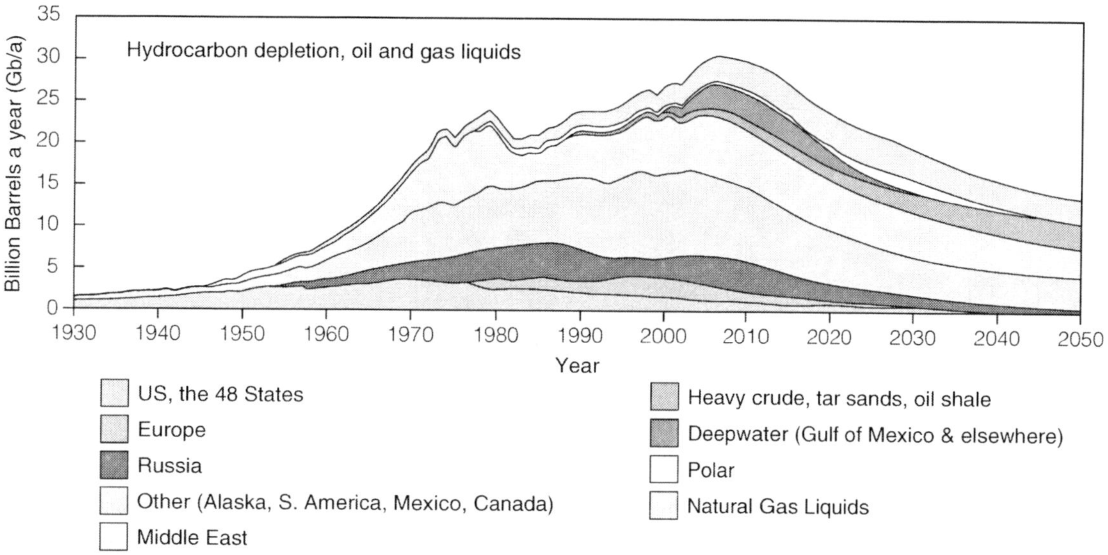

◆ FIGURE 14.12 Hydrocarbon depletion, oil and gas liquids. Association for the Study of Peak Oil (ASPO)

Energy Gases and the Future

Some experts predict that the decline in the supply of affordable oil will begin as early as 2010, some claim it has already happened, and others say the crunch will not start until 2040 or 2050. They refer to the time as *the end of cheap oil*. When it will happen, and the speculation about when the world will run completely out of oil, is not relevant. What does matter is when production will fall off as demand inexorably continues to rise. In other words, price depends on the well-known supply-and-demand curves of the economists. ◆ Figure 14.14 reflects this. It shows the predicted decline in the use of oil in the twenty-first century and the increasing use of natural gases and alternative energy sources. Natural gas is currently reasonably plentiful, and it burns cleaner than coal, gasoline, or heating oil. It currently accounts for 75 percent of the total U.S. energy used in space heating and cooking, and it is in wide use as fuel for electric power plants. Natural gas is measured as cubic feet of gas at an atmospheric pressure of 14.7 pounds per square inch and a temperature of 60°F. A trillion cubic feet (Tcf) of gas would fill a cube 2 miles on a side (8 cubic miles), and the United States consumes about 17 Tcf per year. We have proven reserves of 177 Tcf, which suggests that we will "run out" in 10.4 years. However, it is estimated that about 2,400 Tcf of natural gas exists in the United States (see Fisher in the For Further Information section). Many new discoveries will be made, with 40 percent of those at depths greater than 4,500 m (14,500 ft, almost 3 mi). About 300 Tcf will come from coal-bed methane and 500 Tcf will come from the growth of existing fields. Fisher calls the approaching transition period from oil to gas the *methane economy*.

By the 1990s **coal-bed methane** was recognized as a world-class commodity by several states, and there was a 95 percent

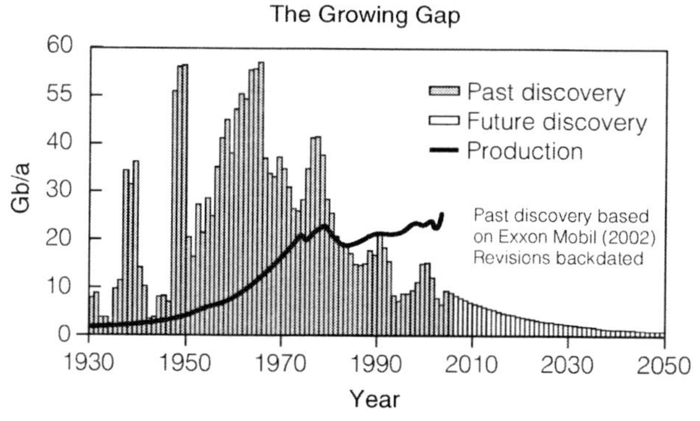

◆ FIGURE 14.13 The growing gap between petroleum discovery and production. Collin Campbell, ASPO

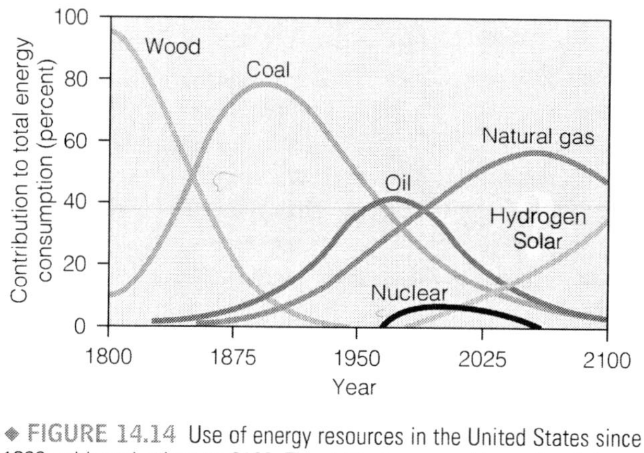

◆ FIGURE 14.14 Use of energy resources in the United States since 1800, with projections to 2100. The shifts from wood to coal and then from coal to oil and natural gas each took about 50 years. Many analysts believe a new shift, to increased use of solar energy and hydrogen gas, will occur over the next 50 years. Note the natural gas peak at about 2050 and the oil peak at about 1970. Department of Energy

Todd Wilkinson

◆ FIGURE 14.15 Drilling for coal-bed methane where the deer and the antelope play, near Pinedale, Wyoming.

probability that 30 Tcf could be found in the coal regions of Wyoming, Colorado, Utah, Montana, and New Mexico. Coal-bed methane is easy to extract by a drilled well (◆ Figure 14.15) as opposed to extracting coal, which involves destructive surface mining. But the gas production, as with all the nonconventional fuels, comes at a price, because it requires disruptive wells, a network of roads, disposal of salty water, pipelines, and methane leakage, all of which degrade the environment. An example of the problems involved is the Powder River Basin, a large drainage basin straddling the Montana-Wyoming border, where coal-bed-methane producers pump salty groundwater out of the coal seams to release the gas. The water is discharged into streams, ponds, and dry washes where the salt becomes a threat to irrigated crops and aquatic ecosystems. Ranchers worry that pumping will deplete the groundwater aquifers upon which they depend. In 2006 most of the 24,200 wells in the basin were on the Wyoming side of the border, but energy companies began moving into Montana. Montana government officials recognize the potential economic benefits to their state from the extraction of coal-bed methane, but they view Wyoming's development as uncontrolled and destructive. Thus, Montana will require producers to re-inject the water into the ground or remove the salt and other pollutants before dumping it into surface water bodies in accordance with the EPA's enforcement of the Clean Water Act. Re-injection and water treatment both require an expenditure of energy, which would cut into the profits of the developers, reduce the net amount of energy produced, and reduce the tax revenue to the state. Montana's tough water-quality standards would require rivers draining north from Wyoming into Montana to meet these standards. Wyoming politicians have accused Montana of "targeting" their state's thriving coal-bed-methane industry. In addition to the two state governments as protagonists, two federal government agencies are also at odds. The EPA wants to enforce the Clean Water Act, but the Department of Energy (DOE) has focused on the potential impact on the national energy supply, stating that re-injection would cut the gas production significantly, resulting in losses of up to $21 billion worth of gas. Montana and the EPA are facing an uphill fight with Wyoming and the DOE; it may be that in this case there is no magic bullet to satisfy both environmental requirements and energy needs.

As we consider methane, don't forget methane hydrates as a potential fuel source (see the chapter opener).

Hydrogen is another gas that has been touted as a potential fuel. Many experts consider it will be the energy of the future as it burns without pollutants, emitting only water vapor. Unfortunately, there are no large reservoirs of hydrogen, and it is not a primary fuel source; it must be manufactured using fossil fuel, which then shifts the burden of pollution to some coal- or gas-fired power plant. It may offer promise as a fuel for airplanes and autos, and it might become economically viable if it can be produced in large quantities using solar, wind, or hydroelectric energy.

Coal

Coal is the carbonaceous residue of plant matter that has been preserved and altered by heat and pressure. Next to oil and oil shales, coal is Earth's most abundant reserve of stored energy. Deposits are known from every geologic period since Devonian time and the appearance of widespread terrestrial plant life some 390 million years ago. Permian coal is found in Antarctica, Australia, and India—pre-continental drift

Gondwanaland (Chapter 3). The large fields of North America, England, and Europe were deposited during the Carboniferous Period, so named because of the extensive coal deposits in rocks of that age. In the United States the Carboniferous Period is divided into the Mississippian and Pennsylvanian Periods, named after the states where the deposits are found. Tertiary coals are found in such diverse locations as Spitsbergen Island in the Arctic Ocean, the western United States, Japan, India, Germany, and Russia.

Coalification and Rank

The first stage in the process of *coalification* is the accumulation of large amounts of plant debris under conditions that will preserve it. This requires high plant production in a low-oxygen depositional environment, the conditions usually found in nonmarine brackish-water swamps. The accumulated plant matter must then be buried to a depth sufficient for the heat and pressure to expel water and volatile matter. The degree of metamorphism (conversion) of plant material to coal is denoted by its **rank.** From lowest to highest rank, the metamorphism of coal follows the sequence *peat,* to *lignite,* to *subbituminous,* to *bituminous,* to *anthracite.* The sequence is accompanied by increasing amounts of fixed carbon and heat (Btu) content, and a decreasing amount of quickly burned volatile material. Once ignited, it is the carbon that burns (oxidizes) and gives off heat, just as wood charcoal does in a barbecue pit.

Bituminous, or soft, coal usually occurs in flat-lying beds at shallow depths that are amenable to surface-mining techniques. Anthracite, the highest rank of coal, is formed when coal-bearing rocks are subjected to intense heat and pressure—a situation that sometimes occurs in areas of plate convergence. In the Appalachian coal basin, high-volatile bituminous coals occur in the western part of the basin and increase in rank to low-volatile bituminous coals to the east. Anthracite formed close to the ancient plate-collision zone of North America and Africa (pre-Pangaea), where the coals were more intensely folded and subject to higher levels of heat. Traveling east from the Appalachian Plateau to the folded Ridge and Valley Province, one goes from flat-lying bituminous terrain to folded anthracite terrain. West Virginia, Kentucky, and Pennsylvania are the leading coal producers of the eastern states. The interior coal basins of the Midwest are characterized by high-sulfur bituminous coal, whereas the younger Western deposits are low-sulfur lignite and subbituminous coal. ⁎ Table 14.5 projects the development of coal-mining activity in the United States. The expansion of mining activity will occur mainly in the Western coal basins, because the coal seams there are very thick and easily mined and the coals are low in sulfur. Wyoming, with its low-sulfur coal, was the leading coal producer in 2005 (587 million short tons), followed in order by West Virginia, Kentucky, Pennsylvania, and Illinois (◆ Figure 14.16).

⁎TABLE 14.5 Active U.S. Coal Mines by Region, 1977 and 2000

	Surface Mines		Underground Mines		Total	
Region	1977	2000	1977	2000	1977	2000
Appalachia*	185	130–175	205	380–500	390	510–675
Midwest	91	95–135	54	120–180	145	215–315
West	141	700–1,005	13	80–110	154	780–1,115

* Encompasses parts of Pennsylvania, eastern Kentucky, West Virginia, Virginia, Maryland, Tennessee, and Ohio.
Source: U.S. Geological Survey

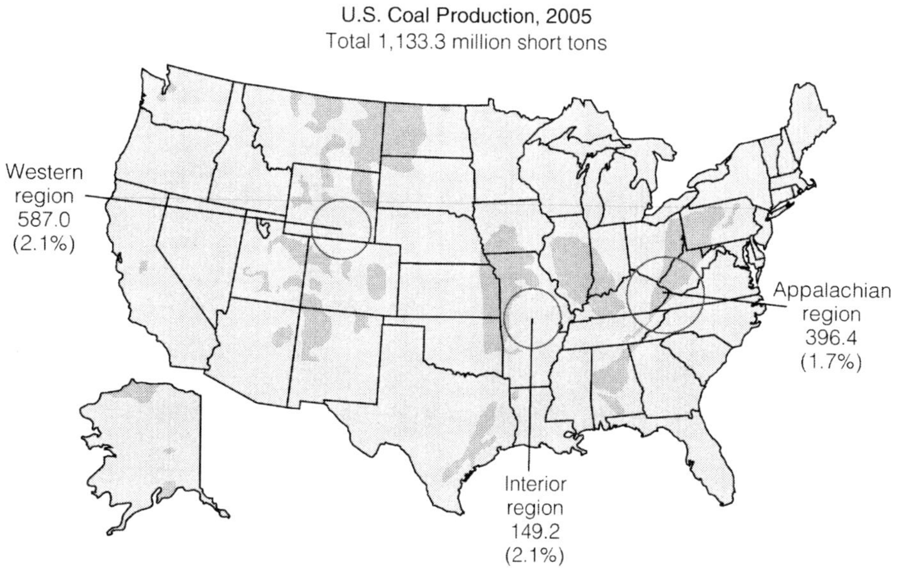

◆ FIGURE 14.16 Production in million short tons (and percent change from 2004) for each of the coal-producing regions and the total production for the United States. Note the shift in production from the Appalachian region to the Western coal states. Energy Information Administration.

Reserves and Production

The future of coal-derived energy in the United States is far more optimistic than the outlook for petroleum. Coal constitutes 80 percent of U.S. energy stores but only 18 percent of present usage. In 1989 electric utilities accounted for 86 percent of the coal consumed, with residential and industrial use accounting for the remainder. It is estimated that there are 84 trillion (84×10^{12}) recoverable tons of coal in the world, the equivalent of 34 trillion barrels of oil. The United States has 283 billion tons of recoverable reserves, which could last 200–300 years at the current rates of production and use. These reserves could meet only 50 years of total U.S. energy demand, however.

It has been estimated that energy demand in the early twenty-first century will be between 100 and 150 quads (a quad is a quadrillion British thermal units, 10^{15} Btu), compared with 73 quads in 1973 and about 80 quads in 1980 (Table 14.6). Coal production will double to meet the demand, and this increase will be mostly in the Western coal basins, where the relatively inexpensive surface-mining techniques can be used. Underground mining costs more, requires a greater capital investment, and takes more time to get into production. Also, miners who work underground face greater risks than those who work at the surface.

The various methods of mining coal are illustrated in ◆ Figure 14.17. A key element for the economic future of coal may lie in its use as a substitute fuel for oil and gas. The technology is in place to make methane from coal—the product, called *synthesis gas*, has been used on a large scale to manufacture petrochemicals, including methanol (methyl alcohol). Direct liquefaction of coal also shows promise. Using coal- or petroleum-based solvents, coal is dissolved and hydrogenated, resulting in about 75 percent gasoline and the remainder propane and butane. About 5.5 barrels of liquid are derived from a ton of coal at a cost less than the most expensive oil during the Iraq War, \$55–\$60 a barrel. A major facet of our energy future lies in converting coal to liquid or gaseous **synthetic fuels**, or *synfuels*. This will be discussed further later in the chapter.

★TABLE 14.6 Energy (Power) Units

Unit	Explanation
British thermal unit (Btu)	The amount of heat required to raise the temperature of 1 pound of water 1° F; ≈ energy released by a burning match
Quad (quadrillion)	10^{15} Btu = 172 million bbl/oil
Barrel of oil	5.8 million Btu = 42 gallons
Bituminous coal (average)	25 million Btu/ton
Natural gas	Variable Btu content, measured in cubic feet (cf)
Megawatt (MW)	1,000 kilowatts (kw), or a million watts
Gigawatt (GW)	1,000 megawatts, or a billion watts

CONSIDER THIS

Your Christmas stocking contains nothing but a lump of anthracite coal. Being an optimist, you rejoice in this gift. Why?

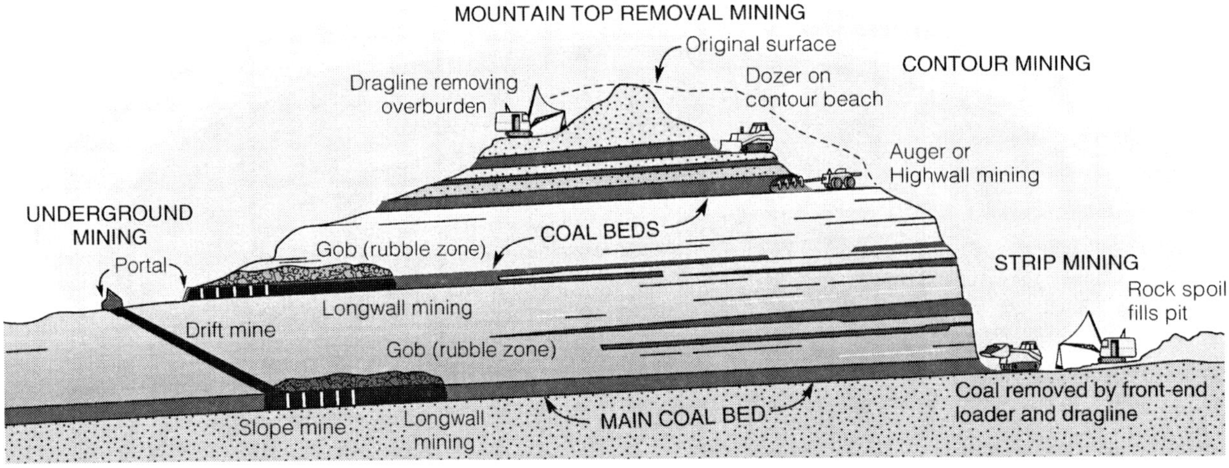

◆ FIGURE 14.17 Schematic cross section illustrating conventional underground and surface coal-mining technology. Adapted from Colorado Geological Survey

Nonconventional Fossil Fuels

Tar Sands

Tar sands contain oil that is too thick to flow at normal temperatures. These deposits, found in Canada, Venezuela, Madagascar, and the United States, are surface mined like coal or drilled using steam to mobilize and recover the oil. Canada's Athabasca sands (◆ Figure 14.18) constitute the largest oil field in the world. Canada claims that more than 1.7 trillion barrels of recoverable oil will come from the deposits.

Tar sands that are too deep (500 m) for surface exploitation underlie Alberta's Cold Lake area. Steam at 300° C (575° F) is injected at high pressure and allowed to "soak" for a month. The oil released from the sand is then brought to the land surface by familiar "rocking-horse" pumps. More than 2,000 steam-injection and oil-production wells deliver 130,000 barrels of oil per day. The area is estimated to contain 220 billion barrels of recoverable oil.

At present only the easily exploited deposits are being worked, and the supply of natural gas necessary to refine the oil is declining precipitously. Apparently, the tar-sand operations are facing shortages of equipment, steel, and workers, numbering 20,000 in 2006 but expected to double by 2008. Operations costs are soaring, with the French company TOTAL delaying its production start-up from 2010 to 2013 and Shell Oil announcing its planned expansion could cost nearly $13 billion, double the original estimate. Clearly, substitutes for conventional oil do not come cheap; nevertheless, production is expected to grow, especially when the world price of crude oil stands at $50 or greater per barrel. Production is expected to reach 30 days of current U.S. consumption by 2030. Currently Canada's tar sands produce less than five percent of what is consumed in the United States, and it is unlikely that it will ever be able to satisfy all of the U.S. demand.

Oil Shales

Oil shales are sedimentary rocks that yield petroleum when they are heated. Found on all continents, including Antarctica, they were originally deposited in lakes, marshes, or the ocean. The original rock oil used in kerosene lamps was produced from black oil shales. The most extensive U.S. deposits, the Green River shales, were formed during Eocene time in huge freshwater lakes in the present-day states of Wyoming, Colorado, and Utah. They are not oily like tar sands, but when heated to 500° C (900° F) they yield oil (◆ Figure 14.19). The source of oil in the rock is **kerogen,** a solid bituminous substance; a ton of oil shale may yield 10–150 gallons of good-quality oil. Potential deposits of economic interest yield 25–50 gallons per ton, and the U.S. Geological Survey includes 3 trillion barrels of oil from shale in their estimates of reserves. Kerogen-rich lake deposits are also found in China, Yugoslavia, and Brazil.

When oil and gas prices were low, even the most accessible oil shale was not economical to mine. Richer deposits are buried under hundreds of feet of barren rock, which would require even more energy to mine. As the end of cheap conventional oil approaches, it is questionable whether exploiting oil shale will ever become economical. Water, too, is a

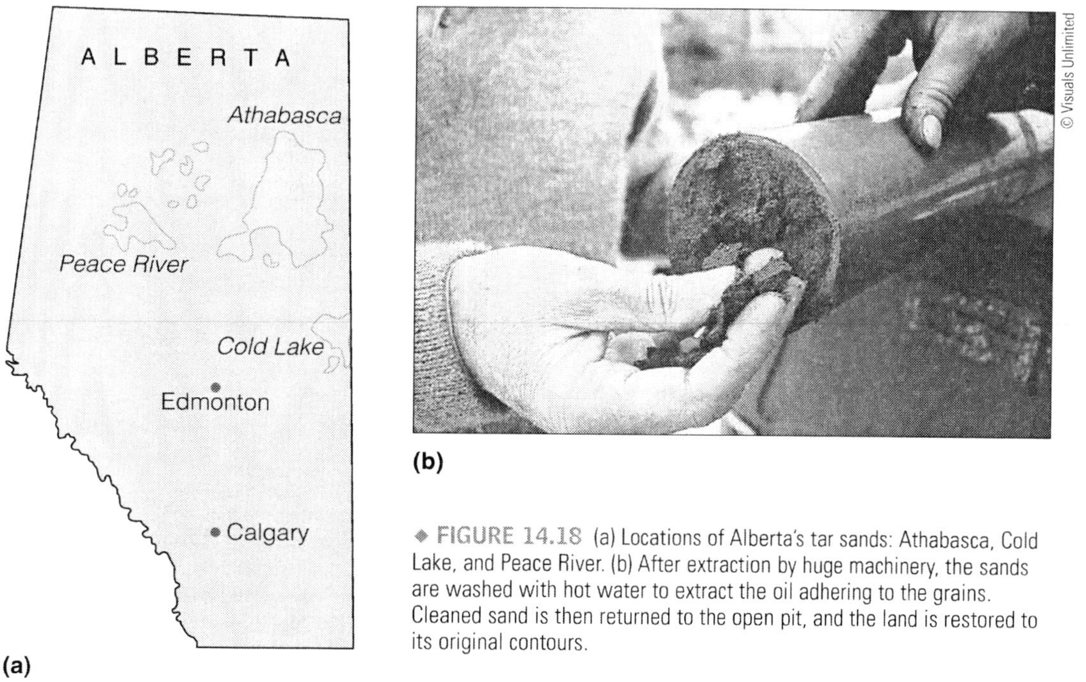

◆ FIGURE 14.18 (a) Locations of Alberta's tar sands: Athabasca, Cold Lake, and Peace River. (b) After extraction by huge machinery, the sands are washed with hot water to extract the oil adhering to the grains. Cleaned sand is then returned to the open pit, and the land is restored to its original contours.

DOE

◆ FIGURE 14.19 Oil shale from the Green River Formation in Wyoming and a beaker of the heavy oil that can be extracted from it.

limiting factor, because processing the shale requires water volumes 3 to 4 times greater than that of the oil extracted. Furthermore, the extraction process destroys the natural landscape, and some of the richest shales are in scenic wilderness areas. At this time the oil-shale industry cannot compete economically in the energy market.

Problems of Fossil Fuel Combustion

Air Pollution

Obtaining energy by burning fossil fuels creates environmental problems of global proportions. It produces oxides of carbon, sulfur, and nitrogen and fine-particulate ash. Carbon monoxide (CO), an oxide produced by combustion of all fossil and plant fuels, is converted to carbon dioxide (CO_2), which contributes to global warming. Burning coal also releases sulfur oxides (SO_x) to the atmosphere, where they form environmentally deleterious compounds. Nitrogen oxides (NO_x), mostly NO and NO_2, are products of combustion in auto engines and are the precursors of the photochemical oxidants, ozone, and peroxyacetyl nitrate (PAN), which we associate with smog. Water and oxygen in the atmosphere combine with SO_2 and NO_2 to form sulfuric acid (H_2SO_4) and nitric acid (HNO_3), the main components of **acid rain**—rain with increased acidity due to environmental factors such as atmospheric pollutants (■ Case Study 14.3 on page 440).

The Clean Air Act of 1963 as amended in 1970 and 1990 specifies standards for pollutant oxides and hydrocarbon emissions. By 1980 new cars were 90 percent cleaner than their 1970 counterparts. The Clean Air Act Amendments of 1990 (CAAA) set goals and timetables that are affecting how and what we drive. Beginning in 1992 fuel suppliers were required to sell only reformulated gasoline in 39 areas where winter air quality was a problem. Reformulations using methyl tertiary butyl ether (MTBE) add as much as 2.5 percent oxygen to the fuel, causing it to burn cleaner and create less ozone, at a cost of about 10 cents more per gallon. Unfortunately, reformulated gasoline (RFG) is stored in underground tanks that sometimes leak and pollute groundwater. As a result, a Blue Ribbon Panel appointed by EPA administrator Barbara Browner has recommended the following:

- Remove the Clean Air Act requirement for 2 percent oxygen in RFG.
- Enhance underground storage tanks, thus improving protection of the nation's drinking water.
- Reduce the use of MTBE nationwide, maintaining current air quality benefits.

So much for MTBE and reformulated fuels.

CONSIDER THIS

Historically, Pennsylvania crude oil (such brands as Quaker State and Pennzoil) has been known for its superior lubricating properties compared with, say, heavier California or Texas crude oils. Why do you suppose this is so? (Table 14.1 can help you deduce a reasonable answer.)

Sulfur Emissions and Acid Rain

Sulfur occurs in coal as tiny particles of iron sulfide, most of which is the mineral pyrite, or "fool's gold" (◆ Figure 14.20). Coal also contains organic sulfur originally contained in the

Hobart King, Mansfield Univ.

◆ FIGURE 14.20 Small pyrite masses in a bituminous coal matrix (gray). The light-gray material is the cell walls of the original plant material, and the darker gray is the solid material that was contained within the cell cavities. Earth pressures compacted the plant material, squeezed out any water, and collapsed the cavities. That pyrite formed within the coal before completion of compaction is evidenced by plant materials bent around solid pyrite masses. The pyrite "blebs" are about 50 microns across.

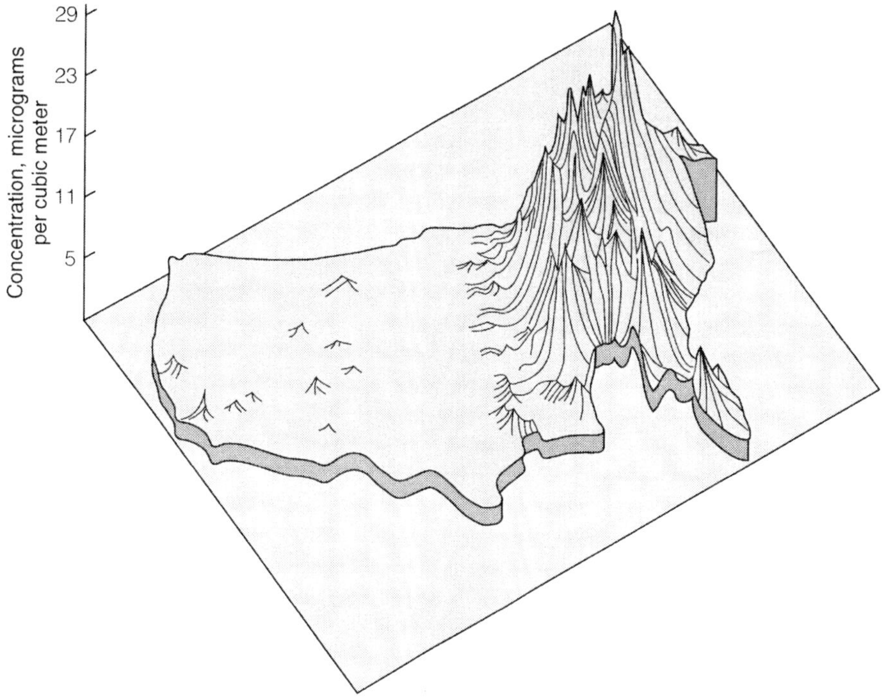

◆ FIGURE 14.21 Atmospheric concentration of sulfate for a 26-quad U.S. coal-use scenario in 1990. The point is that the heaviest concentrations are in the Northeast, where most coal is burned. After A. C. Stern, ed., *Air Pollution*, Vol. 1, San Diego, Academic Press.

coal-forming vegetation. Upon combustion, mostly in coal-burning electrical generating plants, the sulfide is oxidized to sulfur dioxide (SO_2) and carried out through smokestacks into the environment. Emissions of SO_2 are known to have a negative impact on human health, particularly lung problems. Although most coal-burning in the United States occurs in the Midwest and the East, significant volumes of SO_x can travel 1,000 kilometers (600 mi) or more from their source. ◆ Figure 14.21 shows typical U.S. SO_2 emission densities, which reflect the pattern of coal usage very closely. Mobile sources using petroleum fuels emit less than 4 percent of the total.

Sulfur dioxide emitted from a smokestack may be deposited dry near the facility, where it will damage plant life and pose a health threat to animals. If the gas plume from the smokestack is transported away from the facility, its SO_2 will react with the atmosphere and eventually fall to Earth as acid rain. Acid rain can damage crops; acidify soils; corrode rocks, buildings, and monuments; and contaminate streams, lakes, and drinking water. Meanwhile, it is estimated that annual U.S. emissions of sulfur gases will increase almost 30 percent from 1990 levels by 2010.

Sulfur oxides can be removed from combustion gases in smokestacks by using a device called a *scrubber*, which selectively reacts with SO_2 and absorbs or neutralizes it. Many coal-burning facilities are switching to a process known as fluidized-bed combustion. Coal and pulverized limestone are burned together, and the heat of combustion converts the limestone into calcium oxide (CaO) and CO_2. CO_2 is emitted, and the CaO combines with sulfur and oxygen to form calcium sulfate ($CaSO_4$, a gypsumlike substance) as a solid waste product. Sulfur emissions are reduced, but CO_2 emissions increase. It is thus a tradeoff between the health benefits of reduced atmospheric sulfur and the environmental effects of increased contributions to global warming.

Nitrogen Oxides and Smog

Nitrogen oxide emissions (NO_x) are significant from both stationary (coal-burning) and mobile (vehicular) sources. Mobile sources are the most difficult to control, and their contribution to total NO_x is increasing as the number of automobiles increases. NO_x forms when engine combustion causes nitrogen in air to oxidize to NO and NO_2, which then react with oxygen, free radicals (incomplete molecules that are highly reactive), and unburned hydrocarbons to produce **photochemical smog.** Photochemical smog requires sunlight for its formation. Although the word smog is a contraction of the words *smoke* and *fog*, the air pollution in Los Angeles, Mexico City, and Denver has little direct relationship to either smoke or fog.

Both NO_2 and ozone (O_3) are reactive oxidants and cause respiratory problems. A level of only 6 parts per million (ppm) of ozone can kill laboratory animals by pulmonary edema (water in the lungs) and hemorrhage within four hours. In humans, alcohol consumption, exercise, and high temperatures have been found to increase adverse reaction to ozone, whereas vitamin C and previous exposure seem to lessen the reaction. One investigator noted that the person most likely to succumb to ozone would be "an alcoholic with a 'snootful' arriving at LAX for the first time on a hot, smoggy day and jogging all the way to Beverly Hills." Such a scenario may not be all that unlikely in southern California.

Energy-efficient cars and new fuels have gone a long way toward decreasing NO_x emissions. Exhaust-system catalytic conversion of nitrogen oxides, carbon monoxide, and hydrocarbons into carbon dioxide, nitrogen, and water are mandatory automotive equipment on new cars in most states.

Domestic Coal Burning

China is the world's largest producer of coal, accounting for 25 percent of global output. It currently produces 10 percent of the global emissions of CO_2, and it is estimated that by 2025 it will emit more CO_2 than the United States, Canada, and Japan combined. Coal provides 76 percent of the huge country's commercial energy, and it is the only fuel millions of people use.

In a typical home coal is burned in unvented stoves for cooking, space heating, and heating water. In many areas, foods are brought indoors in the fall for drying with coal fires, as the climate is too cool and damp to dry them outdoors. At least 3,000 people in Guizhou Province in southwest China suffer from severe arsenic poisoning caused by consuming chili peppers dried over high-arsenic coal fires (◆Figure 14.22). Although fresh chili peppers have less than 1 ppm arsenic, chili peppers dried over high-arsenic coal have up to 500 ppm arsenic. More than ten million people in the same province suffer dental and skeletal fluorosis due to eating corn that has been cooked over coal containing large amounts of fluorine. About 3.5 billion people worldwide are exposed to toxic fumes indoors caused by coal burning. Each year, an estimated 178,000 people die prematurely due to indoor air pollution in China alone. To combat these health hazards, Chinese and U.S. Geological Survey geologists are working together to identify coal deposits with high toxic-element concentrations that make them unsuitable for domestic use.

Robert Finkelman, USGS

◆ FIGURE 14.22 Drying chili peppers over a coal fire that emits arsenic fumes increases the arsenic content of the chilies up to 500 times, and eating them causes arsenic poisoning. Also, as families gather about coal fires at night, they are exposed to polycyclic aromatic hydrocarbons from incomplete combustion of the coal. Breathing these substances can lead to lung and esophageal cancer.

Mine Collapse

Whenever coal is extracted from shallow underground seams, there is a risk of collapse of the overburden into the mine opening. Lignite and bituminous (soft) coals are most susceptible to collapse because they are exploited from near-horizontal beds at relatively shallow depths. The "room and pillar" method recovers about 50 percent of the coal, leaving the remainder as pillars for supporting the "roof" of the mine. As a mine is being abandoned, the pillars are sometimes removed or reduced in size, increasing the yield from the mine but also increasing the possibility of subsidence at some time in the future. Even when the pillars are left, failure is possible. The weight of the overlying strata can cause the pillars to rupture, and sometimes it physically drives the pillars into the underlying shale beds. Subsidence has been documented in Pittsburgh, Scranton, Wilkes-Barre, and many other areas in Pennsylvania (see Figure 13.24). In 1982 an entire parking lot disappeared into a hole above a mined-out coal seam 88 meters (290 ft) below the surface. Wyoming, Montana, South Dakota, and other western "lignite" states are pockmarked with subsidence features, but human activities and human lives are less likely to be affected by subsidence, or collapse, in those areas.

Energy for the Future

Our oil-based economy is "mature," and certainly the energy scenario of the twenty-first century will be different and not carbon-based as it is now. The primary energy source will be, as always, the Sun. The primary fuel may very well be hydrogen, the most common element in the universe and one that burns readily. This section examines the alternative energy sources that are available to us at this time.

Beyond Petroleum—Alternatives for the Future

The race is on to find a suitable alternative for gasoline for use in automobiles. The contenders are **ethanol, methanol, biodiesel,** electricity, hydrogen, **liquefied natural gas,** and **compressed natural gas.**

Leading the **biofuel** pack is ethanol (ethyl alcohol) derived from corn. Biofuels are any fuels derived from **biomass** (straw; agricultural waste; sugar cane; food leftovers; waste from the wood, paper and forestry industrial sectors; corn; switchgrass; any organic waste; cow manure; etc.). Ethanol made from sugar cane has been used in Brazil for many years as an auto fuel, and at one time almost 90 percent of the cars there burned it.

As with petroleum and coal, biomass is a form of stored solar energy that has been captured by growing plants by the process of photosynthesis. Ethanol derived from corn has received the most attention, the costs being closely competitive to gasoline. A big plus is that as a high percentage blend in automobile fuel, it releases less CO_2 than gasoline, but only in high percentage blends; when used as a low percent additive in a blend with gasoline it worsens CO_2 emissions. Ethanol fuel cuts ozone-forming gases, and it is a safe substitute for the gasoline additive MTBE. A drawback is that ethanol has only about 67 percent of the energy content of an equal volume of gasoline. One reason it is competitive is that the corn-ethanol industry is subsidized by the federal government as a result of policies promoted by the politically powerful corn-growing states. Corn ethanol has three major problems:

- It takes considerable energy to make corn ethanol, and that energy from coal-fired ethanol plants would completely undermine the reduced greenhouse gas emission of ethanol. Furthermore, it requires energy to plant, fertilize, harvest, and transport the corn, all of which produces considerable greenhouse gases that outweigh the advantages.
- The supply problem. If the entire U.S. corn production were devoted to producing E85 (a blend of 85 percent ethanol and 15 percent unleaded gasoline), it would meet only about 12 percent of the nation's current demands. Consequently, for ethanol to win the race, it will need feedstock other than corn.
- It is corrosive, thus not suitable for sending through pipelines. Instead it must be shipped in tank cars constructed of special materials.

The complex question of energy consumed in converting corn to ethanol compared with the energy content of the ethanol has created controversy with estimates ranging widely from a net loss of energy of 20 to 29 percent to a net gain of 67 percent. The range of estimates is due to differences in energy input values, changes in ethanol production processes, and accreditation of the energy value of production by-products (which are used mainly as animal feed). In 2006 the United States had 100 ethanol plants in operation. An additional 200 more are under construction or in the planning stage, which may use coal for power because coal is much less expensive than their current fuel, natural gas. As mentioned above, using coal would offset the greenhouse gains of ethanol.

Methanol (methyl alcohol), running a close second, uses natural gas (methane) as a feedstock, but the United States has vast coal reserves that can be used as feedstock. One company, Norcal Waste Systems in San Francisco, California, is even developing a process for making methane feedstock from dog droppings. (For many years, San Francisco has been a major producer of the stuff, wastefully compelling visitors to the city to carry it away stuck to their shoes.) West Virginia and Montana are developing systems for making coal-derived methanol. Pennsylvania already has a system that turns coal waste into diesel fuel. Methanol can be used directly as fuel; Indy race cars and model airplanes have used it for years. Methanol is an interesting fuel: its combustion generates little nitric oxide (NO) and ozone and it has a high octane number—about 110, which has made it attractive for race cars and high-compression engines. Although it has lower heat (Btu) content than gasoline, it has only about 50 percent of the energy content of an equal volume of gasoline in contrast to ethanol's 67 percent, and it burns more efficiently, which may even things out. On the negative side, methanol releases greater emissions of CO_2. Methanol can also be used in the process of making biodiesel, although biodiesel is currently manufactured from soybeans and other feedstocks, with its use amounting to 75 million gallons in 2005. A challenge facing all the biofuel contenders in the race is the ability to cost-effectively store, collect, and transport biomass feedstocks, and the development of a sustainable and safe supply system.

Direct Solar Energy

Although the amount of solar energy that reaches Earth far exceeds all human energy needs, it is very diffuse. For example, the amount that strikes the atmosphere above the British Isles is 80 times Great Britain's energy needs; it averages about 1 kilowatt per square meter. Cloud cover reduces the amount by 80 percent, but there is still more than enough. So why aren't there solar-electrical plants in every city and hamlet? The reason is cost. The current technology requires a great deal of space, which makes installations expensive. There are essentially two ways to put the sun to work for us:

1. Solar-thermal methods utilize panels or "collectors" to warm a fluid, which is then used for heating or for generating electricity.
2. Photovoltaic cells use the second most common element on Earth, silicon, to convert light directly into electricity.

◆TABLE 14.7 Cost of Electricity by Energy Source

Energy Source	Cost, cents/kwh
Sun	15.8*
Geothermal	13.3*
Biomass	13.0*
Wind	7.5**
Oil	4.4†
Gas	2.9†
Coal	1.2†
Nuclear	0.93†

* The rates that Southern California Edison Company was paying for purchased and manufactured energy in December 1995.
** 1998 value.
† The rates that Southern California Edison Company was paying for purchased and manufactured energy in December 1991 (more recent data not available).

In 1995 the electricity produced from photovoltaic cells cost about 16 cents per kilowatt hour (kwh), many times that produced by fossil fuels (◆ Table 14.7). However, it is estimated that photovoltaic electricity will soon cost less than 10 cents per kilowatt hour and will compete with conventional electrical generation in this century.

Solar-thermal facilities called *power towers* illustrate a creative use of solar energy. The "Solar One" station in the Mojave Desert near Barstow, California, generated electricity between 1982 and 1988 for the Southern California Edison Company. It employed 1,818 suntracking mirrors focused on a water-filled receiver atop a tower where temperatures reached 900° C (◆ Figure 14.23). The super-heated water was transferred to a plant where it flashed to steam and drove turbines that operated an electrical generator. Because of the success of Solar One, a 10-megawatt facility was built on the same site and began operating in mid-1996. Solar Two uses molten nitrate salt as the heat-transfer medium. Molten salt has excellent heat-storing properties; the heat can be used immediately to generate steam, or it can be stored and used later, during cloudy periods or after sundown. Solar Two generated enough power for 10,000 homes and operated from 1996 to 2000, when government funding ended. San Francisco has installed a $5.2 million solar energy project at the Moscone Convention Center. It was dedicated in late 2002, and it is estimated that it will cut energy costs as much as 38 percent and, after debt service, save the taxpayers $200,000.

Photovoltaic cells (PV cells) absorb pulses of light energy on semiconductor materials and turn that energy into an electric current. They are used widely as a power source for satellites, calculators, watches, and remote communication and instrumentation systems. In recent years the cost of PV cells has been reduced by 80 percent, and it is now possible to apply thin films of these materials to shingles, tiles, and window glass, enabling buildings to generate their own electricity. Costs need to fall another 50–75 percent to make them competitive with fossil-fuel generating plants. Several U.S. utility companies have installed PV systems to supply small users in out-of-the-way places without building costly power-line extensions.

Indirect Solar Energy

Wind Energy

Wind power is the biggest success story in renewable energy. It is the fastest growing energy segment in the world, and it has been for the past 25 years. During the same time interval, the cost of wind power has dropped approximately 90 percent, but the present production pales in comparison to what electrical wind generation will be in 20 years. Costs depend on the wind velocity, turbine design and the diameter of the propeller. First, the energy that can be tapped from the wind

Richard Keeler, Southern Calif. Edison Co.

◆ FIGURE 14.23 Solar One, a direct solar-energy power conversion facility in the Mojave Desert near Barstow, California. The heliostat (reflector) field directs sunlight to the central tower.

is proportional to the cube of the wind velocity. For this reason a slight increase in the wind speed results in a large increase in the electricity produced. For instance, an increase of the wind speed from 14 to 16 miles/hour results in an increase in electrical generation of nearly 50 percent. Second, a larger blade size requiring a taller tower yields a good increase in production. Increasing the blade diameter from 10 meters (1980s technology) to 50 meters (2004 technology) gives a 55-fold increase in yearly electrical production. Part of this is because a taller tower is required, and wind speeds normally increase with distance from the ground. Finally, a bigger wind farm is less costly to operate than a small one and thus the electricity is cheaper. Nearly forty percent is the difference between a small 3-MW installation (0.059 ¢/kwh) and a larger 51-MW plant (0.036 ¢/kwh).

California leads the nation in wind power and produces enough electricity to light a city the size of San Francisco. Ninety-five percent of California wind turbines are in three locations: Tehachapi Pass southeast of Bakersfield, Altamont Pass east of San Jose, and Banning Pass near Palm Springs (◆Figure 14.24). These narrow passes through the mountains separate hot, desert-like valleys from areas with cooler maritime climates. As the heated air rises in the valleys, cooler air from the maritime regions is drawn through the narrow passes to replace it. Modern, "smart" windmills with larger props and programmed turbines produce electricity at a price competitive with other sources (Table 14.8).

The wind potential of the lower 48 states is shown in ◆Figure 14.25. One of the most attractive features of wind is that it can produce electricity during peak demand periods, on hot summer days and early in the evening in winter, thus avoiding construction of additional fossil-fuel plants to meet the need. There are environmental problems that accompany wind generation, such as noise, land acquisition, TV interference, and most important, visual blight. Modern generators have largely solved the problem of noise and TV reception, but a couple of hundred windmills are hard to ignore. The feeling is that they destroy the visual aesthetics of a region. On the other hand, ranchers at Altamont love the windmills. They can still graze their cattle and at the same time receive land use royalties that have increased their prewindmill land value.

Europe is the leader in wind-energy development with its wind farms producing almost 35,000 megawatts (MW) (the equivalent of 35 coal-fired power plants) in 2005. One MW is enough electricity to power 250 to 300 homes. The United States, with 2,750 operable wind turbines, generated just over 7,000 MW, and Denmark produced over 3,000 MW. It is estimated that more than 90,000 MW of new wind-generating capacity will be installed annually around the

***TABLE 14.8 Electrical Capacities of Selected U.S. Geothermal Fields**

State	Field	Electrical Capacity, MW
California	The Geysers*	1,970[1]
	Coso	256[2]
	Salton Sea area*	214[2]
	East Mesa	10[3]
	Long Valley	43[3]
Hawaii	Puna	25[2]
Nevada	Dixie Valley	50[2]
	Steamboat Springs	18[2,3]
Utah	Cove Fort	13[2,3]

* 1993 data.
[1] Steam is piped directly to a power plant.
[2] Flash wells produce a mixture of steam and water that must be separated.
[3] Binary wells produce hot water that is piped to a power plant, where heat is transferred to another fluid.
Source: U.S. Geological Survey circular 1125, 1994.

D. D. Trent

◆ FIGURE 14.24 Wind "farm" in San Gorgonio Pass near Palm Springs, California. The San Jacinto Mountains are in the background.

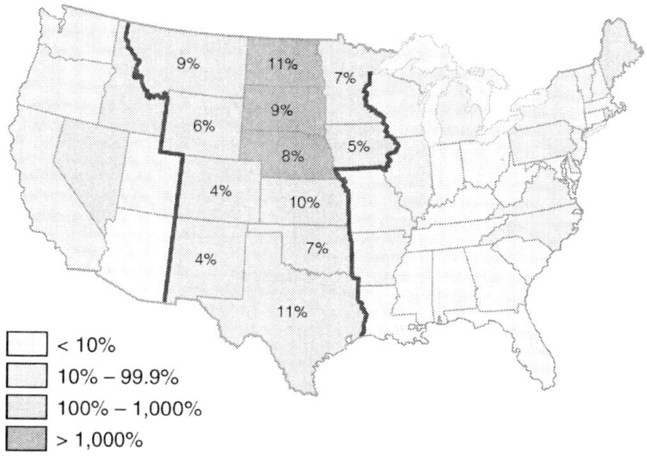

◆ FIGURE 14.25 Wind-electric potential of the 48 contiguous states as a percentage of each state's energy need. Twelve states in the central part of the country could provide about 90 percent of the total U.S. need. From "Energy from the Sun," by C. J. Weinberg and R. Williams, *Scientific American*, 1990.

world by about 2015. Five countries are anticipated to set the pace for adding new capacity through 2007: the United States, 5,285 MW; Spain, 5,200 MW; Germany, 5,000 MW; China, 4,725 MW; and India, 4,000 MW. The Global Wind Energy Council estimates that 12 percent of the world's energy could be provided by wind energy by 2030, resulting in the reduction of 1,832 million tons of CO_2 emissions.

Hydroelectric Energy

Falling water, our largest renewable resource next to wood, has been used as an energy source for thousands of years. First used to generate electricity about 100 years ago on the Fox River near Appleton, Wisconsin, today it provides a fourth of the world's electricity. In Norway 99 percent of the country's electricity and 50 percent of its total energy is produced by falling water. The principle is relatively simple: impound water with a dam and then cause the water to fall through a system of turbines and generators to produce electricity. Because of this simplicity, electricity coming from existing facilities is also the cheapest source of this power. Aitupu Dam on the Parana River between Paraguay and Brazil, completed in 1982, is the largest producing electrical complex in the world; its potential is 12,600 megawatts. It provides much more electricity than the present demand in this part of South America.

Because of land costs and environmental considerations, it is doubtful that any more large dams will be built in the United States (see Chapter 9, also Case Study 9.4). Therefore, dams that were built for other purposes have become attractive for retrofitting with generators. Hydroelectrical facilities are particularly useful for providing power during times of peak demand in areas where coal, oil, or nuclear generation provides the base load. Hydroelectricity can be turned on or off at will to provide peak power, and many utilities have built pumped-water-storage facilities for just that reason. During off-peak hours, when plenty of power is available, water is pumped from an aqueduct or other source to a reservoir at a higher level. Then during peak demand the water is allowed to fall to its original level, fulfilling the temporary need for added electricity.

Although hydroelectric energy is clean, dams and reservoirs change natural ecological systems into ones that require extensive management. One of the social consequences of dam building is displaced persons. This displacement is unacceptable in most societies today. Also unacceptable are recurring attempts to dam areas of great aesthetic value, such as the Grand Canyon. Finally, dam failures—most often geological failures of dams' foundations—result in floods that can take enormous human and economic tolls.

Geothermal Energy

Earth's interior is an enormous reservoir of heat produced by the decay of small amounts of naturally radioactive elements that occur in all rocks. When the deep heat rises to shallower depths, this **geothermal energy** can be tapped for human use. Furthermore, this natural heat is cleaner than many other energy sources, and there are fewer negative environmental consequences of using it. Not surprisingly, the prospects for using geothermal energy are best at or near plate boundaries where active volcanoes and high heat flow are found. The Pacific Rim (Ring of Fire), Iceland on the Mid-Atlantic Ridge, and the Mediterranean belt offer the most promise. Today the United States, Japan, New Zealand, Mexico, and countries of the former USSR are utilizing energy from Earth heat. Iceland uses geothermal energy directly for space heating (one geothermal well there actually began spewing lava), and many other countries have geothermal potential. The versatility of Earth heat ranges from using it to grow mushrooms to driving steam-turbine generators (◆ Figure 14.26).

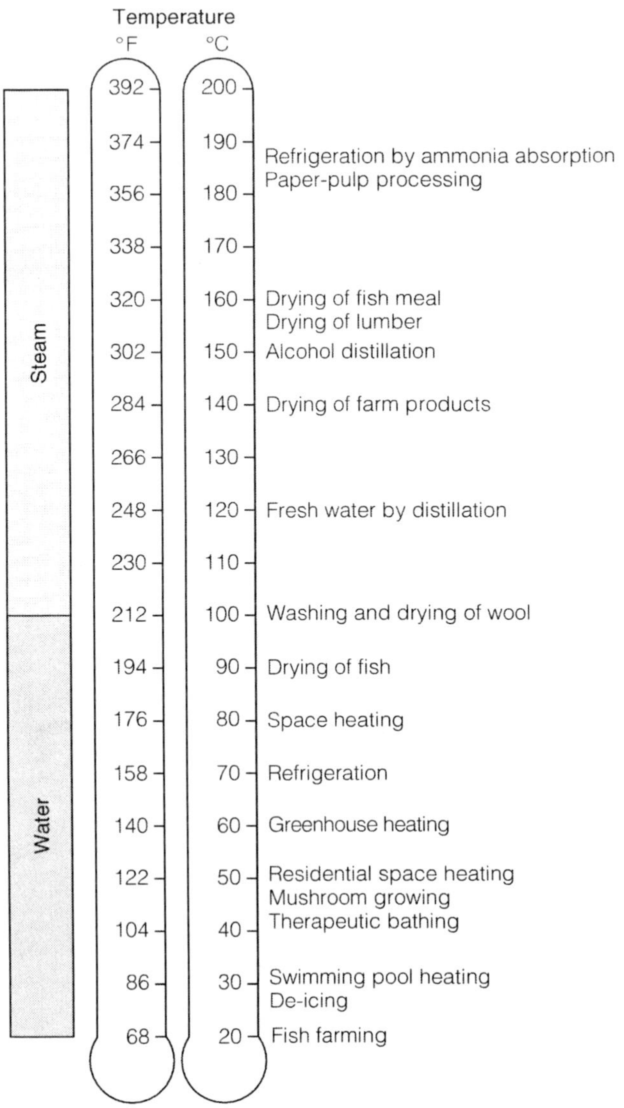

◆ FIGURE 14.26 Some established commercial uses of geothermal energy.

◆ FIGURE 14.27 The Geysers geothermal field in Sonoma County, California, is steam-dominated. It is the largest geothermal-energy producer in the world.

Geothermal energy fields may be found and exploited where magma exists at shallow depths and where there is sufficient underground water to form steam. Geothermal fields are classified as either *steam-dominated* systems or *hot-water-dominated* systems. Larderello, Italy, and The Geysers near Napa Valley in northern California (◆ Figure 14.27) are examples of steam-dominated fields. Such fields are the rarest and most efficient energy producers. They occur where water temperatures are high and discharge is low so that steam forms. They are much like an oil reservoir in that porous rocks that hold steam or very hot water are overlain by an impermeable layer that prevents the upward escape of the steam or water. Table 14.8 shows the electrical generating and direct-use capacities of selected U.S. geothermal fields. The Geysers is the largest producer of geothermal power in the world; its production is sufficient to supply electricity to a city of over one million people. More than 600 wells have been drilled there, some of them as deep as 3.2 kilometers (10,400 ft).

Geothermal energy is clean. The amount of CO_2 it emits is about one-tenth of the CO_2 emitted by coal-burning plants per megawatt of electricity produced, and about one-sixth of relatively clean-burning natural gas. Similar contrasts exist for emissions of sulfur and sulfur oxides.

The problems associated with geothermal energy production are related to water withdrawal and water quality. Some geothermal waters contain toxic elements such as arsenic and selenium and heavy metals such as silver, gold, and copper. Technology is in place to transfer the heat of the toxic brine deep within the wells to a clean, working fluid that can be brought safely to the surface. In some geothermal areas removal of underground water can cause surface subsidence and perhaps even earthquakes. At The Geysers, subsidence of 13 centimeters (5 in) has been measured without noticeable impact. Reinjecting the cooled water back into the geothermal reservoir decreases the subsidence risk and helps ensure a continued supply of steam to the electrical generators. All geothermal fields in California are being monitored for possible seismic activity related to steam production. Small quakes have been reported from The Geysers, but none of significant magnitude.

The future for geothermal energy looks good. About 20 geothermal fields in the United States are generating electricity, and direct-use hydrothermal systems are being developed at more than 30 sites. Electricity was first produced from geothermal energy in 1904. By 2004 the production had grown to about 8,000 megawatts in 21 countries around the world. The United States alone produces 2,700 megawatts of electricity from geothermal energy, equivalent to burning 60 million barrels of oil each year to produce the same amount of electricity.

Nuclear Energy

At one time nuclear energy was considered *the* solution to the world's future energy needs—being clean, no CO_2 emissions, no blots on the landscape other than an occasional cooling tower and containment dome, and the power generated "too cheap to meter." More than 400 nuclear power plants were built worldwide, 109 of them in the United States (five U.S. nuclear plants are now shut down).

Globally, there are now 318 operating nuclear plants generating 16% of the planet's electrical power (◆ Figure 14.28). The United States currently gets 20 percent of its electricity from 104 operating nuclear plants. Some countries have gone heavily into nuclear: France with 59 reactors obtains 79 percent of its power from nuclear fission, and China, facing a shortage of electric power, is building new reactors at a rapid pace—one or two a year.

But the allure of nuclear power began to tarnish with the Chernobyl accident in April 1986, following the 1979 Three Mile Island event in the United States. In addition, the challenge of radioactive waste disposal, and the poor economics compared with fossil-fuel-powered plants, helped reinforce people's anxieties regarding nuclear power. Collectively, these concerns resulted in the cancellation of all orders for new "nuke" plants in the United States. The current threat of terrorist attacks has added to those fears.

Nuclear reactors and fossil-fuel plants generate electricity by similar processes. Both heat a fluid, which then directly or indirectly makes steam that spins turbine blades that drive an electrical generator. Uranium is the fuel of atomic reactors, because its nuclei are so packed with protons and neutrons that they are capable of sustained nuclear reactions. All uranium nuclei have 92 protons and between 142 and 146 neutrons. The common isotope uranium-238 (^{238}U) has 146 neutrons and is barely stable. The next most common isotope, ^{235}U, is so unstable that a stray so-called slow neutron penetrating its nucleus can cause it to split apart completely, a process known as **fission**—a term particle physicists borrowed from the biological sciences. Nuclei that are split

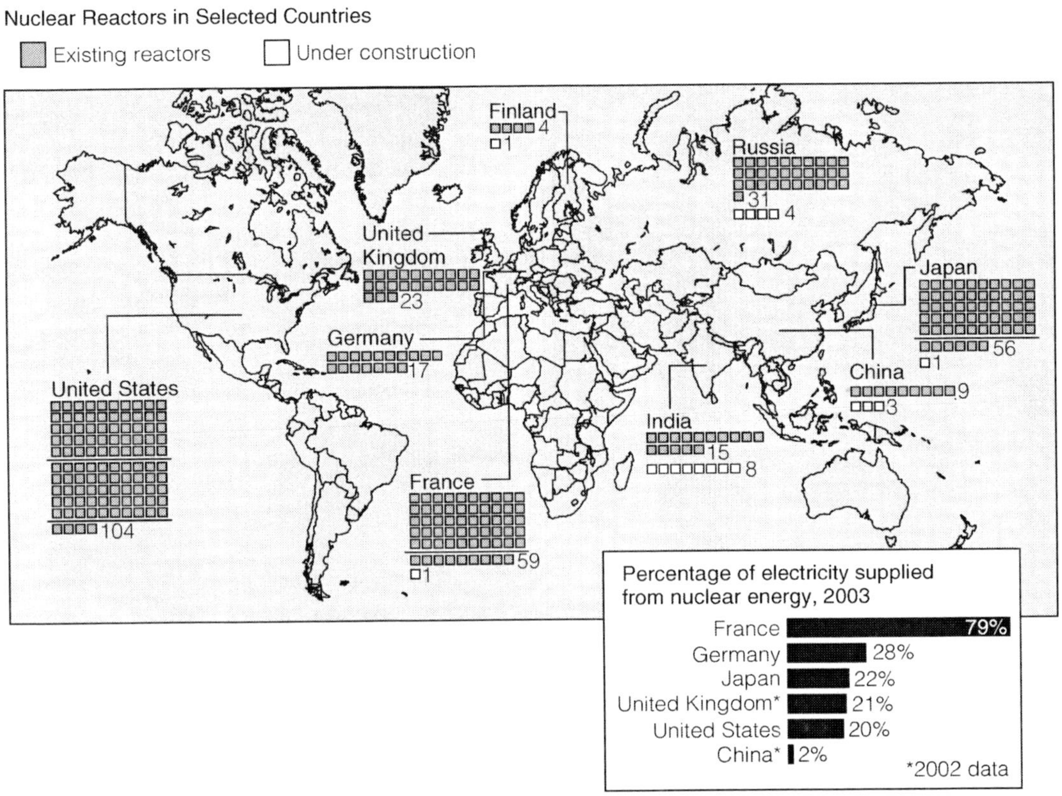

◆ FIGURE 14.28 The percentage of electricity supplied by nuclear reactors and the number of reactors for selected countries, 2003. Note France is the world's most nuclear energy dependent nation with 59 reactors generating nearly 80 percent of its electricity needs. International Energy Agency

easily, such as those of ^{235}U, ^{233}U, and plutonium-239 (^{239}Pu), are called **fissile isotopes.** Fission occurs when a fissile nucleus absorbs a neutron and splits into lighter elements, called *fission products,* while at the same time emitting several "fast" neutrons and energy, 90 percent of which is heat (◆ Figure 14.29). The lighter fission products are elements that recoil from the split nucleus at high speeds. This energy of motion and subsequent collisions with other atoms and molecules creates heat, which raises the temperature of the surrounding medium.

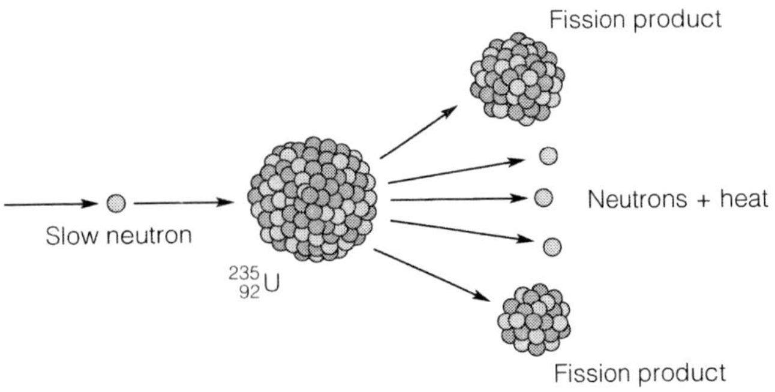

◆ FIGURE 14.29 Nuclear fission. A slow neutron penetrates the nucleus of a U-235 atom, creating fission products (also called *daughter isotopes*), 2 or 3 fast neutrons, and heat.

Meanwhile, the stray fast neutrons collide with other nuclei, repeating the process in what is called a **chain reaction.** A controlled chain reaction occurs when one free neutron (on average) from each fission event goes on to split another nucleus. If more than one nucleus is split by each emitted neutron (on average), the rate of reaction increases rapidly, and the reaction eventually goes out of control. Because ^{235}U makes up only 0.7 percent of all naturally occurring uranium, the rest being ^{238}U, it must be "enriched" for use in commercial nuclear reactors.

The major risk associated with nuclear energy is that of contamination during processing, transportation, and disposal of ^{235}U and high-level nuclear waste products. The locations of these risks in the nuclear fuel cycle are apparent in ◆ Figure 14.30.

Geological Considerations

One of the many criteria in the siting of nuclear reactors is geologic stability. The selected site must be free from landsliding, tsunamis, volcanic activity, flooding, and the like. The Nuclear Regulatory Commission (NRC), the federal

◈ FIGURE 14.30 The nuclear fuel cycle. The steps include mining, enrichment, transportation, power generation, reclamation, and disposal.

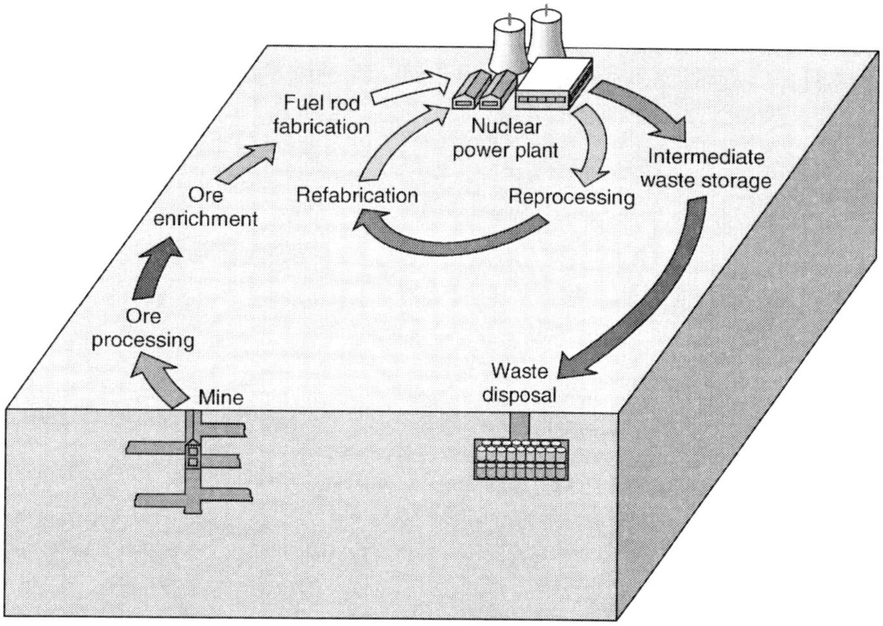

government body that licenses nuclear reactors for public utilities, requires that all active and potentially active faults within a distance of 320 kilometers (200 mi) of a nuclear power plant be located and described. Reactors must be built a distance from an active fault determined by the fault's earthquake-generating potential. Further, the NRC defines an active fault as one that has moved once within the last 30,000 years or twice in 500,000 years. This is a very strict definition, and sites that meet these criteria are difficult to find along active continental margins. To provide additional safety and allow for geological uncertainty, reactors are programmed to shut down immediately at a seismic acceleration of 0.05 *g*. This is conservative indeed, and the shutdown would prevent overheating or meltdown should the earthquake be a damaging one.

And finally, another geological consideration is that nuclear power is far from renewable, because the readily mineable uranium mineral deposits won't last much more than 50 years.

Energy from the Sea

Ocean Thermal Energy Conversion

Ocean thermal energy conversion (OTEC) utilizes the temperature difference between warm surface water and colder deep waters, at least 22° C (72° F), to vaporize a low-boiling-point fluid and operate a generator. A potential OTEC site must be in a coastal area where water depth increases rapidly enough to obtain the needed temperature differential and close enough to shore for power transmission. The Island of Oahu in Hawaii has the population, need, and oceanographic conditions for this form of energy conversion. A mini-OTEC of about 50 kilowatts was operated in Hawaii in the 1980s. The electricity produced was not transmitted to the shore, but a continuously burning lightbulb on a float near the village of Mokapu testified to OTEC capability.

Wave Energy

Capturing energy and converting it to electricity is not a new idea, but until recently its practicality has been little improved since 1900. Research in wave-powered technology by a renewable-energy company based in New Jersey has resulted in their PowerBuoy system that appears successful at extracting the energy of ocean waves. Ocean trials have successfully captured wave energy and converted it to electricity, and the firm now has 40kW-rated PowerBuoys in service in New Jersey and Hawaii. Beginning in 2006 the company began installation of a 1.25-MW PowerBuoy wave "farm" off the northern coast of Spain. An application is pending for placing one of its generating buoys off Gardiner, Oregon, and a full-size demonstration plant of 10 MW capacity is planned for installation in waters of the United Kingdom.

ThomsonNOW™
Click **Geology Interactive** to work through activities on Wave Properties, Wave Progression, and Wind–Wave Relationships through Waves, Tides, and Currents.

Tidal Currents

Tidal currents have been used by coastal dwellers to power mills for at least a thousand years. Restored working tidal mills in New England and Europe are popular tourist attractions. Both flood (landward-moving) and ebb (seaward-moving) tidal currents are used. Modern designs utilize basins that fill with seawater at high tide and empty at low tide, generating electricity in the process. A potential site must have a large tidal range—the height difference between high and low tides—and adequate area for storing the elevated water. It is estimated that there are more than 100 suitable sites in the world that, together, are capable of generating 5 percent of the world's present generating capacity.

The first and best-known large-scale tidal power plant is on the Rance River near St. Malo on the English Channel in northwest France. The tidal range here varies from 9 to 14 meters (30–46 ft), and the power output is a reversible operation; that is, the plant generates power at both flood and ebb tides. Because the hydrostatic head (water drop) is only a few meters, 24 small 10-MW generators with a total capacity of 240 MW are used. On balance, the operation has been successful.

Annapolis Royal in Nova Scotia, Canada, is the only modern tidal-generating plant in North America (◆ Figure 14.31). It generates over 30 million kilowatts per year, enough electricity to support 4,500 homes. An island in the mouth of the Annapolis River was selected as the site for the powerhouse that opened in 1984 after four years of construction (◆ Figure 14.32). It is an ideal location because the Bay of Fundy has the highest tides in the world, and there was an existing causeway on the river with sluice gates that dammed a pond in the river. On the incoming tide the sluice gates are opened and the incoming seawater fills the pond. The sluice gates are then closed, which traps the seawater in the pond upstream of the generating turbine. As the tide recedes, the level of the water below the pond drops and a hydraulic head develops. When the difference is 1.6 meters or more the gates are opened and water flows through the turbine to generate electricity. Tidal power is clean and renewable, and in Canada, what started as a demonstration project turned into a successful power-generating facility.

D. D. Trent

◆ FIGURE 14.31 Annapolis Royal tidal electricity-generating station in Nova Scotia, Canada, the only modern tidal generator in North America. The plant generates more than 30 million kilowatts per year, enough electricity to support 4,500 homes.

CONSIDER THIS

What nonfossil-fuel energy sources are likely to be expanded in the United States in the coming century? In other words, if you wanted to make a long-term investment in one of them, which one would you choose?

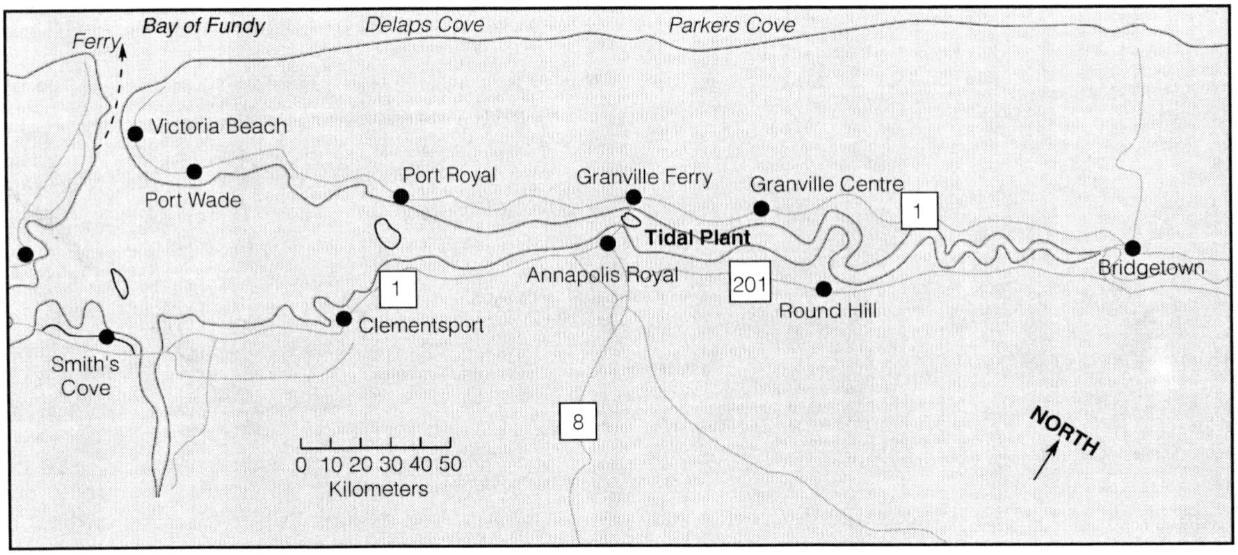

◆ FIGURE 14.32 Map of the site of the Annapolis Royal, Canada, tidal generating station. The Bay of Fundy has the highest tides in the world, which are put to work at the station to generate electricity. Nova Scotia Power, Inc.

CASE STUDY 14.1

Rock Oil and the Colonel

The history of the petroleum industry is full of interesting characters, and probably none was more charismatic than the self-styled Colonel of Titusville, Pennsylvania, Edwin L. Drake (◆ Figure 1). In the 1840s the demand for whale oil and lard had outstripped the supply and was being supplemented by "rock oil" from seeps and tar pits. The oil on northwest Pennsylvania's Oil Creek, so-named because oil bubbled to its surface, had long been skimmed by the Seneca Indians and other Native Americans and used to waterproof their homes and boats. Later, Europeans skimmed it for use as lamp oil, paraffin, liniments, and lubrication.

Drake was a railway conductor in New Haven, Connecticut, who had an interest in petroleum exploration but no experience. There are several versions of why Drake went to Titusville. The most credible one is that he had bought some shares in the Seneca Oil Company in New Haven and was sent by one of its founders, a friend of his, to stimulate the seepage at Oil Creek, which the company had leased from the Pennsylvania Rock Oil Company. Drake had a "Lincolnesque" appearance, and he frequently wore a knee-length topcoat and stovepipe hat. The company bestowed upon him the honorary title of *Colonel* to enhance his credibility in Pennsylvania.

Drake Well Museum, Titusville, Pa.

◆ FIGURE 1 Colonel Edwin L. Drake (right) in front of his well in 1861 with his friend, Titusville druggist Peter Wilson. The men in the background are probably drillers.

Colonel Drake tried digging trenches to stimulate the flow, and he installed a series of skimmers to collect the oil. The process was so inefficient that he persuaded the company to lease the land and attempt drilling. Drake enlisted the help of local "salt borers," who drilled for brine from which salt was extracted. Often these drillers had struck oil instead of water, and they had been run out of town because of the terrible mess it created. The Colonel finally found an experienced driller named Uncle Billy Smith, who used percussion tools featuring a chisel on a cable to chip the rock and a bailer to remove the cuttings. Trouble with cave-ins ensued, and Drake ingeniously drove a large-diameter pipe into the caving zone and drilled the hole through it. On Friday, August 27, 1859, they struck oil, although they didn't know it until they checked the well on Sunday. A black goo had filled the 21-meter (69-ft)-deep hole, and the world had its first successful oil well.

Drake's well produced 10–35 barrels per day, almost doubling the previous world output of rock oil. The oil price jumped to $20/bbl, and a thousand oil pits sprang up around Titusville. Within a year the price had dropped to $0.20/bbl, and the Seneca Oil Company sold the Drake Well. Nearby towns such as Pithole City flourished for a few years and then became ghost towns. At Oil City a few miles south of Titusville, a Cleveland hay and grain dealer named John D. Rockefeller built the first commercial oil refinery in 1863. Seven years later he established the Standard (now Exxon/Mobil) Oil Company.

Because Edwin Drake had filed no patents and established no lease claims, he did not benefit from his vision and labor. He later lost his money speculating on oil stocks and became impoverished. He was forced to live on the generosity of friends. Learning of his dire need in 1873, Titusville citizens took up a collection for him. Three years later the Pennsylvania legislature voted to give him an annual income of $1,500. Drake was buried in Bethlehem, Pennsylvania, in 1880, but his body was reinterred in Titusville in 1901, ironically, the same year that a gusher in Texas, "Old Spindletop," drastically changed the oil supply of the world.

Today, Oil Creek between Titusville and Oil City is the site of the Oil Creek State Park. The ghost town of Pithole is just east of the park, and Drake Well Memorial Park is at the park's northern boundary.

CASE STUDY 14.2

Industry Cracks Open a New Oil Source in the Gulf of Mexico

In what could become the biggest new source of domestic oil in the United States since the discovery of Alaska's North Slope in the 1970s, the first successful oil production from a part of the Gulf of Mexico that lies in very deepwater was announced in September, 2006. The Jack field, 435 kilometers (270 mi) southwest of New Orleans, and a dozen comparable discoveries in the same region, may hold as much as 15 billion barrels of oil and gas reserves (◆ Figure 1)—a barrel of oil is 42 gallons. If successfully exploited, it would boost the U.S. reserves by about 50 percent to more than 29 billion barrels. Production from these deep-water discoveries may well exceed the output of Alaska's giant Prudhoe Bay field, the largest of all U.S. oil fields. True, the new discoveries are sizable, but the Gulf's deep water oil doesn't come close to Mexico's huge Cantarell oil field, or the enormous fields in the Middle East.

The Jack well, one of the world's deepest producing wells at a total depth of five miles, taps a reservoir in rocks of Early Tertiary age. The well was drilled at an estimated cost of more than $100 million. Further development will cost several billion dollars for building platforms (◆ Figure 2), drilling, and laying pipelines to the mainland. Such deep-water discoveries are now possible by advances in seismic exploration and by today's large floating drilling platforms. These technological innovations have made it possible for geologists to search out and drill below the massive salt beds that typically overlie and obscure petroleum reservoirs in these deep-water Early Tertiary sedimentary rocks.

A consortium of three oil companies discovered the Jack field, which lies under 7,000 feet of water, by drilling two wells from a floating platform, the *Cajun Express.* The first well was completed in 2004 at the incredible depth of 29,066 feet, and the second, in 2005, reached a depth of 28,175 feet. The oil produced is light and sweet; that is, low in sulfur, the kind of petroleum that is in demand and realizes the best prices. Nevertheless, the discovery is somewhat provocative as stated by international oil expert Colin Campbell: the discovery "represents a remarkable technological achievement but at the same time reflects the extreme desperation of the industry, evidently having little easier [remaining] to test."

Courtesy of Shell Oil Company

◆ FIGURE 1 Shell Oil Company's tension-leg semisubmersible platform *Ursa* in the Gulf of Mexico, about 265 kilometers (164 mi) south of New Orleans, anchored in water with a depth of 1160 meters (3805 ft). One platform such as this may drill and provide for as many as 30 wells. It is designed to withstand waves up to 22 meters (72 ft) high and hurricane winds up to 225 km/hour (139 mi/hour).

CONTINUED

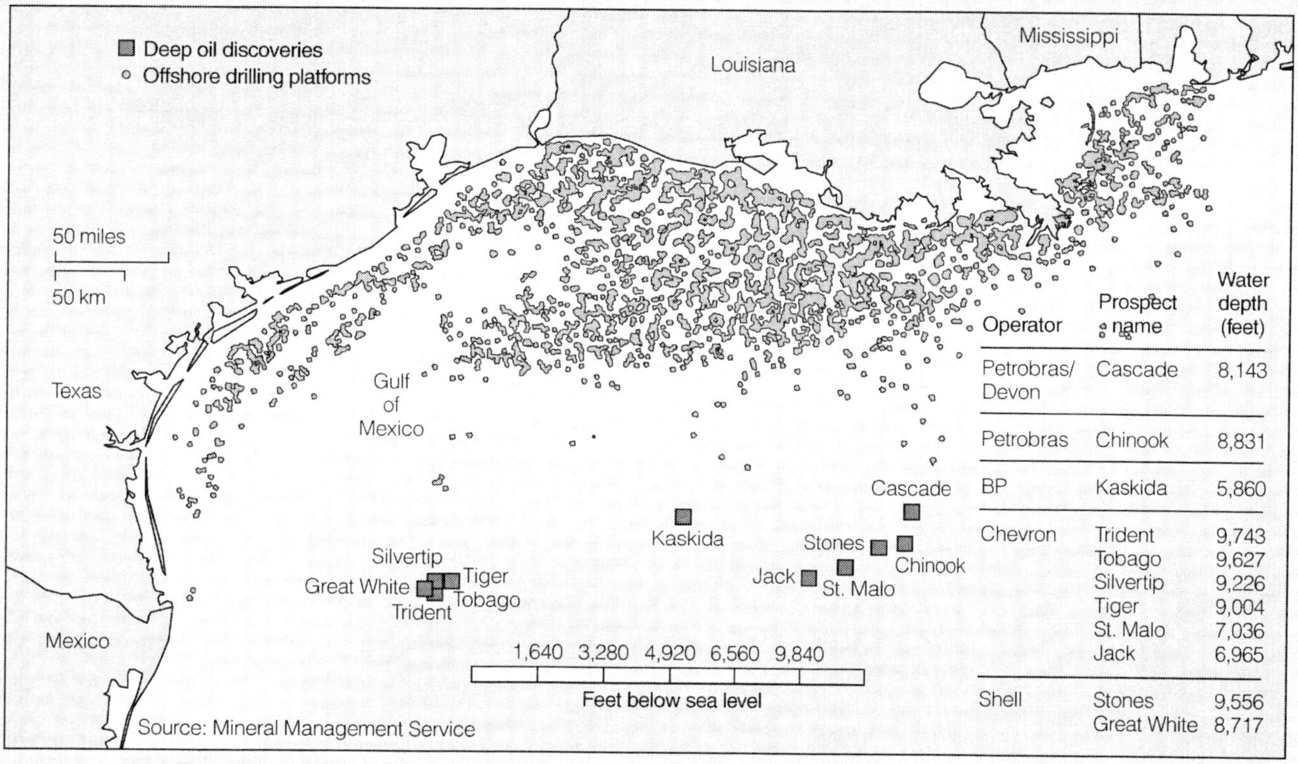

◆ FIGURE 2 Ultradeep water discoveries in the Gulf of Mexico. Oil companies drilling in waters up to 3 kilometers (2 mi) deep have made some significant discoveries. The first test of a well in Chevron's Jack field, which is under 2.1 kilometers (7,000 ft) of water, was drilled in 2004 to a total depth of 8.7 kilometers (28,543 ft or 5.4 mi), and was successful. Ten other ultradeep prospects in the Gulf offer promise. Department of the Interior, Minerals Management Service

CASE STUDY 14.3

Baking Soda, Vinegar, and Acid Rain

The term *acid rain* refers to the atmospheric deposition of acidic substances including rain, snow, fog, dew, particles, and certain gases. Although volcanic activity (see Chapter 5) is the greatest source of materials that may form acids—sulfur, carbon dioxide, and chlorine—sulfur and nitrogen oxides introduced by human activities are approaching the amount of nature's contributions. Combustion of fossil fuels and the refining of sulfide ores are the major human sources of these contaminants. Acids form when these gases come in contact with water in the atmosphere or on the ground. Whereas carbon dioxide forms carbonic acid, a weak acid (see Chapter 6), chloride ion and the oxides of sulfur and nitrogen form the strong hydrochloric, sulfuric, and nitric acids, respectively.

The acidity or basicity (the chemical opposite of acidity) of an aqueous solution is referenced to the pH scale (◆ Figure 1), a measure of the solution's hydrogen-ion activity. A neutral solution will have a pH value of 7.0. The more hydrogen ions floating around in the solution, the more acidic it is, and the lower its pH will be. Note that the scale is reversed, so to speak; a solution with a pH below 7.0 is acidic, and one with a pH above 7.0 is basic, or "alkaline." Acid-rain events with acidities below pH 2.8, the pH of vinegar, have been reported in large cities and heavily industrialized areas. In contrast, lakes without outlets, such as Lake Natron in Africa, may become extremely alkaline and have pHs greater than 11.0—well in excess of the pH of a concentrated solution of baking soda and water.

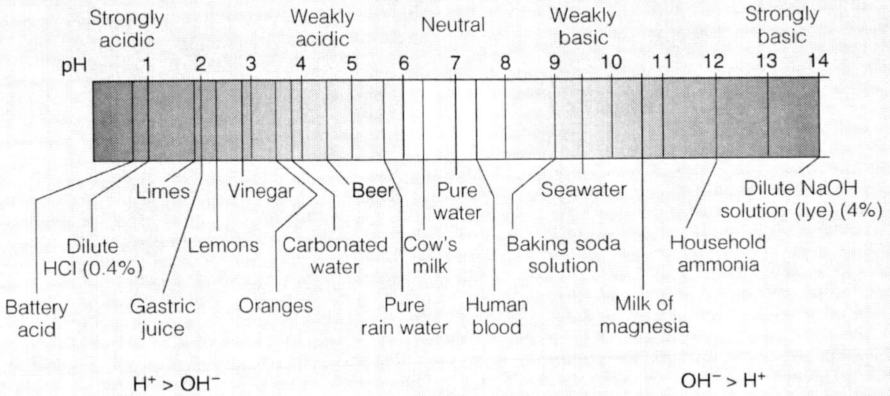

◆ FIGURE 1 The pH scale. A neutral solution has a pH of 7.0. A solution whose pH is less than 7.0 is acidic, and a solution whose pH is above 7.0 is basic, or alkaline.

Energy Is Where You Find It

Humans devote a great deal of mental and physical energy to harnessing Earth's energy. Energy, like fresh water, is invaluable, and when it is in short supply, a family, a tribe, or a country will do almost anything to obtain it. The four photos in this gallery illustrate some of Earth's nonconventional energy sources and some relics of fossil energy. Keep in mind that all these energy resources derive from our Sun.

◆ FIGURE 1 A community gathers around its solar-cell-powered television set; the Republic of Niger. More than likely they are watching a cricket or soccer match, not MTV or *Desperate Housewives*.

John ChiAsson/Gamma Liaison

Peter Bentham, BP Exploration

◆ FIGURE 2 This could be a scene in West Texas, Oklahoma, or Tampico, Mexico, in the 1920s and 1930s. Little regulation of the impact of oil production on the environment existed at the time. In reality, this is the Balakhany Oil Field in Northern Baku, Azerbaijan, in 2003.

Terra Moya Aqua, Inc.

◆ FIGURE 3 Three versions of a new vertical-axis wind turbine (VAWT) that went into commercial deployment in 2006. The inner blades, each shaped as a half-cylinder, rotate around a central vertical axis, while the outer blades, shaped like airplane wings, are fixed. The result is that the VAWTs are 43–45 percent efficient in contrast to conventional propeller-type turbines with only 25–40 percent efficiencies. The VAWTs can function at wind speeds up to 110 kilometers per hour (68 mph) that would shut down conventional turbines. Furthermore, in eight years of experience, the VAWT have proven to be bird and bat friendly.

D. D. Trent

◆ FIGURE 4 The Blue Lagoon, a great hot-water pool 40 kilometers (25 mi) from Reykjavík, Iceland. It uses geothermal power-plant outlet water, which is salty and grows algae, giving it its blue color. Open every day, the lagoon attracts about 100,000 visitors annually.

Summary

Energy

Defined

Capacity or ability to do work. A society's quality of life depends in large part on the availability of energy.

Types

1. Renewable energy is replaced at least as fast as it is consumed; e.g., solar, wind, and hydrologic energy.
2. Nonrenewable energy is replenished much more slowly than it is utilized; e.g., energy derived from coal, oil, and natural gas (fossil fuels).

Petroleum

Defined

Volatile hydrocarbons composed mostly of hydrogen and carbon. Crude oil contains many different hydrocarbon compounds.

Origin and Accumulation

The necessary conditions for an oil field:

1. Source rock—origin as organic matter;
2. Reservoir rock—porous and permeable;
3. Caprock—impermeable overlying stratum; and
4. Geologic trap—such as an anticline, faulted anticline, reef, or stratigraphic trap.

Production

Drilled wells are either pumped or they flow if under gas and water pressure. Even with the best drilling and producing techniques, primary production may recover only half the oil.

Secondary Recovery

Oil remaining in pore spaces underground is stimulated to flow to a recovery well by injecting water, chemicals, steam, or CO_2. Much of the "stuck" oil can be stimulated to flow to the extraction well by these methods.

Value

Varies with weight. Light crude oil is the most valuable.

Reserves

The world's oil production may peak as early as 2010—the exact date is unknown. Alternate energy sources will become increasingly important.

Energy Gases

May be the fuel of the first half of the twenty-first century.

1. Methane hydrates, ice balls of natural gas
2. Natural gas from unconventional sources
3. Hydrogen gas

Coal

Defined

Carbonaceous residue of plants that has been preserved and altered by heat and pressure.

Rank

As coal matures, it increases in heat content and decreases in the amount of water and volatile matter it contains. The sequence is plant matter to peat, to lignite, to subbituminous, to bituminous, to anthracite.

Reserves

The United States may have enough coal to last 200–300 years.

Synfuels

The manufacture of combustible gas, alcohol, and gasoline from coal has a promising future.

Other Fossil Fuels

Tar Sands

Sands containing oil that is too thick to flow and that can be surface-mined. Oil is then washed from the sand. One place the sands are mined is the Athabasca Field in Alberta, Canada.

Oil Shale

Eocene lake deposits containing light kerogen oil that can be removed by heating. Deposits are found in Wyoming, Utah, and Colorado.

Problems with Fossil-Fuel Combustion

Air Pollution

Products of combustion contribute to global warming, smog, and acid rain. Acid rain is a hazard in the U.S. East and Southeast and in eastern Canada, which receives some of the U.S. emissions. This is a major international air-pollution problem.

Mitigation

Sulfur content must be reduced below 1 percent. Methods include fluidized-bed combustion, neutralizing sulfur dioxide in smokestacks with limestone scrubbers, cleaning coals by gravity separation, and chemical leaching methods.

Subsidence and Collapse

Areas underlain by shallow lignite are subject to subsidence.

Alternative Energy Sources

Direct Solar

1. Solar heat collectors warm a fluid, which is then put to work.
2. Photovoltaic cells convert light directly into electricity.

Indirect Solar

Wind is produced by unequal heating of Earth's surface.

Synfuel

A fuel manufactured from organic materials, corn, farm waste, or coal.

Geothermal

Facilities for generating electricity with geothermal energy currently are in place and may supply as much as 2 percent of U.S. electrical energy needs. Geothermal heat is also used worldwide for heating interior spaces, keeping sidewalks free of winter snow and ice, etc.

Hydroelectric

Hydroelectric dams harness the power of water distributed over Earth by the hydrologic cycle.

Oceans

The energy in waves, currents, and tides can be converted to do work for humans. Ocean thermal energy conversion (OTEC) shows some promise, but little has been done with waves and currents. Tidal power is being used successfully only at the Rance River, France, and in Nova Scotia, Canada, at present.

Fuel Cells

Fuel cells generate electricity fueled by hydrogen. Many think this is the fuel–motor combination of the future.

Nuclear Energy

Provides 16 percent of world and 20 percent of U.S. energy needs. Heat generated by fission of uranium isotopes and their daughter-isotopes such as plutonium is converted to electricity. There are 104 commercial nuclear power plants operating in the United States.

1. *Problems*—Mining, transportation, and disposal of high-level nuclear materials are the most dangerous aspects of the nuclear energy cycle. If the chain reaction runs out of control, the heat generated may lead to a central-core meltdown such as occurred at Chernobyl.
2. *Geological considerations*—Very careful geological site studies and investigations are performed to assess hazards from earthquakes, mass-wasting, subsidence, and coastal and riverine floods.

Key Terms

acid rain
anticline
biodiesel
biomass fuel
caprock
chain reaction
coal-bed gas
ethanol
fissile isotopes
fission
geothermal energy
hydrocarbon compound
kerogen
methanol
miscible
nonrenewable resource
oil shale
photochemical smog
rank (coal)
renewable resource
reserves
reservoir rock
salt dome
secondary recovery
slant drilling
source rock
stratigraphic trap
synthetic fuels (synfuels)
tar sands
wildcat well

Study Questions

1. What four conditions are required for an oil deposit to form?
2. What oceanographic conditions favor the formation of a good source rock for oil?
3. Draw a geologic cross section that shows an anticline and a faulted anticline. Assume your cross section is looking north. What are the directions of the dips and strikes of the two limbs of the anticline? Dip and strike determination is explained in Appendix 3.
4. What nonrenewable energy resources exist in the United States, and which one is most abundant?
5. Describe atomic fission. How does it create heat for electrical generation?
6. What are methane hydrates, and what is their potential as fuel for the twenty-first century?
7. What are the environmental consequences of burning fossil fuels? What will be the impact upon the global environment of continued dependence upon fossil fuels?
8. How can solar energy be collected and used for heating, generating electricity, and producing fuels?
9. What sector of the U.S. economy uses the most fossil fuel, and what international problem is connected with this use?
10. Explain the meaning of renewable energy. Which renewable energy sources have been developed, and which undeveloped renewable sources show the most promise?

For Further Information

Fact sheets may be obtained from the USGS free of charge. Write to U.S. Geological Survey (Information Services), P. O. Box 25286, Denver Federal Center, Denver, CO 80225.

Books and Periodicals

Collett, T. S. 2004. Gas hydrates as a future energy resource. *Geotimes 49,* no. 11 November: 24–27.

Cooper, G. A. 1994. Directional drilling. *Scientific American*, May: 82–87.

Deffeyes, Kenneth S. 2003 (Rev. ed.). *Hubbert's Peak: The impending world oil shortage*. Princeton Univ. Press.

Duffield, W. A., J. H. Sass, and M. Sorrey. 1994. *Tapping the Earth's natural heat*. U.S. Geological Survey circular 1125.

Fellows, Larry D. 2001. Energy resources in Arizona. *Arizona geology 31*, no. 2.

Finch, Warren I. 1997. Uranium, its impact on the national and global energy mix. U.S. Geological Survey circular 1141.

Fisher, William L. 2002. The coming methane economy. *Geotimes 47*, no. 11, November.

Goldstein, David. 2004. *Out of Gas: The end of the age of oil*. New York. W. W. Norton and Company.

Haq, Bilal U. 1998. Gas hydrates: Greenhouse nightmare? Energy panacea or pipe dream? *GSA today 8*, no. 11, November.

Heinberg, Richard. 2003. *The party's over: Oil, war and the fate of industrial societies*. Gabriola Island, British Columbia. New Society Publishers.

Hirsch, Robert L. 1987. Impending United States energy crisis. *Science 235*:1467–1473.

Hubbard, N. King. 1971. The energy resources of the Earth. *Scientific American* 225, no. 3, p. 60–70.

McCabe, Peter J., and others. 1993. *The future of energy gases*. U.S. Geological Survey circular 1115, Public issues in Earth science series.

Mossop, Grant D. 1980. Geology of the Athabasca oil sands. *Science 207*, p. 145.

Reese, April. 1999. Bad air days. *E: The environmental magazine*, November–December: 28–35.

Ross, David. 1987. *Energy from waves* (2nd ed.). New York: Pergamon Press.

Schmandt, Jurgen, J. Clarkson, and Hilliard Roderick (Eds.) 1988. *Acid rain and friendly neighbors: The policy dispute between Canada and the United States*. Durham, N.C.: Duke University Press.

Scientific American. 1990. Energy for planet Earth. *Scientific American* special issue, September.

Spencer, Jon E., and Steven L. Rauzi. 2001. Crude oil supply and demand: Long-term trends. *Arizona geology 31*, no. 4.

Sperling, Daniel. 1996. The case for electric vehicles. *Scientific American*, November: 54–59.

Suess, Erwin, Gerhard Bohrmann, Jens Greinert, and Erwin Lausch. 1999. Flammable ice. *Scientific American*, November: 76–83.

U.S. Geological Survey. 1998. Arctic National Wildlife Refuge (ANWR), 1002 Area, petroleum assessment 1998. Fact sheet FS-040-98, May.

U.S. Geological Survey. 2000. USGS world petroleum assessment 2000. Fact sheet FS-070-00, April.

U.S. Geological Survey. 2001. Coal-bed gas resources of the Rocky Mountain region. Fact sheet FS-110-01, November.

U.S. Geological Survey. 2002. Natural gas production in the United States. Fact sheet FS-113-01, January.

ThomsonNOW™

Assess your understanding of this chapter's topics with additional comprehensive interactivities at http://www.thomson.com/login, which also has up-to-date web links, additional readings, and exercises.

index

H

P

Q

T